This b

CH

Fire retardant materials

Fire retardant materials

Edited by

A R Horrocks and D Price

CRC Press
Boca Raton Boston New York Washington, DC

WOODHEAD PUBLISHING LIMITED

Cambridge England

Published by Woodhead Publishing Limited, Abington Hall, Abington
Cambridge CB1 6AH, England
www.woodhead-publishing.com

Published in North and South America by CRC Press LLC
2000 Corporate Blvd, NW
Boca Raton FL 33431, USA

First published 2001, Woodhead Publishing Ltd and CRC Press LLC
© 2001, Woodhead Publishing Ltd
The authors have asserted their moral rights.

British Library Cataloguing in Publication Data
A catalogue record for this book is available from the British Library.

Library of Congress Cataloging in Publication Data
A catalog record for this book is available from the Library of Congress.

Woodhead Publishing ISBN 1 85573 419 2 ✓
CRC Press ISBN 0-8493-3883-2
CRC Press order number: WP3883

Cover design by The ColourStudio
Typeset by Best-set Typesetter Ltd., Hong Kong
Printed by TJ International Ltd, Cornwall, England

Contents

Preface

Public demand for increased safety has led to greater interest in fire retardant materials in the last 30 years. Legislation relating to safety in the home, in work locations, on transport facilities and in public places continues to produce new regulations. The period 1960–80 saw the development of many of the now well established flame retardant materials in which the property of fire retardancy or resistance is conferred by the use of chemical treatments or additives, or where it is an inherent feature of the chemical and physical structure of the material. During this period a number of key and classic texts appeared including *The Chemistry and Uses of Flame Retardants* by J W Lyons (1970) and the subsequent sets of edited texts, *Flame-retardant Polymeric Materials* by M Lewin, S M Atlas and E M Pearce (1975–87) and *Flame Retardance of Polymeric Materials* by W C Kuryla and A J Papa (1973–75). Since then, the developments in science and technology of fire retardancy have been reported and discussed regularly. The UK Fire Chemistry Group of the Society of Chemical Industry, London have meetings in the spring and autumn of each year as do the Fire Retardant Chemical Association (Lancaster, Pa, USA). In 1990 the annual series of conferences commenced at Stamford entitled *Recent Advances in Flame Retardancy of Polymeric Materials*, the related proceedings being edited by M Lewin. There are also biennial European conferences on *Fire Retardant Polymers*.

Within the UK, no authoritative text covering the fire retardancy of materials has been published within the last 10 years apart from D Drysdale's book *Fire Dynamics*, now in its second edition (1998), which covers the broader aspects of fires and their underlying scientific principles. It is timely that, in the USA, the topic of *Fire Retardancy of Polymeric Materials* has been revisited (2000) by A F Grand and C A Wilkie. This book discusses the recent advances in fire retardancy and retardant systems as they are applied to polymers, with emphasis being on the former and not the latter.

Our text considers the material properties first; why materials may need to be fire retarded; how this may be undertaken and the consequences of so doing. This last is particularly important given that the same society that is demanding increased safety (Chapter 13) is questioning the risks to health and the environment by using flame retardants and fire retardant materials (Chapter 3). The book is therefore structured to discuss the fundamental issues which determine whether or not a material is flammable (Chapter 1) and how flame retardancy may be conferred both mechanistically (Chapter 2) and by means of established flame retardant systems (Chapters 7, 8, and 10). In addition, the means of reducing fire hazards of real materials such as textiles (Chapter 4), composites (Chapter 5) and the large group of natural polymers (Chapter 9) are addressed.

The need to anticipate the future must be an essential feature of any study of this type given the external pressures in, for example, requiring the increased environmental sustainability of all materials. Therefore, novel methods of rendering materials fire retardant are explored (Chapters 6 and 11) as well as the anticipated changes for performance-based test regimes (Chapter 12). However, the increasing costs of developing new fire retardant materials is such that mathematical modelling and simulation are increasingly becoming a part of the underpinning science; these topics are also explored (Chapter 14). We have thus attempted to produce a balanced text which addresses not only the advances, which have brought us to our present understanding and the application of these materials, but also the concerns and needs of the future. This book, while being able to stand alone, may be read alongside the earlier texts quoted above and should be seen as a synergistic companion to the recently published *Fire Retardancy of Polymeric Materials (2000)* by Grand and Wilkie.

We would like to take this opportunity of thanking our co-authors for collaborating in the production of this exciting project; the support of our respective research teams and the tolerance and fortitude shown by our respective wives for once again accepting the time taken away from them and our families.

Richard Horrocks
Dennis Price

Editors:

Professor A Richard Horrocks
Dean, Faculty of Technology
Bolton Institute
Deane Road
Bolton BL3 5AB
UK
Tel: 01204 903831
Fax: 01204 381107
E-mail: arh1@bolton.ac.uk

Professor Dennis Price
Fire Chemistry Research
Group
School of Sciences
The University of Salford
Salford
Manchester M5 4WT
UK
Tel: 0161 295 4262
Fax: 0161 295 5222
E-mail: d.price@salford.ac.uk

Chapter 1: Introduction
  Professor D Price, Fire Chemistry Research Group, School of Sciences, University of Salford, Salford, Manchester M5 4WT, UK
  Dr Geoff Anthony, Great Lakes Chemicals Tenax Road, Trafford Park, Manchester M17 1WT, UK
  Dr P Carty, Department of Chemical and Life Sciences, University of Northumbria at Newcastle, Newcastle upon Tyne, NE1 8ST, UK

Chapter 2: Mechanisms and modes of action in flame retardancy of polymers
  Professor Edward D Weil and **Professor Menachem Lewin**, Polytechnic University, Six MetroTech Center, Brooklyn, New York NY 11201, USA
  E-mail: eweil@poly.edu

Chapter 11: Graft copolymerisation as a tool for flame retardancy
Dr Charles A Wilkie, Department of Chemistry, Wehr
Chemistry Building, Marquette University, POB 1881,
Milwaukee, Wisconsin 53201-1881, USA
Fax: 001 414 288 7066
E-mail: Charles.wilkie@marquette.edu

Chapter 12: Performance-based test methods for material flammability
Dr Björn Karlsson, Department of Fire Safety Engineering,
Lund University, Box No. 118, S-22100, Lund, Sweden
Tel: 0046 46222 7363 Fax: 0046 46 222 4612
E-mail: bjorn.karlsson@brand.lth.se

Chapter 13: Fire safety design requirements of flame-retarded materials
Professor Dougal Drysdale, Fire Research Group,
Department of Civil and Environmental Engineering,
University of Edinburgh, Edinburgh EH9 9JL, Scotland
Tel: 0131 650 5724 Fax: 0131 667 9238
E-mail: Dougal@s.lvo.edu.ac.uk

Chapter 14: Mathematical modelling
Dr J Staggs, Fuel and Energy Department, University of
Leeds, Leeds LS2 9JT, UK
Tel: 0113 243 1751 Fax: 0113 244 0572
E-mail: j.e.j.staggs@leeds.ac.uk

1

Introduction: polymer combustion, condensed phase pyrolysis and smoke formation

DENNIS PRICE

Fire Chemistry Research Group,
School of Sciences
University of Salford
Salford, UK

GEOFFREY ANTHONY

Great Lakes Chemical Corporation
Trafford Park
Manchester, UK

PETER CARTY

School of Applied Molecular Sciences, University of
Northumbria, Newcastle upon Tyne, UK

Plastics and textiles find many uses and add greatly to the quality of modern-day life. However, a major problem arises because most of the polymers on which these materials are based are organic and thus flammable. In the UK alone some 800–900 deaths and roughly 15 000 injuries result from fire each year.[1] Most of the deaths are caused by inhalation of smoke and toxic combustion gases, carbon monoxide being the most common cause, whilst the injuries result from exposure to the heat evolved from fires. In addition, the annual cost of damage to buildings and loss of goods varies between £0.5 billion and £1.0 billion. Thus, there are great economic, sociological and legislative pressures on the polymer industries to produce materials with greatly reduced fire risk. To facilitate such developments this book aims to be an authoritative reference source for the highly diverse field of flame retardant materials. This introductory chapter gives an overview of the various interacting stages of the complex phenomenon of polymer combustion and flame retardance together with a more detailed consideration of condensed-phase processes and smoke. Thus it provides the background for understanding the many and varied aspects of flame retarded materials which are considered in much greater detail in the following chapters.

1.1 Polymers

1.1.1 Classification of polymers

Polymers can be classified in a variety of ways, several of which are worth considering.[2] Firstly, they have often simply been classified as natural or synthetic (and sometimes as synthetic modifications of natural polymers). However, a classification based on their physical/mechanical properties can also be used, in particular their elasticity and degree of elongation. Under these criteria, polymers can be classified into elastomers, plastics and fibres. Elastomers (rubbers) are characterised by having a high extensibility and recovery; plastics have intermediate properties, while fibres can have very high tensile strength but low extensibility. Plastics are often further sub-divided into thermoplastics (whose deformation at elevated temperature is reversible) and thermosets (which undergo irreversible changes when heated).

1.1.2 Chemical classes of polymers

Polymers can also be classified in terms of their chemical structure, and this gives an important indication of their reactivity, including their fire perfor-mance and their tendency to produce smoke when they burn.

The main carbon-containing polymers with no heteroatoms present are the polyalkenes (polyolefins) and the aromatic hydrocarbon polymers. The main polyolefins are thermoplastics: polyethylene (repeating unit: $-(CH_2-CH_2)-$) and polypropylene (repeating unit: $-(CH(CH_3)-CH_2-)$, which are the most widely used synthetic polymers. The most important aro-matic hydrocarbon polymers are based essentially on polystyrene (repeat-ing unit: $-(CH(Phenyl)-CH_2)-$). Polystyrene is extensively used as a foam and as a plastic for injection-moulded articles. A number of styrenic copoly-mers including acrylonitrile-butadiene-styrene terpolymers (ABS), styrene-acrylonitrile polymers (SAN) and methyl methacrylate butadiene styrene terpolymers (MBS) are also important.

The most important and widely used oxygen-containing polymers are cellulosics, polyacrylics and polyesters. Polyacrylics (not to be confused with polymers based on acrylonitrile) are the only major oxygen-containing polymers which contain carbon-carbon chains. The most important cellu-losics are used in the timber industry and in the manufacture of paper and textiles. The main polyacrylic is poly(methyl methacrylate) (repeating unit: $-(CH_2-C(CH_3)-CO-OCH_3)-$; PMMA), widely used as a substitute for glass. The most important polyesters are manufactured from glycols (such as polyethylene terephthalate (PET), or polybutylene terephthalate (PBT)), or from bisphenol A (polycarbonate). They are used as engineer-ing thermoplastics, in applications such as soft drink bottles (PET), as fibres (PET), for injection-moulded articles and as unbreakable replacements for glass (polycarbonate). Other oxygenated polymers include phenolic resins, polyethers, such as polyphenylene oxide (PPO), a very thermally stable engineering polymer and polyacetals (such as polyformaldehyde), used for their intense hardness and resistance to solvents.

Nitrogen-containing materials include nylons (polyamides), polyurethanes and polyacrylonitrile. Nylons, having repeating units containing the characteristic group $-CO-NH-$, are made into fibres and also into a number of injection-moulded articles, as well as for specialist uses in the wire and cable industry. Nylons are synthetic aliphatic polyamides, but there also exist natural polyamides (wool, silk, leather) and synthetic aromatic polyamides (of exceptionally high thermal stability) which are used as fibres in protective clothing. Polyurethanes (with repeating units containing the characteristic group $-NH-COO-$), are normally manufactured from the condensation of polyisocyanates and polyols. Their principal area of application is as foams (flexible, for use in furniture or as filling materials and rigid, for use in packaging or as thermal insulation). Both these types of polymers have carbon-nitrogen chains, but nitrogen can also be contained in materials with carbon-carbon chains, the main example being polyacrylonitrile (repeating unit $-(CH_2-CH-CN-)$). It is used mostly to make into fibres and as a constituent of the engineering copolymers SAN and ABS.

The most important chlorine-containing polymer is poly(vinyl chloride) (PVC, repeating unit: $-(CH_2-CHCl)-$). (Together with polyethylene and polypropylene, it is the most widely used synthetic polymer.) PVC is unique in the sense that it is used both as a rigid material (unplasticised, as pipes, sheets, rods, bottles, siding, injection-moulded appliance housings, etc.) and as flexible material (plasticised, as wire and cable coatings, wall coverings, furniture fabrics, foams, inflatable toys, protective clothing, etc.). Flexibility is achieved by adding plasticisers. Semi-flexible materials can also be made; they are manufactured into pipes and wire and cable materials. The further chlorination of PVC leads to another member of the family of chlorinated materials: chlorinated poly(vinyl chloride) (CPVC), which has very different physical and fire properties from PVC. An additional chlorinated material of commercial interest is poly(vinylidene chloride) (PVDC, with a repeating unit: $-(CH_2-CCl_2)-$) used for making films and fibres.

Fluorine-containing polymers are characterised by high thermal and chemical stability and low coefficients of friction. The most important material is polytetrafluoroethylene (PTFE), while others include poly(vinylidene fluoride) (PVDF), poly(vinyl fluoride) (PVF) and fluorinated ethylene polymers (FEP). They are used as insulators, particularly in the wire and cable industry, in printed circuits and in gaskets, diaphragms and as metal coatings for 'non-stick' surfaces.

1.1.2.1 Polymer combustion

Natural and synthetic polymers, when exposed to a source of sufficient heat, will decompose or 'pyrolyse' evolving flammable volatiles. These mix with the air and, if the temperature is high enough, ignite. Table 1.1 provides a

Table 1.1 Decomposition and ignition temperatures* together with heats of combustion of some common thermoplastic polymers and cellulose (cotton)

Polymer	Decomposition range/°C	Flash ignition temperature/°C	Autoignition temperature/°C	ΔH_c/kJ kg^{-1}
LDPE	340–440	340	350	46.5
Polypropylene	330–410	350–370	390–410	46.0
Polystyrene	300–400	345–360	490	42.0
PVC (rigid)	200–300	390	455	20.0
PMMA	170–300	300	450	26.0
Cellulose (cotton)	280–380	210	400	17.0

*determined by ASTM D 1929.
Key: LDPE is low density (non-linear) polyethylene; PMMA is poly(methyl methacrylate).

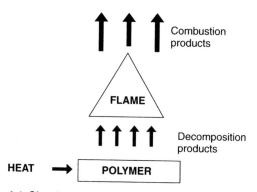

1.1 Simple representation of polymer combustion processes

listing of the decomposition and ignition temperatures for a range of common polymers. Ignition occurs either spontaneously (autoignition) or due to the presence of an external source such as a spark or a flame (flash ignition). If the heat evolved by this ignited flame is sufficient to keep the decomposition rate of the polymer above that required to maintain the concentration of the combustible volatiles, i.e. the 'fuel', within the flammability limits for the system, then a self-sustaining combustion cycle will be established. Figure 1.1 is a simple representation of this behaviour.

1.1.3 The simple flame

A flame is a gas-phase combustion process. Two types occur. Firstly, there is the 'premixed' flame in which the gaseous fuel and oxygen are mixed

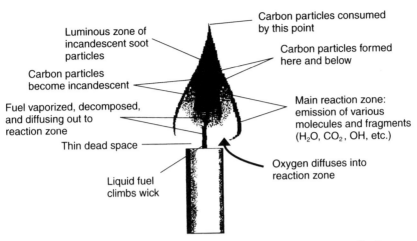

1.2 Principal processes occurring in each region of a candle flame

prior to combustion. The best examples are the bunsen burner and the gas/air flames of the domestic cooking stove. Secondly, there is the 'diffusion' flame, so-called because the oxygen necessary for combustion diffuses into the gas mixture from the surrounding atmosphere. The best known example of a diffusion flame is that of the candle, shown schematically in Fig. 1.2.[3] Wax melted by the heat radiated from the flame migrates up the wick by capillary action and is pyrolysed on its surface at temperatures between 600 and 800 °C. The gaseous products from this pyrolysis migrate further and either remain within the inner part of the flame or reach the outer flame mantle. The inner region of the flame is deficient in oxygen so that reducing conditions exist there. Hydrocarbon fragments from the pyrolysis migrate to regions in which temperatures reach 1000 °C. Under these conditions, cyclisation and aromatisation lead to the formation of carbon particles, i.e. soot. These are transported further and start to glow, causing luminescence of the flame. Soot particles are consumed in this luminescent region by reaction with water to form carbon monoxide. The pyrolysis gases are carried to the exterior and encounter oxygen diffusing inwards. In this outer flame reaction zone, high energy, primary oxygen-containing free radicals are generated at temperatures around 1400 °C. These maintain the combustion reaction. If the process is uninterrupted and an adequate supply of oxygen is available, the end products of combustion of the candle flame are carbon dioxide and water.

The processes which take place during the combustion of plastics are, in principle, similar to those of the candle flame. However, before discussing polymer flames in detail, some important terms need to be defined. Combustion is a catalytic exothermic reaction maintained by internally generated free radicals and heat. Provided the supply of radicals and heat exceeds the energy required for combustion, the combustion proceeds at an increasing rate until an explosion occurs. If the energy supply is constant and equals the demand, a stationary equilibrium will be established, i.e. a steady flame occurs. If the available energy is below that required to maintain this equilibrium, then the rate of combustion will decrease until the flame extinguishes. The radicals, oxygen and heat necessary to sustain the combustion reach the site by various transport mechanisms:

Mass transfer
- due to turbulent flow, e.g. flow processes such as eddy diffusivity
- due to concentration gradients, i.e. molecular diffusion
- due to temperature gradients, i.e. thermal diffusion

Energy transfer
- due to temperature gradients, i.e. thermal conduction
- due to radiation

As previously stated, a flame is a gas-phase combustion process. Combustion when both solid and gas phases occur together is also known. If the volatilisation temperature of a solid is higher than its combustion temperature, combustion occurs directly on the surface. At low temperatures in the presence of sufficient oxygen, incandescence occurs, i.e. flameless combustion takes place. With an insufficient oxygen supply, smouldering occurs and neither flames nor incandescence appear.

1.1.4 Polymer flames

Combustion reactions liberate the energy stored in the chemical bonds of the molecules of the fuel. A fuel is any substance that will release energy during its reaction with oxygen, usually in air, generally initiated by an external heat source. Typical fuels are wood, hydrocarbons, coal and animal fats. In the main, they are organic materials as are most synthetic polymers. Polymer combustion is a complex process involving a multitude of steps and is best described in qualitative terms. Figure 1.3 is a schematic diagram of the various steps which combine to establish the polymer combustion process. The three essential stages required to initiate the combustion are heating, thermal decomposition or pyrolysis and ignition. Ignition is normally caused by the presence of an external heat source such as a flame or a spark or, if the temperature is high enough, occurs spontaneously.

The temperature of the solid polymer is raised either due to an external heat source such as radiation or a flame, or by thermal 'feedback' as indicated in Fig. 1.3. During the initial exposure to heat thermoplastics, which have a linear chain structure, soften or melt and start to flow. On the other

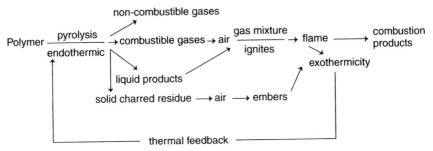

1.3 Schematic representation of many processes involved in polymer combustion (after Troitzsch, ref. 4, p. 16)[4]

hand, thermosetting plastics have a three-dimensional cross-linked molecular structure which prevents softening or melting. Additional heat causes both types of polymer to pyrolyse and evolve smaller volatile molecular species. Because of their structure this occurs at higher temperatures for thermosetting as opposed to thermoplastic polymers. Since most plastics are organic in nature the evolved species will also be organic and thus flammable. Such flammable evolved species provide the fuel to sustain the flame. Thus we see that the mechanism of combustion contains both a condensed phase and a vapour phase contribution.

Pyrolysis is an endothermic process which requires the input of sufficient energy to satisfy the dissociation energies of any bonds to be broken ($200–400\,kJ\,mol^{-1}$) plus any activation energy requirements of the process. As individual polymers differ in structure, their decomposition temperature-ranges vary within certain limits. The limits will again change somewhat when a polymer is compounded with various additives and subsequently processed to produce what are commonly known as 'plastics'. Table 1.1 collates the decomposition ranges of some common thermoplastics polymers together with that for the natural polymer cellulose. The fuel generating pyrolysis reactions that control polymer combustion occur in the condensed phase and are considered in detail in section 2.

1.1.4.1 Ignition

Flammable products, i.e. 'fuel', evolved from the decomposing polymer/plastic mix with oxygen from the surrounding air. When the lower flammability limit is reached the mixture will either 'flash' ignite due to the presence of an external flame or spark, or autoignite if the temperature is above the autoignition temperature. The flash and autoignition temperatures of some common polymers are given in Table 1.1. Ignition depends on numerous variables like oxygen availability, temperature, and physical and

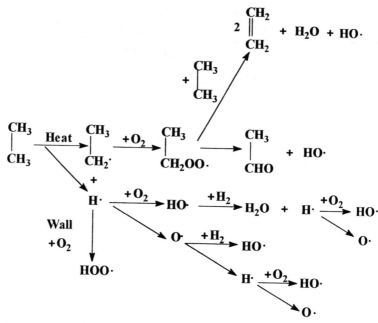

1.4 Free radical generation during the combustion of ethane[5]

chemical properties of the polymer/plastic. Once ignited the burning process is exothermic and, if sufficient energy is evolved this will override the endothermicity required for the polymer pyrolysis. Thus flame spread will be initiated.

1.1.4.2 Flame spread

The heat generated via the burning process sustains the polymer pyrolysis process as shown by the self-sustaining cycle depicted in Fig. 1.3. The rate of pyrolysis will be accelerating leading to an increasing supply of fuel to the flame which then spreads over the polymer surface. As a simplified model of the flame chemistry, consider the reactions occurring in a hydrocarbon diffusion flame. The important step is the chain branching step propagated by the highly reactive H· and OH· radicals. These confer a high velocity on the flame front resulting in rapid flame spread. In the case of the OH· radicals, their avalanche-like proliferation can be illustrated by the combustion of ethane as shown in Fig. 1.4. It should be remembered that a similar contribution will also be made by the H· radicals.

A schematic description of flame spread along a surface is given in Fig. 1.5. The diffusion flame advances over the decomposing polymer surface.

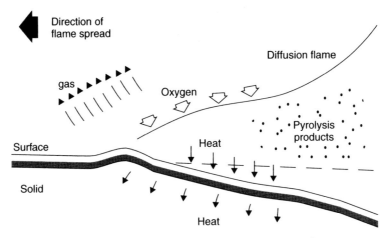

1.5 Schematic representation of flame spread [after Atka[6]]

As with the candle flame, the surface temperature of the polymer surface (500 °C) is lower than that of the diffusion flame and of the edge of the flame, where the reaction with oxygen occurs (1200 °C). The extent of flame spread will also be affected by the heat of combustion of the polymer. The greater the heat of combustion the greater the amount of heat liberated into the flame to sustain the burning cycle. Table 1.1 gives values of heats of combustion for some common polymers. There is not a simple relationship between heat of combustion and combustibility of a polymer. For example, cotton has a low heat of combustion, 17.0 kJ kg^{-1}, but is extremely flammable.

Concurrent with the extremely rapid gas-phase reactions controlled by diffusion flames, slower oxygen-dependent reactions also take place. These give rise to smoke, soot and carbon-like residues. Some can occur in the condensed phase resulting in glow or incandescence.

1.1.5 Flame retardation

Of major interest in the plastics and textiles industries is not the fact that their products burn but how to render them less likely to ignite and, if they are ignited, to burn much less efficiently. The phenomenon is termed 'flame retardance'. This book is intended to provided a detailed account of the flame retarded materials developed to meet these objectives.

A simple schematic representation of the self-sustaining polymer combustion cycle is shown in Fig. 1.6. Flame retardants act to break this cycle, and thus extinguish the flame or reduce the burning rate, in a number of possible ways:

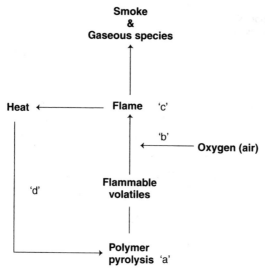

1.6 Schematic representation of the self-sustaining polymer combustion cycle; a–d represent potential modes of flame retardant action

- by reducing the heat evolved to below that required to sustain combustion
- by modifying the pyrolysis process to reduce the amount of flammable volatiles evolved in favour of increasing the formation of less flammable char which also acts as a barrier between the polymer and the flame ('a')
- by isolating the flame from the oxygen/air supply ('b')
- by introducing into the plastic formulations compounds which will release chlorine or bromine atoms if the polymer is heated to near the ignition temperature. Chlorine and particularly bromine atoms are very efficient flame inhibitors ('c')
- by reducing the heat flow back to the polymer to prevent further pyrolysis. This can be achieved by the introduction of a heat sink, e.g. aluminium oxide trihydrate (ATH, $Al(OH)_3$) which decomposes endothermically or by arranging that a barrier, e.g. char or intumescent coating, is formed when the polymer is exposed to fire conditions ('d')
- by developing inherently flame retarded polymer systems

Most flame retardant systems in use today have been developed empirically. Current interest in obtaining a better understanding of polymer combustion and interaction of flame retardants therewith is motivated by the requirement to develop environmentally friendly flame retardant systems with enhanced performance. This work needs to be based on sound scientific principles.

A much more detailed account of flame retardants will be given in Chapter 2.

1.2   Condensed-phase processes

As indicated earlier, when exposed to heat such as a source of ignition, or the combustion flame, the surface temperature of the polymer can rise to a point at which its structure will break down and it will release volatile material. Physical properties, which can influence this, are thermal conductivity, heat capacity and the ability to melt back away from an ignition source. As part of the degradation mechanism some polymers will also produce carbonaceous char.

The behaviour of a polymer in a fire risk situation is therefore the result of a combination of many different physical and chemical processes, which take place in the condensed phase. The kinetics of these processes are particularly important both as a function of temperature and relative to each other. This section will consider the processes in detail and the way that they affect polymer combustion. Because it is such an important aspect of fire safety, emphasis will be given to the physics and chemistry of char formation.

1.2.1   Bond dissociation

Thermal decomposition of a polymer is often initiated by dissociation of covalent bonds to form radicals. Bond dissociation energies (BDE) will depend on the nature of the atoms making up the bond and also the precise structural environment in which the bond occurs. Bond dissociation values can often be used to explain why one bond dissociates in preference to another, and are of particular importance for polymers, which degrade by free radical mechanisms. Table 1.2 collates the most important BDEs of relevance to polymer chemistry.

As will be shown in subsequent sections, some polymers degrade by concerted mechanisms, in which bonds are broken and formed simultaneously. These are usually lower energy processes, and take place at relatively low temperatures. In the following section examples will be given of both free radical and concerted mechanisms which take place in the condensed phase.

1.2.2   Chemistry of polymer degradation

The mechanisms of polymer degradation, and the temperatures at which they occur will depend very much on the polymer's structure. Mechanisms

Table 1.2 Dissociation energies of some covalent bonds[7]

Bond	Dissociation energy kJ mol^{-1}	Bond	Dissociation energy kJ mol^{-1}
C–H	340	H–I	297
C–C	607	C–F	553
C–O	1076	C–Cl	398
H–H	435	C–Br	280
H–F	569	C–I	209
H–Cl	431	C–P	515
H–Br	386	P–O	600

are usually hypothesised after the analysis of degradation products, and there is still often considerable debate over the precise path taken to reach these structures. It is not the purpose of this introduction to review these mechanisms comprehensively, but to present a few examples which show how the structure of a polymer can influence the way it degrades thermally.

To gain a general picture of the possible degradation paths, consider the hypothetical polymer below:

$$\left[A - \underset{\underset{C}{\overset{|}{B}}}{} \right]_n$$

This generalised structure is of oligomer units joined by atoms A and B containing a pendant side group (C), which is not attached to any other oligomer unit. Typical of this type of polymer are polyvinyl chloride [–CH$_2$–CH(Cl)–]n, and polystyrene [–CH$_2$–CH(C$_6$H$_5$)–]n, which differ only in their side group, yet behave very differently when pyrolysed.

As discussed earlier, polymers at their degradation temperature can form radicals due to bond scission. For polystyrene, degradation to volatile products initiates around 300 °C and has been explained by hydrogen loss followed by C–C bond scission to form a chain terminating carbon radical.[8] This radical can then further degrade by stepwise elimination of the styrene monomer until the polymer molecule is completely degraded, or the radical is stabilised. This is a low energy process, which is often referred to as 'unzipping', and accounts for about 50% of the volatiles formed from polystyrene pyrolysis.

Scheme for polystyrene

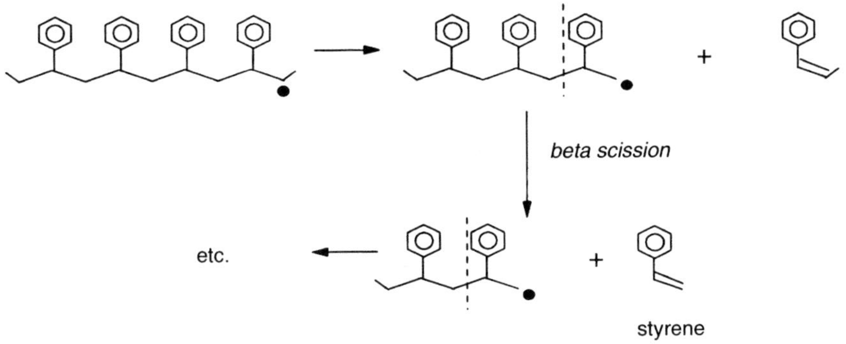

beta scission

etc.   +

styrene

It can be seen that each elimination of styrene results from the breaking of the C–C bond which is beta to the carbon bearing the radical, and produces a similar radical structure which can undergo further styrene elimination.

For poly(vinylchloride) however an alternative initial degradation path requiring less activation energy than the radical formation in polystyrene is the concerted elimination of hydrogen chloride. This takes place at the lower temperature of 250 °C. The resulting condensed-phase structure is a polyolefin. This new structure offers several alternative and sequential degradation pathways, which have only recently been fully evaluated.[9] One of these pathways is the elimination of benzene. This also appears to be an unzipping reaction, although it is not the simple beta scission process seen with styrene. An alternative and competing pathway is cross-linking of the polyolefin to produce a more thermally stable structure. This degrades at higher temperatures with skeletal rearrangement to produce volatile materials such as toluene and xylenes, and also undergoes extensive proton migration to produce aliphatic hydrocarbons and char.

The reason that these two polymers degrade so differently is due to the energetically favourable elimination of hydrogen chloride in poly(vinylchloride). Once this has taken place, the molecule cannot undergo simple C–C bond scission, and is forced to take alternative degradation pathways. As the temperature rises other mechanisms become energetically and sterically favourable, each one producing a more thermally stable condensed phase and resulting ultimately in char.

Scheme for poly(vinylchloride)

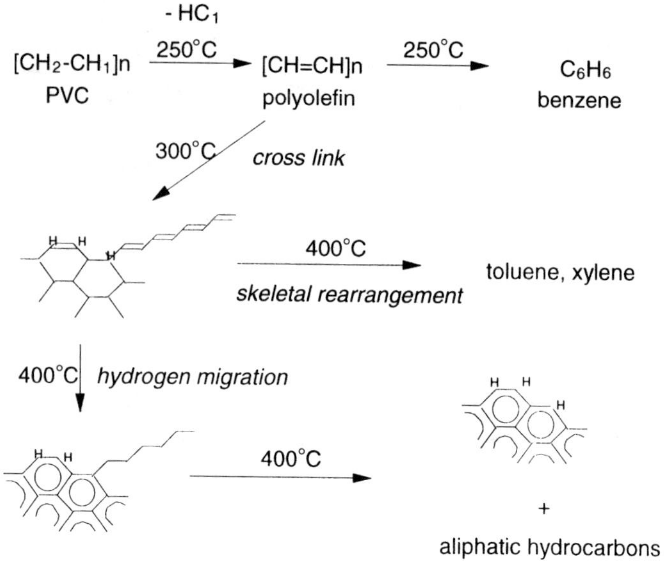

$- HC_1$

$[CH_2-CH_1]n$ $\xrightarrow{250°C}$ $[CH=CH]n$ $\xrightarrow{250°C}$ C_6H_6

PVC   polyolefin   benzene

$300°C$ *cross link*

$\xrightarrow[\text{skeletal rearrangement}]{400°C}$ toluene, xylene

$400°C$ *hydrogen migration*

$\xrightarrow{400°C}$

+

aliphatic hydrocarbons

1.2.3 Char-forming polymers

Char formation is probably the most important condensed-phase mechanism for modifying the combustion process. It serves as a barrier to heat and mass flow, and as a means of stabilising carbon, thus preventing its conversion to combustible gases. The efficiency of a char as a barrier in these processes depends greatly on its chemical and physical structure. The ability of char formation to prevent sustained ignition will also depend on its rate of formation in relation to other degradation mechanisms, especially the release of combustible gases. Polymers such as polycarbonate, novolaks and polyphenylene oxide all burn with the formation of a carbon rich residue called char. This char forming property is also reflected during thermogravimetric (TG) experiments which show initial degradation producing a more thermally stable material. Consider the following examples.

1.2.3.1 Bisphenol A-polycarbonate

For bisphenol A-polycarbonate, degradation starts at 450°C to produce a 25% residue, which is stable to over 650°C. The shape of the TG curve (Fig. 1.7), is due to a combination of the processes:

polymer → volatiles
polymer → cross-linking (char)
char → volatiles

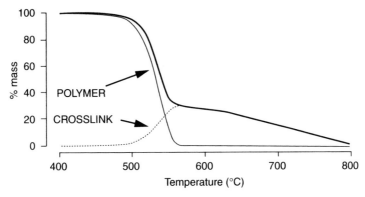

1.7 Thermogravimetric of bisphenol A-polycarbonate in nitrogen showing underlying processes[10]

Although thermogravimetry shows that mass loss is not observed until about 450 °C it is known that structural changes can take place at lower temperatures to produce molecular weight loss. This will certainly occur if traces of water are present, since the carbonate ester group is susceptible to hydrolysis. Even under anhydrous conditions such as the high vacuum of a mass spectrometer, however, it has been shown that polycarbonate can degrade to low molecular weight cyclic oligomer units at temperatures of 410 °C.[10]

The thermal degradation of bisphenol A-polycarbonate has been studied by several research groups, often under quite different experimental conditions. Not surprisingly these have resulted in different degradation products and different proposed mechanisms. The high vacuum conditions of mass spectrometry are particularly suited to the study of primary degradation processes, since larger molecules are sufficiently volatile to leave the condensed phase, and the analysis of volatile materials is sufficiently rapid to prevent further degradation. At atmospheric pressure however, degradation is likely to continue, until products are formed with sufficient volatility to escape into the vapour phase. In sealed containers, volatile degradation products can continue to react with each other and with materials in the condensed phase.

Using TG-GLC-MS, the volatile materials from bisphenol A-polycarbonate during the main mass loss region have been shown to be phenol, p-cresol, p-ethylphenol and isopropylphenol, Fig. 1.8. The same materials have been reported as degradation products formed by flash pyrolysis of polycarbonate at 850 °C.[11]

An explanation of the cross-linking process was proposed based on a skeletal rearrangement[11] leading to benzoate esters.

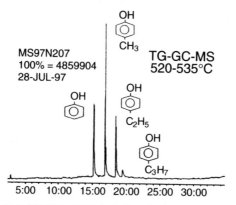

1.8 GC-MS of TG volatiles from polycarbonate degradation

Skeletal rearrangement of BPAPC

It can be reasoned that although this mechanism will explain stabilisation of the polymer by the formation of more covalent bonds, it does not explain the process of carbon enrichment which is also an essential part of a char forming mechanism. Although there has been considerable debate in the literature concerning the mechanism of cross-linking and char formation, we believe it to be based on free radical scission of C–C and C–O linkages followed by proton abstraction as proposed by McNeill.[12]

Thus phenolic groups are formed by scission of the C–O link followed by hydrogen abstraction from the condensed phase and isopropyl and phenyl groups are formed by scission of the C–C link followed by hydrogen abstraction from the condensed phase.

Methyl and ethyl substituted phenols indicate some skeletal rearrangement is also taking place. All of these volatile molecules will however require hydrogen radical abstraction from the condensed phase to enable their formation. The evolution of volatile materials, which are richer in hydrogen than the original polymer, means that the condensed phase becomes depleted in hydrogen and richer in carbon. Recombination of the condensed-phase radicals (R) formed by hydrogen abstraction is a means of cross-linking and a route to stabilisation and char formation.

1.2.3.2 Polyphenylene oxide and novolak polymers

Poly (2,6 -dimethyl-1,4-phenylene oxide) (PPO) is a char-forming polymer sometimes blended with other polymers to reduce flammability. Despite the dissimilarity in structure to novolak polymers, which are condensation products of phenols and formaldehyde, these two materials both pyrolise to produce cresols and xylenols, all of which are hydrogen rich when compared with the starting material. This is because PPO undergoes a skeletal rearrangement during thermal degradation to form a novolak-type structure.[13]

1.2.4 Char structure

The degree of protection provided by a char during combustion depends on both its chemical and physical structure. Whereas pure graphite is highly stable to heat and oxygen, chars from polymer combustion do not have this property. Although chars are richer in carbon than the original polymer, they are rarely all carbon. Char formed after combustion of bisphenol A-polycarbonate was found to contain only 90% carbon, with about 3% hydrogen, and 7% nitrogen remaining.[14] In the same study Raman spectroscopy was used to detect some graphitic material in chars from both pyrolysis and burn experiments, and infrared spectroscopy used to recognise residual hydrogen, and polar groups. Other analytical methods used to study char structure are X-ray photoelectron spectroscopy (XPS) and solid phase NMR. The latter is particularly useful for studying the incorporation of hetero-elements such as phosphorus in chars.[15] A review of char formation has recently been published by Levchik and Wilkie.[16]

To illustrate the importance of the physical structure of char on fire retardant properties it is useful to describe an ideal and non-ideal char. These are depicted schematically in Fig. 1.9. The ideal char for fire retardant properties is an intact structure of closed cells containing pockets of gas. For this to happen the bubbles of gas must become frozen into the expanding and thickening polymer melt, which ultimately solidifies to produce the honeycombed structure. This prevents the flow of volatile liquids or vapours into the flame and provides sufficient thermal gradient to keep the remaining polymer or polymer melt below its decomposition temperature.

The non-ideal or poor char structure, does not contain closed cells but channels or fissures through which gaseous decomposition products or

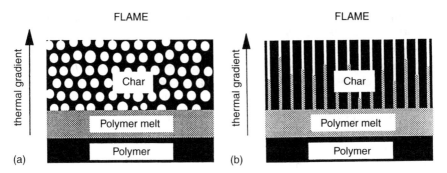

1.9 (a) Ideal char structure, (b) Poor char structure

polymer melt can escape. Of these two effects the more important is the movement of liquid products which can be drawn by capillary action into hotter regions where they are more likely to decompose.[17] This negates any heat insulating effects that the char may have on the virgin polymer beneath.

Factors which influence the type of char formed are still not properly understood, but will include melt viscosity, the surface tension of the melt-gas interface and the kinetics of gasification and polymer cross-linking.

1.3.   Smoke

Most of the work done on smoke generation during polymer burning has taken place over the last 20 years or so beginning with the development of test methods for measuring smoke density from burning materials. Even now no single smoke test is universally recognised for its predictive ability or high correlation with real fire situations. The term *smoke* has a vague definition. In general use, smoke is considered to be a cloud of particles, individually invisible, which is opaque as a result of scattering and/or absorption of visible light. *Fumes* are considered to be less opaque forms of smoke. It is necessary to distinguish between 'combustion gases' and 'visible smoke' since the two can have different effects, as well as different methods of measurement and significance in fires.

Among the combustion gases, carbon monoxide, CO, is of chief concern.[18] Other toxic gases can be formed in fires include hydrogen cyanide, nitrogen oxides, hydrogen chloride, sulphur oxides and some very toxic organics. The problems associated with combustion gases will be considered in detail in Chapter 3. In certain fire situations both aspects can be of comparable concern, since the loss of visibility due to heavy smoke can hinder escape until toxic gas concentrations and temperatures become critical. Presumably, reductions in the rate or intensity of visible smoke development will help to increase escape time, hence the need to develop effective smoke suppressants for polymers.

Visible smoke from burning polymers is generally a result of incomplete combustion. Since polymer flames are diffusion flames (*vide-infra*), proper mixing for complete burning does not readily occur.

Recent investigations using pyrolysis/gas chromatography/mass spectroscopy and special kinetic methods, have shown that within a flame, unsaturated hydrocarbon molecules formed by thermal cracking of the fuel will polymerise and dehydrogenate to form carbon, or soot. During these processes intermediate molecules can form unsaturated species or they can cyclise to form polybenzenoid structures, both of which will lead to soot formation. These polybenzenoid structures take on more importance as intermediates when they are formed directly from aromatic fuels. More detailed reviews of the chemistry of flames and soot formation have been published.[19,20]

In the presence of a sufficiently intense heat source a polymer will pyrolyse, breaking down to low molecular weight species. These species diffuse from the solid phase into the gas phase, where they form the smoke observed in the absence of flame. At high heating rates and with ignition, these low molecular weight species fuel the polymer flame. A much simplified picture of this is shown in Fig. 1.10. Aliphatic fuels are cracked to smaller alkyl radicals which, in the absence of oxygen grow to form

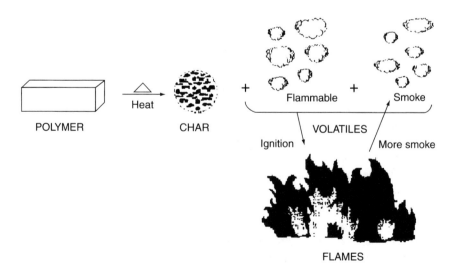

1.10 Polymer pyrolysis, burning and smoke evolution

conjugated polyenes or polybenzenoids which may be radical, ionic, or neutral. Ultimately the intermediates, which are highly reactive, react with other unsaturated species and condense to form soot. Aromatic fuels are thought to proceed directly to polybenzenoid intermediates. In these cases, heavy soot formation occurs rapidly. Since oxidation to oxides of carbon is competitive with soot formation, oxygen-containing fuels generally show a decreased tendency for soot formation. On the other hand, halogens may promote soot formation through dehydrohalogenation, assisting ring closure and the formation of olefins and polyenes.[21]

The nature of the cracked species and pyrolyzates generated is thus a major factor in determining smoke formation, given similar conditions of polymer combustion. Pyrolyzates of some common polymers are listed below:

- polyethylene – ethylene, propylene, higher olefins
- poly(methyl methacrylate) – methyl methacrylate (monomer)
- polystyrene – styrene, styrene oligomers, aromatics
- polyurethanes – aromatic isocyanates and amines, aldehydes
- poly(vinyl chloride) – HCl, benzene, other aromatics and low molecular weight alkenes and alkanes
- poly(ethylene terephthalate) – acetaldehyde, unsaturated esters, carboxylated aromatics

The relative distribution of pyrolysis products from an individual polymer is dependent on the pyrolysis temperature, the heating rate and the pyrolysis atmosphere. For example, it has been shown in a series of polyesters that yields of styrene and toluene pass through maxima at 600°–700°C[22] and naphthalene is found among the pyrolyzates at 700°C and above. The amount of smoke generated in a nitrogen atmosphere passes through maxima with increasing temperature in several of the polyesters whereas from others the smoke increases steadily with temperature. A study of the variation of smoke intensity with temperature for a group of 8 natural and 12 synthetic polymers showed that smoke density passes through a maximum in the region of 400°–600°C apart from poly(methyl methacrylate).[23]

1.3.1 Smoke measurement

Smoke formation during burning or smouldering is not an inherent property of a material as are heats of combustion, melting points or densities. The level of smoke actually measured in a test depends on the burning conditions (i.e. heat flux, oxidant supply, specimen geometry, the presence or absence of flame) as well as the test conditions (ambient temperature, test chamber volume, ventilation, etc.). As a result, no single smoke test or even a set of smoke measurements from different tests is likely to provide a comprehensive definition of smoke behaviour in a real fire.

A number of tests to measure smoke generation have been developed.[4] The names and features of some of the more common tests are summarised in Table 1.3. Full experimental and procedural details are available in the appropriate test literature. As the table shows, two types of smoke measurement are used: light attenuation (optical) or smoke mass (weight). Optical methods are the more common. The scale of the test methods also varies widely, for example, in the Steiner tunnel test (ASTM E-84) the sample is 7.6 metres long and 49.5 cm wide. In most cases, the smoke test is combined with some type of flammability test, and the smoke test is secondary to the fire test. Examples of test methods which were designed principally for smoke measurement include the NBS smoke density chamber, the Rohm and Haas XP-2 chamber and the Arapahoe smoke chamber. The cone calorimeter is the most recent innovation in the fire testing scenario. It was designed initially to measure heat release rate but since its inception it has been modified so that it is capable of measuring other fire parameters such as smoke release rate and soot production, heat of combustion, mass loss rate, ignitability and toxic gas formation.

Table 1.3* Commonly used smoke tests

Name	Feature	Principle
Underwriters' Laboratories tunnel (Steiner tunnel) (ASTM E-84)	7.6 m flame-spread chamber, monitored at exhaust	Optical
Rohm and Haas Chamber (XP-2) (ASTM D-2843)	Specimen variability, flaming combustion, accumulated smoke monitored in 0.07 m³ chamber	Optical
NBS Chamber (ASTM E-662)	76 × 76 mm specimen, radiant heating with or without flame, accumulated products monitored within 0.51 m³ chamber	Optical
Arapahoe smoke chamber	Soot collected on glass filter	Weight
Ohio State calorimeter (OSU) (ASTM E-906)	Variable heat input smoke monitored at top of exhaust stack	Optical
Cone calorimeter (ASTM E-1354)	100 mm × 100 mm sample thickness 6 to 50 mm horizontal sample 0–110 kW m⁻² heat flux	Optical

* Based on a table taken from Lawson.[24]

The amount of smoke produced in full-scale fires is a function of both the smoke-producing tendency and the amount of material burnt. It has been mentioned earlier that direct measures of smoke formation in laboratory tests do not correlate well with actual fire performance. However using the cone calorimeter, the parameter 'smoke factor' (the product of peak heat release and the total smoke produced at 5 minutes into the test) does give a better indication of the tendency of a material to produce smoke in a full-scale fire test. Hirschler has recently determined the fire performance (including smoke formation) of 35 widely used polymer formulations.[25] It must be borne in mind that the greater majority of commercial materials used are rarely in the form of pure polymer and often the additives in the polymer formulations, e.g. plasticisers, can themselves greatly influence the flammability properties of a material.[26]

There has been some interest in comparing and correlating various large- and small-scale laboratory smoke tests, and also in validating smoke density tests with full-scale fires.[27] In these tests major interest has been in measuring the total amount of smoke formed. However, since full-scale tests indicate that fire hazard from smoke is associated with its appearance early in the fire time scale, the rate of generation of smoke is also an important consideration. It is now agreed that the major fire hazards presented by visible smoke are fear and a slowing down in the rate of escape from the building where the fire is taking place.

1.3.2   Effect of polymer structure on smoke formation

The structure of a polymer influences both flammability and smoke formation.[28] Polymers with aliphatic backbones, or those that are largely aliphatic and oxygenated, have a tendency toward low smoke generation, while polyenic polymers and those with pendant aromatic groups generally produce more smoke. There are two important exceptions to this simple rule, the saturated polymers poly(vinylchloride) and poly(vinylidene chloride). In poly(vinylchloride), dehydrohalogenation gives a polyenic structure which cyclises to form aromatic products, while poly(vinylidene chloride) forms more char and gives different volatile pyrolyzates. Polymers with high thermal stability or which form small amounts of flammable pyrolyzates generally produce little visible smoke. Increasing char formation is one way of minimising the yield of pyrolyzates and hence smoke reduction.

Structural factors in smoke generation are thus important, insofar as they contribute to the inherent stability of the polymer and largely determine the nature of the pyrolyzates which form the combustion fuels. The following much simplified generalisations about structural factors can be made:

- Aromatic and polyenic polymers have greater tendencies to produce smoke than aliphatic or oxygenated polymers.
- Polymers with aromatic units in the backbone have lower tendencies to reduce smoke than polymers with pendant aromatic groups.
- Halogenation to low or intermediate levels tends to increase the amount of smoke formed, but highly halogenated polymers have reduced smoking tendencies.
- The amount of smoke produced is related to the types of fuels formed on degradation and to the degree of thermal stability of the polymers.

Many other factors such as sample size, sample orientation, ventilation, heat flux etc. also contribute to the amount of smoke formed.

1.3.3 Smoke suppressant technology

The combustion of polymers involves a variety of processes (both physical and chemical) occurring in several phases. Thus, polymer melting and degradation, heat transfer in both solid and liquid phases and diffusion of the breakdown products through the degrading polymer into the gas phase accompany the various combustion reactions which occur. As a result, polymers and polymer formulations can be modified so that additive smoke-suppressing compounds are effective in reducing smoke during burning. Several of the smoke suppressant additives known to be effective in burner fuels are also effective in polymers. Approaches used for reducing smoke during burning have included the use of fillers, additives, surface treatments, and structural modification of the polymers themselves. Certain chemical reactions occurring during combustion processes affect the generation of visible smoke. These include the following:

Gas phase:
- oxidation and hindered nucleation of soot
- flocculation and growth of soot particles
- dilution of the fuel
- modification of the type and composition of pyrolyzates
- modification of flame temperatures, etc.

Solid and liquid phases
- dilution of combustible polymer content
- dissipation of heat
- surface insulation and protection of combustible substrate
- promotion of char formation
- alteration of pyrolysis reactions and energy flow
- reduction of polymer mass burning rate, etc.

1.3.4 Smoke suppressants for polymers

1.3.4.1 Fillers

Fillers are non-polymeric compounding materials used at concentrations greater than about 20% of the polymer mass and often at concentrations as high as 40%. Two classes of fillers, based on their apparent smoke suppressant functions are known: they are 'inert' and 'active' smoke suppressant fillers.

'Inert' fillers reduce the amount of smoke generated from a given mass or volume of a polymer simply by diluting or reducing the amount of combustible substrate present and also by absorbing heat to reduce the burning rate. Examples of such fillers are silica (SiO_2), clays, $CaCO_3$, and carbon black. It is possible that a filler may be inert in one polymer system, but active in another. For example, $CaCO_3$ often remains unchanged during the combustion of polypropylene, but will react with HCl formed during combustion of chlorinated polymers. Inert fillers usually give only marginal improvements in flame retardancy, unless present in very high concentrations.

'Active' fillers promote the same diluent and heat absorption functions as inert fillers, but they absorb more heat per unit weight by endothermic processes. Gases such as water, carbon dioxide, or ammonia, released during heating may also dilute the fuel volatiles and modify flame reactions. Examples of currently used active fillers include $Al(OH)_3$,[29,30] basic magnesium carbonate, $Mg(OH)_2$[31,32] and talcs. In these the water of hydration and/or carbon dioxide is released at temperatures approximating to those of polymer decomposition, producing both flame-retardant and smoke-suppressant effects.

The physical properties of polymers are often badly affected by the addition of fillers. High-impact resistant polymers, such as ABS, suffer massive reductions in impact strength in the presence of fillers, often making filler treatments impractical in some applications. Loss of impact and tensile properties in other polymers can be somewhat reduced by the addition of impact modifiers.

1.3.4.2 Additive compounds

Additives are generally non-polymeric compounding substances used at levels below about 20% of the polymer resin. A number of different types of additives have been shown or are claimed to have smoke-suppressant properties. By far the largest class of smoke suppressants are metal compounds (mainly oxides or hydroxides) used in poly(vinylchloride) (PVC) and other halogen-containing polymers. Non-metallic additives include some dicarboxylic acids, sulphur, and various plasticisers and surfactants.

A brief discussion of some of the more effective, and most frequently used metal based additives concludes this section.

Antimony compounds. Antimony oxides, e.g. Sb_2O_3 and Sb_2O_5, although widely used flame-retarding additives in halogen-containing polymers such as PVC, are not considered as smoke suppressants because their effect on smoke production is variable. Some commercial modified antimony oxides have been introduced which claim to reduce smoke generation in PVC. Antimony(V) oxide is more effective in reducing smoke than antimony(III) oxide.[33]

Iron compounds. Ferrocene (dicyclopentadienyl iron) was one of the early additives for which synergistic flame retardancy and smoke suppression was claimed.[34,35] Other, less volatile, organo-iron compounds have been investigated as potential smoke-suppressants for PVC (rigid and plasticised).[36] Basic iron(III) oxide (FeOOH) is especially effective and has been shown to have excellent smoke suppressing effects in a wide range of chlorine containing polymers and blends of these with other polymers, especially ABS.[37] Recently FeOCl has been identified as the active char forming/smoke suppressing compound formed by reaction of FeOOH with HCl at low temperatures, while iron(III) oxide and iron(III) chloride are formed at higher temperatures. Iron(III) and ferricenium chlorides have also been shown to suppress the formation of benzene and smoke in PVC and CPVC.[38]

Carty and White have investigated the reactivity of basic iron(III) oxide (together with other flame retarding/smoke-suppressing compounds) in a wide range of blends of PVC and CPVC with ABS and other polymers and plasticisers.[39] In all the polymer blends examined, it was shown that FeOOH and its synergistic interaction with HCl had the most significant smoke-suppressing activity among the compounds studied. A mechanism showing how FeOOH/HCl reduces smoke by increasing char formation has been proposed.[40] Detailed mechanistic studies using FeOOH in PVC (rigid and plasticised), PVC/ABS, CPVC/ABS and CPVC plasticised have recently been reported.[33,41] The flame-retarding activity of FeOOH/PVC/CPVC combinations is not yet fully understood.

Molybdenum compounds. Molybdenum compounds have been used in flame-retardant treatments for some time.[42] By using MoO_3 as a partial replacement for Sb_2O_3, smoke and flammability properties can be optimised for PVC. Other molybdenum compounds have been reported to give smoke suppressant effects in polymers other than PVC, but halogen is a necessary co-ingredient.

Zinc compounds. Zinc compounds have also been shown to impart smoke-suppressant characteristics, largely in chlorinated polymers. Smoke-suppressant effects appear to be optimum at 1 phr or lower. Synergistic flame-retardant and smoke-suppressant effects are found in PVC formulations when used in combination with antimony oxide. Some examples of synergistic effects on flammability and smoke generation have also been recognised with other zinc compounds such as zinc borate in chlorine-containing polymers.

Mixed-metal additives. In most of the examples discussed above, metal compounds (mainly oxides) are added alone or in combination with antimony oxide or alumina trihydrate. However, strong smoke-suppressant effects have been achieved in a number of cases by combining two or more metal oxides in an additive system. Kroenke has reviewed the smoke suppressing activity of a wide range of metal compounds (mainly oxides) separately and in combination.[43] The first examples of this were in patents claiming the use of iron powder in combination with copper(I) oxide or molybdenum trioxide.[44]

Zinc and molybdenum oxides form a particularly effective and interesting two-metal smoke-suppressant system for PVC. Carty and White have also shown that combinations of FeOOH and Sb_2O_3 and Sb_2O_5 have excellent flame-retarding/smoke-suppressing effects in PVC, CPVC and blends of these with ABS.

A series of two-metal additive combinations for rigid PVC which significantly reduce smoke generation has been reported and in some cases smoke reductions in the order of 60–100% are obtained. Synergism has been shown to occur in some of these systems. The components may suppress smoke individually and some are synergistic when used in combination. There is, as yet, no detailed explanation for the chemical behaviour of these mixed-metal systems. However, all these metal additives were found to increase char, and additive concentration is a particularly important factor with MoO_3, ZnO, and FeOOH.

It should be noted that some metal compounds which have smoke-suppressant activity have been used for some time in polymer formulations, although for different reasons. Stabilisers for PVC based on Ba, Cd and Zn compounds are well known, and tin-based compounds have also been used as stabilisers in PVC systems. Interference with compounding ingredients may also reduce the effects of some smoke-suppressant additives. For instance, ferrocene is reportedly ineffective as a smoke suppressant in PVC plasticised with phosphate plasticisers.[34]

There is current interest in developing smoke suppressants based on reductive-coupling chemistry using mainly Cu(I) compounds[41] and in the use of nanocomposite systems (see Chapter 5) which produce robust chars with improved barrier and structural properties.[45]

1.4 Conclusions

This chapter has provided a concise account of the phenomenon of polymer combustion and flame retardance. Because polymer combustion is always preceded by decomposition of the solid polymer, condensed-phase processes were considered in more depth. This aspect is particularly relevant to the requirement to develop environmentally friendly flame retardant systems. In this respect, char formation is particularly important. Smoke generation is a major hazard in the event of a fire involving polymer materials. Methods for smoke suppression were also considered.

For a more in depth treatment of the subject the reader is referred to the bibliography.

1.5 Bibliography

The Combustion of Organic Polymers, C F Cullis and M M Hirschler, Clarendon Press, Oxford, 1981: ISBN 0-19-851351-8.

International Plastics Flammability Handbook, J Troitzsch, 2ⁿᵈ edition, Hanser Publishers, Munich, 1990: ISBN 3-446-15156-7 and 0-19-520797-1; 3ʳᵈ edition, J Troitzsch, S Bourbigot and M le Bras, eds, Hanser Publishers, New York, in press, due 2001.

Heat Release in Fires, V Brabauskas and S J Grayson (eds), Elsevier Applied Science, London (1992).

Polymer Degradation & Stabilisation, N Grassie and G Scott, Cambridge University Press, 1985: ISBN 0-521-249619.

R M Fristrom, *J. Fire & Flamm.*, 5 (1974) 289–320.

F J Martin, *Combustion and Flame*, 12 (1968) 125–135.

M Elomaa, L Sarvaranta, E Mikkola, R Kallonen, A Zitting, C A P Zevenhoven and M Hupa, *Critical Reviews in Biochem. and Molec. Biol.*, 27 (1997) 137–197.

Various articles in *Chemistry in Britain*, (1987) 213–249 and (1998) 20–24.

References

1 UK Fire Statistics, Home Office, London, 1998.

2 Hirschler M M in *Heat Release in Fires*. (eds Brabauskas, V and Grayson, S J), Elsevier Applied Science, London (1992) pp 375–8.

3 Lyons J W, 'FIRE', Scientific American Box, New York (1985).

4 Troitzsch J, *International Plastics Flammability Handbook, Principles – Regulations-Testing and Approval*. (2nd ed), Hanser Publications, Munich (1990).

5 Schmidt W G, *Inst. Plastics Ind., Trans.*, (1965) 12, p 247.

6 Atika K, in H H G Jellinek (ed) *Aspects of degradation and stabilisation of polymers*, Chap. 10, p 514, Elsevier Scientific Publishing Co., Amsterdam, 1978.

7 *Handbook of Chemistry and Physics*, 73rd edition, edited by David R Lide, 1993.

8 Carniti P, Beltrame P L, Armada M, Gervasini A and Audisio G, Ind. Eng. Chem. Res. 30, (1991) 1624; A. Goyot, *Polym Degrad Stab*, 15 (1986) 219.

9 Anthony G M, *Polym Degrad Stab*, 64 (1999) 353.

10 Montaudo G, Puglisi C and Samperi F, *Polymer Degradation and Stability*, 26 (1989) 285.

11 Ballistreri A, Montaudo G, Scamporrino E, Puglisi C, Vitalini D and Cucinella S, *J. Polym. Sci. A*, 26 (1988) 2113.

12 McNeill I C and Rincon A, *Polym Degrad Stab*, 31 (1991) 163.

13 Jachowicz J, Kryszewski M and Kowalski P, *J. App Polym. Sci.*, 22 (1978) 2891.

14 Factor A, *in: Fire and polymers* ACS Symp. Ser. 425 (1990) 274.

15 Bourbigot S, le Bras M, Delobel R, Breant P and Tremillon J M, *Carbon*, 33 (1995) 283.

16 Levchik S V and Wilkie CA, Chapter 6 in: Fire retardancy of polymeric materials, A F Grand and C A Wilkie (eds), Marcel Dekker Inc, New York, 2000, 171–215.

17 Gibov K M, Shapovalova L N and Zhubanov B A, *Fire Mater.*, 10 (1986) 133.

18 Hirschler M M (Editor in Chief), *Carbon Monoxide and Human Lethality: Fire and Non-Fire Studies*, Elsevier Applied Science, London (1993).

19 Calcote H F, *Comb Flame*, 42 (1981) 215.

20 See ref. 2, pp 233–50.

21 Ahlstrom D A, Liebman S A and Quinn E J, *Poly. Prep. Amer. Chem. Soc. Div. Poly. Chem.*, 14 (1973) 1025.

22 Calcraft A M, Green R J S and McRoberts J S, *Plast. Polym.*, 42 (1974) 200.

23 Edgerley R G and Pettett K, *Fire Mat*, 2 (1978) 11.

24 Lawson D F and Kay E L, *J. Fire Ret Chem.*, 2 (1975) 132.

25 ref. 2, pp 375–442.

26 Carty P and White S, *Engineering Plastics*, 8 (1995) 287.

27 Costa L, Camino G, Bertelli G and Borsini G, *Fire and Materials*, 19 (1995) 133.

28 van Krevelen D W, *Polymer*, 16 (1975) 615.

29 Brown S C and Herbert M J, '*Flame Retardants 92*' Conference Paper pp 100–19, Elsevier Applied Science, London (1992).

30 Antia F K, Cullis C F and Hirschler M M, *Eur Polym. J.*, 18 (1982) 95.

31 Hornsby P, 'Inorganic Fire Retardants – All Change', RSC Conference Paper, Fire Retardancy and Smoke Suppression in Polymers Modified with Magnesium Hydroxide, London (1993).

32 Lawson D F, Kay E L and Roberts D J, *Rubber Chem. Technology*, 48 (1975) 124.

33 White S, PhD Thesis, University of Northumbria (1998).

34 Kracklauer J J and Sparks C J, *Plast. Eng.*, 11 (1974) 57.

35 Lawson D F, *J. Appl. Poly. Sci.*, 20 (1976) 2183.

36 Carty P and White S, *Appl. Organometallic Chem.*, 10 (1996) 101.

37 Carty P, White P and S, *Polymer Networks and Blends*, 5 (1995) 205.

38 Carty P and White S, *Polymer*, 38 (1997) 1111.

39 Carty P and White S, *Polymer Networks and Blends*, 7 (1997) 121.

40 Carty P and White S, *Polymers and Composites*, 6 (1998) 33–8.

41 Starnes W H Jr. and Huang O, Proc. 2nd. Beijing Int. Symposium on Fire Retardants, 1–5, (1993).

42 Starnes W H, Wescott L D, Reents W D, Cais R E, Vollacorta G M, Plitz I M and Anthony L J in *Polymer Additives* (ed. J E Cresta), Plenum Press, New York (1984) pp 237–48.
43 Kroenke W J, *J. Appl. Poly. Sci.*, 26 (1981) 1167.
44 Mitchell L C, US Patent 3845001 (1974); US Patent 3903028 (1975); US Patent 3870679 (1975).
45 Porter P, Metcalfe E and Thomas M J K, *Fire Mat*, 24 (2000) 45–52.

2

Mechanisms and modes of action in flame retardancy of polymers

MENACHEM LEWIN AND
EDWARD D WEIL

Polymer Research Institute
Polytechnic University, Brooklyn, New York

2.1 Introduction

Some basic mechanisms of flame retardancy were recognised as early as 1947 when several primary principles were put forward.[1] These included the effect of the additive on the mode of the thermal degradation of the polymer in order to produce fuel-poor pyrolytic paths, external flame retardant coatings to exclude oxygen from the surface of the polymer, internal barrier formation to prevent evolution of combustible gases, inert gas evolution to dilute fuel formed in pyrolysis and dissipation of heat away from the flame front. Discovery of the flame-inhibiting effect of volatile halogen derivatives subsequently led to the postulation of the radical trap-gas-phase mechanism.[2] The gas-phase and the condensed-phase proposals have long been generally considered as the primary, though not the only, effective mechanism of flame retardancy. This situation is now being modified as new mechanisms of new flame-retarding systems, especially those based on physical principles, evolve and as new insights into the performance of flame retardants is being gained. In many cases several mechanistic principles operate simultaneously and consequently it is difficult to identify one dominant mechanism. In such cases *modes of action* of particular flame-retarding formulation may be defined and described.

This paper attempts to review some of the principles, mechanisms and modes of action which prevail at present in the field of flame retardancy of polymers.

2.2 General considerations

Pyrolysis and combustion of polymers occur in several stages. The polymeric substrate heated by an external heat source is pyrolysed with the generation of combustible fuel. Usually, only a part of this fuel is fully combusted in the flame by combining with the stoichiometric amount of atmospheric oxygen. The other part remains and can be combusted by drastic means, e.g. in the presence of a catalyst and by an excess of oxygen. A part of the released heat is fed back to the substrate and causes its continued pyrolysis, perpetuating the combustion cycle. Another part is lost to the environment. The energy needed to heat the polymer to the pyrolysis temperature and to decompose and gasify or volatilise the combustibles and the amount and character of the gaseous products determines the flammability of the substrate. A flame retardant acting via a condensed-phase chemical mechanism alters the pyrolytic path of the substrate and reduces substantially the amount of gaseous combustibles, usually by favouring the formation of carbonaceous char and water.[3] In this case the heat released in the combustion decreases with an increase in the amount of the flame-retarding agent.

In the gas-phase mechanism the amount of combustible matter remains constant but the heat released in the combustion usually decreases with an increase in the amount of the flame-retarding agent. The amount of heat returned to the polymer surface is therefore also diminished and the pyrolysis is retarded or halted as the temperature of the surface decreases. The flame-retarding moiety has to be volatile and reach the flame in the gaseous form. Alternatively it has to decompose and furnish the active fraction of its molecule to the gaseous phase. The char remaining in the substrate will contain less of the active agent. The pyrolysis of the polymer should, in the limiting case, proceed as if there would have been no flame-retarding agent incorporated in it. In addition presence of the gas-phase active agent should not influence the composition of the volatiles reaching the flame.[3]

2.3   Gas-phase mechanisms

The gas-phase activity of the active flame retardant consists in its interference in the combustion train of the polymer. Polymers, like other fuels, produce upon pyrolysis species capable of reaction with atmospheric oxygen and produce the H_2-O_2 scheme which propagates the fuel combustion by the branching reaction:[4]

$$H^\bullet + O_2 = OH^\bullet + O^\bullet \tag{2.1}$$

$$O^\bullet + H_2 = OH^\bullet + H^\bullet \tag{2.2}$$

The main exothermic reaction which provides most of the energy maintaining the flame, is:

$$OH^\bullet + CO = CO_2 + H^\bullet \tag{2.3}$$

To slow down or stop the combustion, it is imperative to hinder the chain-branching reactions [2.1] and [2.2]. The inhibiting effects of halogen derivatives, usually chlorine and bromine, is considered to operate via the gas-phase mechanism. This effect in the first instance occurs either by releasing a halogen atom, if the flame-retardant molecule does not contain hydrogen, or by releasing a hydrogen halide:

$$MX = M^\bullet + X^\bullet \qquad\qquad [2.4]$$

$$MX = HX + M^\bullet \qquad\qquad [2.5]$$

where $M^\bullet$ is the residue of the flame-retardant molecule. The halogen atom reacts with the fuel, producing hydrogen halide:

$$RH + X^\bullet = HX + R^\bullet \qquad\qquad [2.6]$$

The hydrogen halide is believed to be the actual flame inhibitor by affecting the chain branching:

$$H^\bullet + HX = H_2 + X^\bullet \qquad\qquad [2.7]$$

$$OH^\bullet + HX = H_2O + X^\bullet \qquad\qquad [2.8]$$

Reaction [2.7] was found to be about twice as fast as [2.8] and the high value of the ratio H_2/OH in the flame front indicates that [2.7] is the main inhibiting reaction.[5] It is believed that the competition between reactions [2.7] and [2.1] determines the inhibiting effect. Reaction [2.1] produces two free radicals for each H atom consumed, whereas reaction [2.7] produces one halogen radical which recombines to become the relatively stable halogen molecule.

2.3.1 Comparing flame-retardant activity of halogen derivatives

Equation [2.7] represents an equilibrium with a forward reaction and a reverse reaction. The equilibrium constants of equation [2.7] for HBr and HCl are:[6]

$$K_{HCl} = 0.583\exp(1097/RT); \quad K_{HBr} = 0.374\exp(16760/RT$$

The equilibrium constants decrease strongly with increase in temperature, which explains the decreasing effectivity of halogen derivatives in large hot fires.[6] Petrella[5] calculated that in the temperature range 500–1500 K the forward reaction predominates and K_{HBr} is much higher than K_{HCl}. Both are highly effective at the ignition temperature range of polymers. The flame-retardant effectivity of the halogens was stated to be directly proportional to their atomic weights, i.e. $F:Cl:Br:I = 1.0:1.9:4.2:6.7$.[7] On a volumetric basis 13% of bromine was found to be as effective as 22% of chlorine when comparing the tetrahalophthalic anhydrides as flame retardants for polyesters.[8,9] A similar effect was found for PP, PS and PAN[3] and when comparing NH_4Cl to NH_4Br in cellulose.[9]

The activity of the halogens is also strongly affected by the strength of the respective carbon-halogen bonds. The low bond strength of I–C and consequently the low stability of the iodine compounds virtually exclude their use. The high stability of the fluorine derivatives and the high reactivity of the fluorine atoms in reactions [2.7] and [2.8] will prevent the radical quenching processes in the flame. The lower bond strength and stability of the aliphatic compounds, their greater ease of dissociation as well as the lower temperature and earlier formation of the HBr molecules are responsible for their higher effectivity as compared to the aromatic halogen compounds. The higher stability of the latter along with their higher volatility allow these compounds to evaporate before they can decompose and furnish the halogen to the flame.

2.3.2 Physical modes of action of halogenated flame retardants

The radical trap activity is not the only activity of the halogenated flame retardants. The physical factors such as the density and mass of the halogen and its heat capacity, have a profound influence on the flame-retarding activity of the agent. In addition, its dilution of the flame which thus decreases the mass concentration of combustible gases present are effective. Larsen[7,199,200] demonstrated the important role of the heat capacity of the flame retardant. In flame retardant polymer systems the halogens appear to work by reducing the heat evolved in the combustion of the gases given off by the decomposing polymer (low or zero fuel value plus action as a heat sink) such that to sustain burning the mass rate of gasification must be increased by the application of an increased external heat flux.[10] Other authors[11] showed by thermochemical computation that most of the action of a wide variety of halocarbon flame inhibitors could be correlated to a combination of heat capacity and endothermic bond dissociation.

A physical effect, often mentioned but rarely demonstrated or evaluated, is the 'blanketing' effect of excluding oxygen from the surface of the pyrolysing polymer. Ignition generally takes place in the vapour phase adjacent to the condensed phase, when an ignitable fuel–air mixture is reached. There is, however, evidence that the rate of pyrolysis may be affected by the oxygen getting to or into the condensed phase, and that in polyolefins surface oxidation may provide energy for pyrolysis.[12] The rate of isothermal pyrolysis of cellulose was found to be higher in the presence of air as compared to vacuum pyrolysis by an order of magnitude.[13,14] The rate of pyrolysis in the presence of air was also found to decrease linearily with increase in orientation of rayon fibres. Increase in chain orientation brings about a decrease in distance between chains and their more compact packing with consequent decreased penetrability and rate of diffusion of oxygen into the polymer resulting in a decreased rate of pyrolysis.[13,14] Although some doubt has been cast on the significance of the blanketing effect,[15] it is self-evident that an outgoing stream of bulky halogen and other non-fuel molecules emitted from the pyrolysing polymer could retard the penetration of the oxygen into the polymer and slow down the pyrolysis.

The proponents of the *physical theory* of the flame-retardant activity of halogenated additives compare the halogen activity to that of inert gases, CO_2 and water.[7] The physical theory takes into consideration the basic parameters of the flame as well as the processes occurring in the solid phase leading to the production of the combustibles, and enables in certain cases an estimate to be made of the amount of flame-retardant agent needed to inhibit a flame. There appears to be no contradiction between the radical trap theory and the physical theory with regard to halogens. Both approaches complement each other. It is difficult to determine in a general way the relative contribution of each of the two modes of activity. This will usually depend on the structure and properties of the polymer and of the flame retardant as well as on the conditions and parameters of the flame and on the size of samples used.

2.4   The condensed-phase mechanism

The condensed-phase mechanism postulates a chemical interaction between the flame-retardant agent, which is usually added in substantial amounts, and the polymer. This interaction occurs at temperatures lower than those of the pyrolytic decomposition. Two principal modes of this interaction were suggested: dehydration and cross-linking. They have been established for a number of polymers including cellulosics and synthetics.[16,17,202]

2.4.1 Principal modes of the condensed phase mechanism: dehydration and char formation

The varying effectivity of phosphorus compounds in different polymers has been related to the polymers susceptibility to dehydration and char formation: this explains the decreasing activity with decreasing oxygen content of the polymer. Whereas cellulosics are adequately flame retarded with around 2% of phosphorus, 5–15% of it is needed for polyolefins.[18] The interaction of phosphorus derivatives with the polymers not containing hydroxyls is slow and has to be preceded by an oxidation. It has been suggested that 50–99% of the phosphorus derivatives may be lost by evaporation, possibly of P_2O_5 or other oxides formed in the pyrolysis of the phosphorus derivatives.[19]

Two alternative mechanisms have been proposed for the condensed phase in cellulose: dehydration of cellulosics with acids and acid-forming agents of phosphorus and sulphur derivatives. Both mechanisms lead to char formation.[20] (a) esterification and subsequent pyrolytic ester decomposition (see Scheme 2.1) and (b) carbonium ion catalysis (Scheme 2.2):

$$R_2CH\text{–}CHR'OH + ZOH(acid) \rightarrow R_2CH\text{–}CHR'OZ + H_2O$$
$$\rightarrow R_2C\text{=}CHR' + ZOH(\text{where } Z = \text{acyl radical of the acid})$$

(Scheme 2.1)

$$R_2CH\text{–}CHR'OH \rightarrow R_2CH\text{–}CHR'OH_2^+ \rightarrow H_2O$$
$$+ R_2CH\text{–}C^+HR'$$

(Scheme 2.2)

Differential thermal analysis (DTA) and oxygen index (OI) data indicated that phosphorus compounds reduce the flammability of cellulosics primarily by the Scheme 2.1 mechanism, which, being relatively slow, is affected by the fine structure of the polymer. Less-ordered regions (LOR) pyrolyse at a lower temperature than the crystalline regions and decompose before all of the phosphate ester can decompose, which decreases the flame-retarding effectivity and necessitates a higher amount of phosphorus. Sulphated celluloses, obtained by sulphation with ammonium sulphamate, are dehydrated by carbonium ion disproportionment (Scheme 2.2) and show a strong acid activity which rapidly decrystallises and hydrolyses the crystalline regions. The fire-retardant activity was accordingly found not to be greatly influenced by the fine structural parameters, and the same amount of sulphur was needed to flame-retard celluloses of different crystallinities.[20]

2.4.2   Cross-linking and char formation

It was early recognised that cross-linking promotes char formation in pyrolysis of celluloses.[21] Cross-linking has been assumed to be operative in P–N synergism.[22] Cross-linking reduces in many cases, albeit not always, the flammability of polymers. Although it increases the OI of phenolics, it does not markedly alter the flammability of epoxides.[23] A drastic increase in char formation is observed when comparing cross-linked polystyrene (PS), obtained by copolymerising it with vinylbenzyl chloride, to uncross-linked PS. PS pyrolyses predominantly to monomer and dimer units almost without char. Cross-linked PS yielded 47% of char.[24] Cross-linking and char formation were recently obtained by an oxidative addition of organometallics to polyester.[25]

Cross-linking promotes the stabilisation of the structure of cellulose by providing additional covalent bonds between the chains, which are stronger than the hydrogen bonds, and which have to be broken before the stepwise degradation of the chain occurs on pyrolysis. However, low degrees of cross-linking can decrease the thermal stability by increasing the distance between the individual chains and consequently weakening and breaking the hydrogen bonds. Thus, although the OI of cotton increases marginally with increasing formaldehyde cross-linking, that of rayon markedly decreases.[26]

The formation of char in celluloses is initiated by rapid *auto-crosslinking* due to the formation of ether oxygen bridges formed from hydroxyl groups on adjacent chains. The auto-crosslinking is evidenced by a rapid initial weight loss, due to evolution of water, in the first stage of pyrolysis at 251 °C, and is linearly related to the amount of char. Formaldehyde cross-linking of rayon interferes with the auto-crosslinking reaction, decreases the initial weight loss and reduces char formation.[26]

It was suggested that cross-linking may increase the viscosity of the molten polymer in the combustion zone, thereby lowering the rate of transport of the combustible pyrolysis products to the flame.[27]

2.4.3 Structural parameters

In addition to bond strength and intermolecular forces, there are several other parameters, such as chain rigidity, resonance stability, aromaticity, crystallinity and orientation, that have a pronounced influence on pyrolysis and combustion. The linear correlations of van Krevelen between OI and char and between the char-forming tendency (CFT) and char residue (CR), are well known.[3] The CFT (equation 2.9) is defined as the amount of residue at 850 °C per structural unit, divided by 12, i.e. the amount of C equivalents per structural unit, where each group has its own CFT. These equations hold only for untreated polymers and for polymers containing condensed-phase flame retardants. They do not hold if halogen is present.[3]

$$CR = 1200 \left\{ \sum_i (CFT)_i \right\} \Big/ M \qquad [2.9]$$

where M is the molecular weight per structural unit

Recent work on the relationship between chemical structures and pyrolysis and on the effects of introducing substituent functionalities into aromatic and heterocyclic structures on the modes of pyrolysis has been reviewed by Pearce.[28,29]

An interesting attempt to develop a generalised kinetic model of polymer pyrolysis was recently made by Lyon.[30] The model is based on some of the mechanisms important in the burning process, i.e. generation of combustible gases and char formation, but can be solved for the overall mass loss history of the specimen; for verification, special thermogravimetric techniques can be used.

2.4.4 Fine structural parameters and pyrolysis of polymer blends

In addition to orientation, crystallinity and degree of polymerization (DP) also have a strong influence on the energy required to melt and degrade polymers, on the rates of vacuum and air pyrolysis and on char formation and yield. That the DP has an effect on the degradation temperature of various polymers is known. Vacuum pyrolysates of purified celluloses were found to increase with increasing orientation and less-ordered regions (LOR) and to be inversely proportional to the square root of the DP.[13,14] The decrease in thermal stability with increasing orientation was ascribed to the straining of the hydrogen bonds. The extent of the auto-crosslinking reaction, discussed earlier, was found to be directly proportional to the percentage of char. The char increases with the increase in LOR of the polymer.

The energy of activation of pyrolysis of cellulose was found to increase strongly with the increase in crystallinity, indicating different mechanisms operating for the crystalline and less-ordered regions.[14]

Little is known of the effect of the fine structural parameters on the pyrolytic behaviour of polymers other than cellulose. The inclusion of these parameters in mechanistic models might prove to be of considerable interest. One such area might be the pyrolysis and flame retardancy of blends, as evidenced in the case of cotton-wool blends.[31] The DSC endotherm of cotton at 350 °C, which is due to the decomposition of the levoglucosan monomer formed on pyrolysis, disappears with the addition of relatively small amounts of wool. Since levoglucosan is formed from the crystalline regions of the cellulose, its disappearance was attributed to the swelling decrystallization of the amino derivatives formed in the pyrolysis of the wool, which occurred at 225 °C, i.e. lower than the 300 °C at which the cotton pyrolysis begins. This is also manifested by a 'synergism' in char production. There is a strong increase in char in these blends, beyond the char amounts predicted by the composition of the blend. This rise stems from the increase in the LOR due to the swelling. Consistent with the above is also the decrease in activation energy of the pyrolysis from 220.1 for cotton to $103.4 \, \text{kJ mol}^{-1}$ for the blend with 18% of wool. It is important to note that the above interaction between the ingredients of the blend is physicochemical in nature and depends on temperature. Pyrolysis-gas chromatography of a series of wool–cotton blends at 1000 °C for 30 s yielded all the peaks in the relative area ratios as expected from a simple additive calculation in the absence of any interaction. The degree of interaction of components in a blend is therefore to be considered as a kinetic process governed by temperature and time.[31] The increase in char does not result in improved flammability. Actually more additive is needed for the blend than for the individual components.[31] A similar situation exists in the case of cotton–polyester blends. In this case more flammable gases, such as ethylene, are evolved from the blend than from the individual components.[9,16]

2.5   Modes of action of halogen-based flame retardants: synergistic systems

Halogen derivatives are used as a rule together with co-additives enhancing their flame inhibiting activity. These co-additives are usually termed *synergists*. There is a considerable number of such co-additives, the most prominent one being antimony trioxide. Their effects are based on widely differing modes of action, embracing both the radical trap and physical effects mechanisms as well as principles of the condensed-phase mechanism and intumescence. The differences between the various co-additive-synergistic systems are not only in the modes of action but also in the extent of the synergistic effect.

The term *synergism*, as currently used in the FR-terminology is poorly defined. Strictly speaking, it refers to the combined effect of two or more additives, which is greater than that predicted on the basis of the additivity of the effect of the components. In order to characterise and compare synergistic systems we introduced the term *synergistic effectivity* (SE), used in this and previous publications of this laboratory.[31-34,201] It is defined as the ratio of the FR-effectivity (EFF) of the flame-retardant additive plus the synergist to the EFF of the additive without synergist, compared at the same additive level. EFF is defined as the increment in OI for 1% of the flame-retardant element, at a given level of the flame-retardant element. The values of EFF and SE cited in this paper were computed from results of work in this laboratory as well as from data published in the literature, and were tabulated and published in previous publications.[35] The SE values are based in most cases on results obtained for additive/synergist ratios yielding the highest effect. More general mathematical definitions of synergism have been proposed.[53]

2.5.1 Halogen–antimony synergism

Formulations based on the halogen–antimony synergism are being widely used and have been described for a variety of polymers: cellulosics,[1] polyesters, polyamides, polyolefins, polyurethanes, polyacrylonitrile and polystyrene.[24]

The mode of action responsible for this synergism appears to depend both on condensed-phase as well as vapour-phase activities.[36,37] It is believed that during the pyrolysis, first, some hydrogen halide is released in the self-decomposition of the halogenated compound or by interaction with antimony trioxide and/or with the polymer. The HX reacts with the Sb_2O_3 producing either SbX_3 or $SbOX$.[36,37] Although some SbX_3 is found in the first stage of the pyrolysis, the weight loss pattern found in one study implied the formation of less volatile Sb-containing moieties, obtained by progressive halogenation of Sb_2O_3.[38] During the transformations gaseous SbX_3 is evolved and released to the gas phase, whereas SbOX, which is a strong Lewis acid, may operate in the condensed phase, facilitating the dissociation of the C-X bonds.[39]

Several cases of condensed-phase activity of Sb_2O_3 are known. Lowering the charring temperature by adding Sb_2O_3 to cellulosic fabrics treated with chlorine compounds has been observed.[16] Adding Sb_2O_3 to polyolefins treated with Dechlorane Plus was found to increase the char yield substantially.[3]

The main effect of Sb_2O_3 is, however, in the gas phase. The antimony halides, after reaching the gas phase, react with atomic hydrogen producing HX, SbX, SbX_2, and Sb. Antimony reacts with atomic oxygen, water and hydroxyl radicals, producing SbOH and SbO, which in turn scavenge H atoms. SbX_3 reacts with water, producing SbOH and HX. A fine dispersion of solid SbO and Sb are also produced in the flame and catalyse the recombination of H•. In addition, it is believed that the antimony halides delay the escape of halogen from the flame, and thus increase its concentration, and at the same time also dilute the flame. The antimony halides may also exert a 'blanketing' effect by hindering the penetration of oxygen into the pyrolysing polymer.[13,14]

Values of EFF and SE of aromatic and aliphatic bromine derivatives with antimony trioxide, computed from data of van Krevelen,[3] show SE of 2.2 and 4.3, respectively. Similarly, for aliphatic chlorine derivatives with antimony trioxide an SE value of 2.2 is computed for polystyrene.

2.5.2 Mode of action of ammonium bromide

Ammonium bromide was recently found to have a high FR-effectivity of bromine, i.e. 1.24 for NH_4Br encapsulated in PP as compared to 0.6 for aliphatic bromine compounds. It has been explained by the low dissociation energy of NH_4Br to HBr and NH_3 which is much lower than the dissociation energy of the C–Br bond. The degree of dissociation is 38.7% at 320 °C, so that sizable amounts of HBr are readily available when PP begins to decompose. The radical trap activity of the HBr as well as the physical effects exerted by the HBr and the ammonia clearly operate here simultaneously. The possibility of synergism between the HBr and NH_3 in the gaseous phase should, however, not be discarded, as both compounds reach the flame at about the same time. Little is known about the behaviour of ammonia in the flame, particularly in the presence of H•, Br•, OH•, and O• radicals.

2.5.3 The mode of action of mixtures of bromine-based and chlorine-based additives

Attention has recently been drawn to the enhancement in the flame retardancy when mixtures of brominated and chlorinated flame retardants are applied to ABS, HIPS and PP and a synergistic interaction was postulated and discussed in several papers.[40–44] In most cases the maximum effect is found with a Br:Cl ratio of 1:1 and with 10–12% of the sum of chlorine and bromine. When using Dechlorane Plus and brominated epoxy resin (51% Br) with ABS in the presence of 5% Sb_2O_3, the FR-EFF was calculated as 0.80. The SE obtained was 1.67. This synergism is *in addition* to the Hal–Sb synergism and is known until now to be effective only in the presence of antimony trioxide. The synergistic effect increases with the amount of antimony and reaches a maximum at 6% level.

Some light was thrown on the Br–Cl synergism in pyrolysis experiments, carried out in the ion source cell of a mass spectrometer with mixtures of polyvinyl bromide (PVB) and polyvinyl chloride (PVC) or polyvinylidene chloride (PVDC).[45] Whereas relatively high concentrations of both HCl and HBr were found, the amounts of $SbCl_3$ were very low compared to those of $SbBr_3$, which points to a much slower interaction of HCl with the oxide than that of HBr, which is not surprising, considering the higher stability and lower reactivity of HCl. This indicates that the Br–Cl synergism operates via the bromine–antimony route,[45] and is supported by the fact that no information is available on the FR-behaviour of Br–Cl systems without antimony. Additionally, it is also conceivable that the radicals Br• and Cl• might recombine not only to Br_2 and Cl_2 but also to BrCl, which is polar and more reactive and will react with the H• radicals to produce HBr and another Cl• radical, thus increasing the effectivity. This may explain the higher effectivity of the formulations containing Br- and Cl-based additives as compared to formulations in which *only* Br-based additives are applied.

The extent of the contribution of this essentially radical trap effect to the overall synergism of the bromine–chlorine system cannot be estimated at present. Experiments on identical formulations with and without antimony trioxide could elucidate the matter.

The Br–Cl synergism has been investigated up to now for a small number of polymers and little systematic work reported on it. The chemical structure and stability of the brominated and chlorinated additives and their concentrations and ratios in the formulations, with different amounts of antimony trioxide or other synergists, may have a profound effect on the synergistic activity in various polymers and provide a new picture of this phenomenon.

2.5.4   Modes of action – synergism of mixtures of bromine- and phosphorus-based additives

Synergistic interactions between bromine- and phosphorus-based derivatives are described in several publications.[46,47] Of particular interest is the case of polyacrylonitrile (PAN) treated with varying ratios of ammonium polyphosphate (APP) and hexabromocyclododecane (HBCD).[46] An SE value of 1.55 was calculated in this case. It was demonstrated in the study that the system acts via an intumescent mechanism. The bromine compound was proven by OI and nitrous oxide index (NOI) tests not to operate in the gas phase in the flame in the radical trap mode, but rather as a *blowing agent* to foam the char. This appears to be the first reported case in which a condensed-phase activity is shown for a bromine-based additive. This finding opens the way for reconsidering the mechanism of operation of bromine compounds as flame retardants in other polymers and systems. A similar phenomenon is observed in flexible polyurethane made from polyols containing bromine and phosphorus; the bromine was found to enhance the formation of a more copious foamed char.[47]

Similar SE values are computed from data given by Roderig et al[26] for a polycarbonate–polyethylene terephthalate (PC–PET) blend treated with varying ratios of triphenyl phosphate (TPP) and brominated PC. An SE value of 1.38 is found. It was suggested that when the bromine and the phosphorus atoms are included in the same additive molecule the synergism is more pronounced.[48,49] This is indeed the case when a brominated phosphate, with the ratio of bromine to phosphorus of 7:3, is added to the same blend; the SE value is 1.58.[48] These SE values are similar to those of the PAN /APP/HBCD system discussed above. There are some indications from the foamed bulky appearance of the char, that in these cases bromine may also serve, at least partly, as a blowing agent in an intumescent process, operative in these Br–P formulations, instead of or together with the radical trapping activity in the gas phase. The Br–P SE values are considerably lower than the bromine–antimony values, i.e. 2.2–4.3, as well as the PP/pentaerythritol/APP intumescent values, i.e. 5.5–11.3,[31–35] pointing to the possibility of a different complex mechanism.

It has recently been suggested that phosphorus compounds may replace antimony as a halogen synergist.[28,49] In the case of oxygen-containing polymers, such as nylon 6 and PET a strong synergism was demonstrated. For PET the total amount of additives (Br-based plus P-based) decreased by over 90% compared to the regular Br-based and Sb additive. A similar synergistic activity of bromine and phosphorus was obtained for PBT, PP, PS, HIPS and ABS. A decrease in the amount of total additive of 40% was obtained for PE.[49] The P–Br synergism was also demontrated in a case when both atoms are part of the same molecule of dialkyl 4-hydroxy-3,5-dibromobenzyl phosphonates, developed as a flame retardant for ABS.[50] The results were based on comparative experiments with related compounds having only the bromine related structures or only phosphorus related structures. A similar synergism was also lately obtained for UV-curable urethane acrylate to which variable amounts of tribromophenyl acrylate and triphenyl phosphate were added. A Br:P ratio of 2:1 was found to yield the maximum synergism.[51] This suggests a stoichiometric interaction, but is not consistent with the occasionally postulated formation of POBr$_3$. This requires a Br:P molar ratio of 3:1 which was actually the case in a red phosphorus-decabromodiphenyl system.[52] The mode of action of these synergistic formulations has not yet been elucidated and needs further research. It has been pointed out that synergism of halogens and phosphorus is not a general phenomenon: additivity is often observed.[53]

2.6   Modes of action of phosphorus-based flame retardants

2.6.1   General comments

Several reviews of phosphorus flame retardants, or of flame retardants more broadly, contain discussions of mode of action of phosphorus compounds.[54–61] A review by Granzow in 1978[62] is still highly useful and contains results not readily found elsewhere. Overviews have been published by Brauman in 1977[63] and by Weil in 1992–4.[55,64,65] More recent results have further emphasised the multiple modes of action of phosphorus.

Various phosphorus-based flame retardants have been shown to exert action in both the condensed phase and in the flame. Physical and chemical actions have been implicated in both phases. Flame inhibition, heat loss due to melt flow, surface obstruction by phosphorus-containing acids, acid-catalysed char accumulation, char enhancement and protection of char from oxidation have all been noted in particular polymer systems containing phosphorus-based flame retardants,[63] although the relative contribution of each mode of action depends on the polymer system and the fire exposure conditions. It is quite likely that in many cases, more than one mode of action is involved.

2.6.2 Condensed-phase modes of action

2.6.2.1 Charring modes of action

There is very convincing evidence, especially in oxygen-containing polymers such as cellulose and rigid polyurethane foam, that phosphorus compounds can increase the char yield. Formation of char means that less material is actually burned. Secondly, char formation is often accompanied by water release, which dilutes the combustible vapours. Moreover, the char can often protect the underlying polymer and the char-forming reactions are sometimes endothermic.

The pyrolysis behaviour of cellulose (cotton, paper, wood) has been discussed in sections 2.4.2–2.4.4, and a great deal of work has been published on the flame-retardant action of phosphorus in cellulose; several detailed reviews are available.[66–68,203] When cellulose is heated to its pyrolysis temperature, it normally depolymerises to a tarry carbohydrate product (mainly levoglucosan) which further breaks down to smaller combustible organic fragments. However, when a phosphorus-containing flame retardant is present in the cellulose, the retardant breaks down to phosphorus acids or anhydrides upon fire exposure. These reactive phosphorus species then phosphorylate the cellulose, generally with release of water (see Scheme 2,1). Phosphorylated cellulose then breaks down and forms char. A flame-retardant effect results from the formation of a non-combustible outward-flowing vapour (water), the reduction in fuel, and in some cases the protective effect of the char. A greater degree of flame retardancy seems likely if the char resists oxidation, although even a transitory char may have some inhibitory effect. Even if the char does undergo oxidation (usually by smouldering), the presence of a phosphorus compound tends to inhibit complete oxidation of the carbon to carbon dioxide, and thus the heat evolution is lessened. Besides its effect in enhancing the amount of char, the phosphorus flame retardant may coat the char and thus help prevent burning and smouldering by obstruction of the surface.

Another mode of action in which phosphorus is important as a char former is in intumescent fire-retardant paints and mastics. These typically have a phosphorus compound such as ammonium polyphosphate and a char-forming polyol such as pentaerythritol, along with a blowing agent such as melamine, and, of course, a binder.[69] Briefly stated, the phosphorus compound provides a phosphorylating agent which reacts with the pentaerythritol to form polyol phosphates which then break down to char through a series of elimination steps.[70–78] In some such formulations, melamine is combined with the phosphorus acid as melamine phosphate or melamine pyrophosphate,[79–80] and the released melamine and/or its breakdown products provide an endothermic action and a blowing action.

It has been known for a long time that certain nitrogen compounds such as melamine, ureas or dicyandiamide will enhance (perhaps synergise) the action of phosphorus in cellulose. This is not a general phenomenon, and it depends on the nitrogen compound[81] and the polymer system. The effect has been attributed to the formation of P–N bonded intermediates which are better phosphorylating agents than are the related phosphorus compounds without the nitrogen.[82] Another theory is that the nitrogen compounds retard the volatility loss of phosphorus from the condensed phase[83] while another study with urea-formaldehyde–diammonium phosphate on polyester–cotton blend fabric suggests just the opposite[84] and proposes that the nitrogen resin somehow enhances the vapour-phase action of the phosphorus. A further study of red phosphorus with melamine and various other nitrogen compounds in several thermoplastics also suggested that the nitrogen compounds enhanced the oxidation of phosphorus and gave off inert gases including ammonia. The only nitrogen compound which did not enhance retardancy was benzotriazole which did not give off ammonia.[85] We could find no study on effects of ammonia, hydrogen cyanide or cyanamide on flames from the burning of other fuels. Even considered as a source of 'inert gas,' ammonia can make a significant physical impact on the fuel value of the evolved fuel gases, and thus enhance flame retardancy, aside from chemical effects.[86] Physical effects are rarely assessed in studies of flame retardancy. A related effect of the nitrogen compounds, noted in a study referenced above,[85] was the reduction of the average rate of production of volatiles (presuming limiting fuel to the flame). This may also be partly a physical effect, shared with endothermic additives.

The topic of phosphorus–nitrogen synergism has been further studied at Polytechnic University.[87,88] As previously believed, it was confirmed not to be general; for example, phospham in nylon-4,6 was no more active than its elemental phosphorus equivalent.[89] In the case of a synergistic mixture of melamine phosphate and a cyclic phosphonate in EVA, enhanced rate of char formation at the optimum ratio was noted and the char was found by X-ray photon spectroscopy (XPS) to contain phosphorus acids with P–NH bonds, which may have contributed to faster charring and/or to better retention of phosphorus.[90]

Besides that quantity of char and its rate of formation, the char quality as a mass-transfer and heat-transfer barrier is also important. If the char is porous, a prevalence of closed cells over open cells is best and freedom from cracks and channels is also important. However, the amount of char rather than its volume seems to be dominant, at least in one Russian study.[91]

Another case where char enhancement by phosphorus is important is in rigid polyurethane foams.[92–95] The analytical evidence shows that phosphorus appears to be largely retained in the char[96–97] which is usually noted to be more coherent and thus better as a protective barrier.[94,95] However, there is also evidence that, besides char formation, part of the mode of action of phosphorus additives in rigid polyurethanes involves vapour-phase action in the flame zone.[98]

In contrast to the situation in rigid foams, char formation is probably not the basis of the action of phosphorus retardants in flexible polyurethane foam, and the formation of a small amount of char, insufficient to protect, can even reduce the flame retardancy of the flexible foam if the char acts as a wick.

In poly(ethylene terephthalate)[99–101] and poly(methyl methacrylate),[102–105] various phosphorus flame retardants cause an increase in the amount of residue and a retardation of the release of volatile fuel. This is probably the result of acid-catalysed cross-linking, perhaps by way of anhydride linkages.

Based on a study in nylon-6,[106] red phosphorus flame retardants may work by a condensed-phase mode of action if the classical interpretation is given to the very similar curves of OI and nitrous oxide index vs concentration: this evidence points away from flame chemistry. The red phosphorus becomes oxidised to phosphoric acids and some of the nylon structure becomes attached to the phosphoric acid in the form of alkyl ester structures suggested by the infrared spectrum.

It appears that phosphorus flame retardants may not produce enough char in nylons to be entirely effective in a non-dripping mode of extinguishment. Recently, studies at Du Pont have shown that melamine pyrophosphate can be further activated by a latent strong heteropoly acid such as phosphotungstic or silicotungstic acid to make sufficiently enhanced char, thus achieving UL 94 V-0 ratings.[107] Levchik et al[108] showed that at sufficient concentration of ammonium polyphosphate in nylon-6, a flame retardant intumescent char could be produced; a 5-aminopentyl polyphosphate was identified as an intermediate. Besides the protective action of the char, a protective P–N-containing cross-linked ultraphosphate coating was postulated on the basis of infrared evidence.

In oxygen-free hydrocarbon polymers such as polyolefins and styrenics, which generally do not char very easily, phosphorus flame retardants are usually not very effective unless they are supplemented by other char-forming additives. In the absence of such a char-forming additive, the flame-retardant action which the phosphorus compound still does exert seems to involve some combination of vapour-phase activity, protective acid coating and melt-drip enhancement. In polymers that do not contain reactive functional groups, the phosphorus flame retardant may react with oxygen groups or double bonds produced in the surface as the polymer burns.[109–113] Some evidence was also adduced by XPS spectroscopy that the ammonia of ammonium polyphosphate may also interact with oxygen groups eventually to form nitrogen heterocycles comprising part of the intumescent char.[114]

An important approach to effective use of the charring mode of action of phosphorus in those polymers such as polyolefins and styrenics, which are poor char formers, is to introduce a char-forming additive. Some of these char-forming additives are the same ones which were earlier found effective as 'carbonifics' (char formers) in intumescent paints and mastics.[69] Others are good char-forming smaller molecular weight additives such as tris(hydroxyethyl)isocyanurate[115] or polymers deliberately synthesised mainly as char-formers such as a polyester made from tris(hydroxyethyl) isocyanurate[116] or a triazine-piperazine-morpholine condensation polymer.[117] Char-forming polymers having, *per se*, useful thermomechanical properties such as poly(2,6-dimethylphenylene oxide) can be used in main ingredients where they contribute to a useful plastic blend as in GE's NORYL, a PPO-HIPS blend flame retarded by aryl phosphates[118] or as minor ingredients where they help the flame retardancy but not the mechanical properties.[119]

In some instances, the phosphorus flame-retardant moiety is built chemically into a char-forming structure. No generic advantage has been found to this approach, but it has been the subject of a great deal of industrial research. One example which reached the market place with limited success is the bis(melaminium) salt of pentaerythritol spirodiphosphate (Great Lakes' CHARGUARD 329).[120] A more recent combination of the catalytic char-forming action of phosphorus with the char-forming pentaerythritol structure is the bicyclic monophosphate of pentaerythritol (Great Lakes' NH-1197).[121]

Cyclic phosphorus ester structures such as neopentylene phosphonates are reasonably good char formers. Monsanto's XPM-1000 is a tris-neopentylene ester of nitrilotrimethylphosphonic acid, and is a fairly good char former, for example in a flame-retardant ethylene-vinyl acetate formulation enhanced by the additional char-forming catalytic action and blowing action of a melamine phosphate, giving the effect of a synergistic combination.[90]

An interesting recent development of a self-intumescing phosphorus–nitrogen compound is Albright & Wilson's ethylenediamine salt of phosphoric acid (Amgard NK, now Antiblaze NK).[122,123] When heated, it gives off a gas (probably water) and forms an expandable mass. Being stable to only about 200 °C, this salt is limited in its use to polymers processed at lower temperatures; it has been, for example, successful in caulks.

Studies have been made of both the physical and the chemical properties of the intumescent shield formed on a combustible polymer by the interaction of a char-forming additive and a phosphorus compound acting as charring catalyst. Useful overviews have been published by researchers at Lille[124] and Turin.[70-78] Evidence from a study of six commercial intumescent systems in polypropylene indicated that the yield of 'residue after transition,' which is seen as a plateau (or the last plateau) after the main weight loss in a TGA curve under air, correlated well (r = 0.99) with the oxygen index.[125]

From a mode of action standpoint, there is support by XPS and elemental analysis for the postulated formation of a protective 'phosphocarbonaceous' structure through phosphorylation of oxygenated functional groups on the 'carboniferous additive' (typically a pentaerythritol) and on the nascent char.[124,126]

2.6.2.2 Coating modes of action

Some researchers have emphasised the surface chemistry of charring in systems containing a charrable polymer such as PPO; endothermic rearrangement of the polyether to a methylene-linked polyphenol followed by a phosphate-accelerated dehydrative and endothermic dehydrogenative charring is indicated by analysis of surface material.[127] Phosphorylation of the phenolic rearrangement product is likely to be an early step in the chemistry of the char formation.

Phosphorus can also inhibit smouldering, also known as glowing combustion of the char;[128-130] the mode of action has long been postulated to involve some sort of polyphosphoric acid coating which is possibly a physical barrier action; besides this a deactivation of oxidation-active centres on the carbon can be demonstrated.[130-132]

It has been shown that incorporation of phosphorus even in amounts as small as 0.1% can inhibit oxidation of graphitic carbon by free oxygen. Hydrophilic phosphorus acid groups and other P=O structures can bond to oxidation-prone sites (the 'armchair' sites) on the surface.

Research at Alma-Ata recently showed that a phosphorus flame retardant can reduce the permeability of char, improving its barrier action.[133] Whether this is due to a coating or to a structural change in the char is not clear.

Brauman postulated that a phosphorus acid acts as a physical barrier to the vaporisation of fuel from a hydrocarbon polymer flame-retarded by ammonium polyphosphate or triphenyl phosphate[134,135] and infrared bands ascribed to the polyphosphoric acid coating were observed.

2.6.2.3   Effects on melt viscosity

In some cases, phosphorus compounds can act under fire-exposure conditions by generating acids which catalyse thermal breakdown of the polymer melt,[136] reducing melt viscosity and encouraging the flow or drip of the molten polymer from the fire zone. In poly(ethylene terephthalate) fabric, as little as 0.15% phosphorus permits the fabric to pass a vertical flame test, presumably by dripping or melting away from the flame.[137]

The melt viscosity depressing effect can be enhanced in thermoplastics prone to free-radical degradation by peroxide 'synergists'; thus, foamed polystyrene flame retarded with the now-obsolete tris(2,3-dibromopropyl) phosphate could be made more easily to pass a test allowing drip extinguishment by including dicumyl peroxide.[138] The effect of peroxides was shown to be, at least in large part, a melt degradation and flow enhancement effect by Gouinlock et al.[139] Small amounts of sulphur in styrenic polymers flame-retarded with phosphates can also produce this effect, as shown in an Albemarle patent.[140]

A melt-flow, melt-viscosity-depressant mode of action can be defeated by the presence of any non-melting solid which can retard the melt flow or which can act like a candle wick: cotton threads in a flame-retardant PET fabric can have such an effect. A particularly impressive example is the antagonistic effect of traces of silicone oil on flame-retarded polyester fabric; the fabric is rendered flammable probably because the silica formed on pyrolysis of the silicone reduces melt flow.[141] It has been shown that pigment-printing with an infusible pigment can spoil the flame retardancy of a phosphorus-containing flame-retarded polyester such as flame-retardant TREVIRA,[142] and this may be a melt flow retarding effect. Materials such as mineral or cellulosic powders or fibres, even if merely present on the surface, can defeat the action of various flame retardants in flexible urethane foams.[143]

2.6.2.4   Condensed-phase free-radical inhibition modes of action

This idea has been proposed by Russian and Czech researchers, who offer some evidence in support of free-radical inhibition, or at least of an antioxidant effect, by non-volatile phosphorus flame retardants.[144-146] Electron spin resonance data indicate that aryl phosphate flame retardants may scavenge alkylperoxy radicals in the polymer surface.[145] The contribution of this action to flame retardancy is not clear.

2.6.2.5   Condensed-phase modes of action based on surface effects on fillers

This relatively unexplored area has two principal aspects. Firstly, phosphorus compounds having characteristics of surfactants, such as alkyl acid phosphates, can aid the dispersion of solid flame retardants such as alumina trihydrate (ATH).[147] Improved dispersion usually results in improved flame-retardant efficiency. Secondly, some char enhancement may be possible as the result of improved binding or possibly from the catalytic action of a surface active agent. Thus, certain alkoxytitanates and alkoxyzirconate coupling agents having alkyl acid pyrophosphate anions seem to enhance the UL 94 flammability ratings of polypropylene with various mineral fillers.[148] Interestingly, on one filler, barium sulphate, the effect of the titanate seemed to reach a maximum at 1% concentration, and the effect was lower at lower and higher concentrations. This needs re-examination and further study.

2.6.3 Vapour-phase modes of action

2.6.3.1 Chemical modes of action

It has been shown that volatile phosphorus compounds are efficient flame inhibitors.[149,150] Mass spectroscopy studies by Hastie at the National Bureau of Standards[151–153] showed that triphenyl phosphate and triphenylphosphine oxide break down in the flame to small molecular species such as P_2, PO, PO_2 and HPO_2. These species cause the hydrogen atom concentration in the flame to be reduced, thus quenching the flame. The step in the flame chemistry which is inhibited is the rate-controlling branching step (equation 2.1) involving the reaction of a hydrogen atom with an oxygen molecule to give a hydroxyl radical and an oxygen atom. This is the same step which is believed to be inhibited by the hydrogen atom scavenging effect of halogens (discussed earlier).

Further studies of volatile phosphorus flame retardants such as trialkyl phosphates and trialkylphosphine oxides show evidence of flame-zone mode of action.[62,154] Phosphine oxides in particular seem most apt to show vapour-phase action; they are quite unreactive in most plausible condensed-phase chemistry. Some empirical evidence has been adduced for vapour-phase flame-retardant action of phosphine oxide: for instance, trime-thylphosphine oxide in rigid polyurethane foam showed very different oxygen index- vs nitrous oxide index-concentration curves.[154] Even with the phosphine oxide structure reacted into the polymer, as in a series of modified nylons made at VPI and tested at NIST,[155] it appears that the greater part of the flame-retardant action was vapour-phase, although there was also a small char yield increase. In a similar manner, having the triphenylphosphine oxide structure reacted in or as an additive gave no significant difference in fire performance. In this rigid polyurethane foam system, the condensed-phase active additive was more efficient on a phosphorus basis than the vapour-phase active additive.

The contrary was found in studies of flame-retardant finishes on wool and wool-polyester blends. The results showed that a demonstrably more volatile phosphonium structure gave better flame-retardant results than a less volatile polymer-bonded phosphine oxide structure; in this instance, vapour phase action seems more efficacious.[156,157] Not surprisingly, the relative efficacy of condensed-phase and vapour-phase action of phosphorus is substrate-dependent, probably dependent on the relative propensity to release volatile fuel and to form char.

Vapour-phase flame-retardant activity appears to be a substantial part of the mode of action of triaryl phosphates in the commercial blends of polyphenylene oxide with high-impact polystyrene; the polyphenylene oxide gives a protective char while the triaryl phosphate provides the flame inhibition needed to suppress the combustion of the polystyrene thermal breakdown products in the vapour phase.[118]

Recent developments in regard to new and superior aryl phosphate additives have emphasised higher molecular weight diphosphates and oligomers. Examples are tetraphenyl resorcinol diphosphate and tetraphenyl bisphenol-A diphosphate, which in the commercial form also contain smaller amounts of mono- and oligo- meric phosphates. Even with these materials, in PC/ABS and in HIPS, the flammability data suggest a correlation between UL rating and volatility; the more volatile phosphates gave the higher ratings.[158] This fact implies that the action of the flame retardant in the vapour phase is an important contributor to the overall action.

Even with the more volatile triphenyl phosphate however, the vapour-phase action seems not to represent the entire mode of action. In poly(2,6-dimethylphenylene oxide) (PPO) with HIPS blends, part of the triphenyl phosphate is retained and promotes the rearrangement of PPO to a benzylic hydroxyphenylene polymer which gives an enhanced char yield.[159] Hydrogen bonding seems to delay the triphenyl phosphate volatilisation.

2.6.3.2 Physical modes of action

Vapour-phase action does not have to involve flame chemistry. A physical mode of flame inhibition, based on heat capacity and heat of vaporisation, and, possibly, endothermic dissociation in the vapour phase, may be important. This physical aspect of vapour-phase flame retardancy has been discussed in connection with halogen systems by Larsen,[7,10,160,161] and, in the same way, phosphorus compounds may contribute at least part of their flame-retardant effect by virtue of their heat of vaporisation and their heat capacity.

Weil attempted to assess the relative contribution of vapour-phase and condensed-phase modes of action with a variety of phosphorus additives in poly(methyl methacrylate) (PMMA).[64] All the additives were compared at equivalent phosphorus loadings by OI measurement. The smallest eleva-tion of OI was found with trimethylphosphine oxide, a stable volatile com-pound, and the largest elevation of OI was found with phosphoric acid or alkyl acid phosphates which are non-volatile. These results suggest that the condensed-phase mode of action is a more efficient one than the vapour-phase mode of action with PMMA, even though PMMA is a polymer which depolymerises thermally to volatile monomer.

In the study by Ravey et al,[143,162] it was found that about 80% of the tris(dichloroisopropyl) phosphate (TDCPP) vaporises from a flexible urethane foam before most of the foam decomposes. But injection of large amounts of TDCPP into the flame of a burning non-flame-retardant foam produced no flame extinguishment. In the bottom-up mode of burning the TDCPP-retarded foam, as in the usual CAL 117 test, there was neither char formation nor acid coating produced on the surface. Likewise, in a miniaturised version of the bottom-ignition CAL 117 test, it was possible to see a self-extinguishing effect from merely having the usual self-extinguishing amount of flame retardant placed only on the surface or even on just the corners of a small bar of foam. Thus a *physical* vapour-phase action is indicated which may be produced by a combination of endothermicity, fuel dilution, and the Damkoehler number effect of the outward flow of a non-combustible vapour. However, with the same foams containing TDCPP, top-down burning showed the slow accumula-tion of a phosphorus-containing carbonaceous barrier layer. Thus, the predominant mode of action was seen to shift depending on the geometry of burning.

2.6.4   Some comments on interaction of phosphorus retardants with other flame retardants

2.6.4.1   *Further aspects of the mode of action of halogen–phosphorus combinations*

Synergism has been discussed earlier in this chapter. Some problems of defining synergism, with particular attention to the question of halogen–phosphorus synergism, were critically reviewed with reference to real and dubious examples in flame retardancy.[163] Phosphorus–halogen synergism, unlike antimony–halogen synergism, does not appear to be general. For-mation of phosphorus oxyhalides, while possible, lacks any direct experi-mental support but might be inferred from a few instances where the optimum Br:P ratio is about 3:1, such as in a red phosphorus–decabromobiphenyl system.[164] Other than a few instances of synergism, good additive results are often obtained with combinations of halogen- and phosphorus based flame retardants. One study with tris(dichloroisopropyl) phosphate in epoxy resin showed evidence of vapour-phase action (deduced from volatility), condensed-phase (acceleration of resin degradation with char enhancement) and even some suggestion of a contribution by physical barrier action.[165] However, there are some cases where phosphorus-halogen synergism seems to have been demonstrated. These were discussed earlier in the halogen section.[48,51,59]

2.6.4.2 Interaction of phosphorus with antimony

There are quite a few published formulations showing the attempted use of antimony oxide in combination with phosphorus and halogen. Results sometimes seem to be favourable, but a number of quantitative studies show convincing evidence of an antagonism between antimony and phosphorus.[55,61] In the most pronounced cases, one element cancels out the flame-retardant effect of the other, and in less drastic cases, the flame-retardant effects of the combination are lower than might be expected from adding the effects of the two separate compounds. Detailed studies of triaryl phosphate and antimony oxide in PVC showed that this antagonism only occurred in a part of the composition range.[166,167] This antagonistic effect probably is the result of the formation of antimony phosphate which is very stable and practically inert as a flame retardant.

2.6.4.3 Interactions with mineral fillers

Good E-84 tunnel results are obtained with polyester resins containing alumina trihydrate (ATH) in combination with dimethyl methylphosphonate or triethyl phosphate.[168] By some interpretations of the data, the combination might be said to be synergistic.

A careful study by Scharf compares the effect of TiO_2 and SnO_2 on the flame-retardant char-forming effect of ammonium polyphosphate in polypropylene, together with an intumescent nitrogenous resin.[169] TiO_2 increased flame retardancy by giving a stronger and more continuous char in higher yield; SnO_2, on the other hand, was antagonistic and made the char flakier and more porous, and did not enhance the char yield. The beneficial action of TiO_2 was considered to be a physical 'bridging' effect; the deleterious action of SnO_2 was attributed to an unfavourable chemical interaction with the phosphorus compound.

Recent studies in Lille[170] have shown remarkable synergism of the ammonium polyphosphate-pentaerthritol intumescent flame-retardant system in an ethylene–butylacrylate–maleic anhydride terpolymer based formulation by low levels of an acid zeolite. The presence of the zeolite appears to enhance the quality of the protective char by decreasing the size of the amorphous domains and preventing the formation of crack-susceptible large domains in the carbon. The formation of aluminophosphates, retention of volatile cracking products, and increased radical concentration in the char are also implicated in the protective mode of action. The relative contribution of these various effects remains to be apportioned.

2.6.4.4 Interactions between different phosphorus flame retardants

What we might call 'phosphorus–phosphorus synergism' has been reported in a few cases where two different phosphorus compounds were used together as flame retardants. Some examples are combinations of a phosphonium bromide or phosphine oxide with ammonium polyphosphate in polypropylene or polystyrene.[171,172] The use of regression analysis provides statistical evidence for, and some measure of, this interaction effect.[119] The reported cases may be instances of a vapour-phase-active phosphorus flame retardant combined with a condensed-phase-active flame retardant, but this is only a hypothesis.

2.6.5 Built-in vs additive phosphorus flame retardants

No general answer can be given as to whether there is any advantage to building a phosphorus flame retardant into a polymer rather than adding it. A review of phosphorus-containing polymers leads to the conclusion that despite the large amount of work done on the synthesis of phosphorus-containing polymers,[173,174] the number of such polymers which are commercial is much smaller than the number of successful phosphorus-containing additives. The same conclusion would certainly be arrived at on the basis of total tonnage. This may be largely because it is more difficult and costly to make a useful phosphorus-containing polymer than to make an additive. However, one study comparing a built-in phosphine oxide structure with an additive phosphine oxide in a polyester showed no advantage for the built-in phosphorus.[175] A study by Stackman[176-178] who compared additive vs co-reacted phosphonate structures in polyesters showed that at a low percentage of phosphorus, the additive was slightly better while at a higher percentage of phosphorus, the co-reactant was slightly better. Flame-retarding polyester fabric by use of built-in phosphorus structures is successful, as is flame-retarding by phosphorus additives; in both cases, less than 1% phosphorus levels are all that are needed to obtain the melt-flow type of extinguishment permitted by tests such as NFPA 701.

2.6.6 Some guidelines from consideration of phosphorus modes of action

To make use of two main categories of phosphorus modes of action, the plastics compounder may find it useful to try to combine a vapour-phase-active (i.e. relatively volatile) phosphorus flame retardant with a condensed-phase-active (i.e. relatively less volatile) phosphorus flame retardant.

To enhance further the condensed-phase mode of action, it may be found helpful to formulate with additional char-forming additives, at least in those cases where the polymer itself does not char very well. Blending good char-forming polymers into poor char-forming polymers along with phosphorus flame retardants may prove helpful. The models for success are PPO-HIPS and PC-ABS blends.

2.7   Modes of action of borates

The use of borax (sodium borate) to flame-retard cellulosics goes back over two centuries.[179] Water-soluble sodium borates as well as borate–boric acid combinations continue to be used in cellulosics and other hydroxyl-containing polymers. The mode of action appears to be a combination of the effect of the formation of a conspicuous glassy inorganic layer, which is often intumescent, and an increase in char formation perhaps through the formation of borate esters as well as through the blocking off of volatile fuel release. Borates and boric acid can also give off water, which provides a heat sink, a fuel diluent, a propellant for the fuel out of the flame zone, and a blowing agent for the glassy intumescent coating.[180]

Zinc borates have become major flame and smoke retardants. Here, a multi-modal action can also be demonstrated. Most of the zinc borates in commercial use are hydrates, with sharply defined endothermic water release temperatures.[181] The use of $2ZnO \cdot 3B_2O_3 \cdot 3.5H_2O$, US Borax's FIRE-BRAKE ZB, provides release of 13.5% water at 290–450 °C, which is a good match to the decomposition temperature of PVC and many other common polymers. The water released from FIREBRAKE ZB absorbs $503 \, J \, g^{-1}$ of heat, serves to blow char to a foam, and dilutes the fuel. A more thermally stable hydrate, FIREBRAKE 415, loses water at 415 °C so it is a good match for the decomposition temperature of high-temperature engineering thermoplastics. Hydrated barium borate is of value as a flame retardant with low water solubility. Hydrated calcium borate has been proposed as a flame retardant.[182] Both these borates share at least the endothermic water-release mode of action with the hydrated zinc borates.

The largest use of FIREBRAKE ZB is in PVC where it can replace part of the usual antimony oxide synergist with good or even superior flame retardancy, greatly reduced smoke, and lower heat and carbon monoxide release. Part of the smoke-reducing action is due to promotion of char, which represents carbon that did not get into the vapour phase. Most of the boron remains in the char, as does the zinc.

The zinc chloride formed in the condensed phase can catalyse dehydrochlorination and cross-linking. The minor part of the zinc chloride which volatilizes may have flame-inhibiting action. The boric oxide released from zinc borate by the action of acid is a glassy melt which can stabilize the char and inhibit afterglow.[183]

In halogen-free systems, FIREBRAKE ZB still can work by the water-release mechanisms. The flame-retardant action of anhydrous zinc borate in non-halogen systems may be due to an improved char (barrier) layer. Moreover, in the presence of alumina trihydrate, zinc borates can form a porous and ceramic-like sintered layer at temperatures above about 550 °C. This layer can act as a barrier for heat and mass transfer.[184,185]

Detailed studies have been done on the way in which boron compounds inhibit the oxidation of the graphitic structures which are present in char along with amorphous carbon.[186] The boron appears to poison specific oxidation-prone sites on the graphite crystal surface whereas phosphorus poisons a different set of sites.[131]

The question arises whether boron chlorides or boron bromides play a role in the action of zinc borate in halogenated polymers. It has been shown that boron halides are flame inhibitors with about the same order of magnitude of radical-scavenging efficacy as hydrogen halides.[187] However, in a recent Chinese study using zinc borate in PVC, it was found that only a small amount of the boron is lost, presumably as volatile boron halides. In fact this small boron halide release action was suggested to be deleterious – it was not enough to contribute much inhibition but enough to break up the integrity of the barrier layer formed.[188]

2.8   Modes of action of metal hydroxides and other hydrated inorganic additives

Alumina trihydrate (aluminium trihydroxide) and magnesium hydroxide actually do not have water of hydration in their structure as such. The hydroxyl groups bonded to the metal have to undergo endothermic decomposition to produce free water, and this starts at about 220 °C and 330 °C for ATH and magnesium hydroxide respectively. The enthalpy of water release is $1.17 \, kJ \, g^{-1}$ for ATH and $1.356 \, kJ \, g^{-1}$ for magnesium hydroxide.[189] This endothermicity is certainly part of the mode of action of ATH and magnesium hydroxide.[190]

It has been suggested that this endothermic fuel-diluting water release is not the total explanation for the action of ATH or magnesium hydroxide.

In fact, it has been shown that at low levels, anhydrous alumina can be a more potent flame retardant than hydrated aluminium; this was the case in an epoxy resin, for instance.[191] Anhydrous alumina is an acid catalyst and may be expected to aid charring of polymers prone to acid-catalysed dehydration. Moreover, a layer of refractory mineral can act as a heat barrier. In a case where catalysis can scarcely be invoked, a layer of silica was shown to have a profound depressing effect on heat release, perhaps as a heat-transmission barrier by poor conduction and by reflection of radiant heat.[192,193,194] Magnesia (MgO) is a good thermal insulator, often used as such, and it may be playing this role in magnesium hydroxide-retarded plastics after the water-release mode of action has been exhausted.

Further enhancement of magnesium hydroxide by certain additives may be due to development of the barrier action; novolac synergists which increase the action of magnesium hydroxide in polypropylene visibly retard melt flow at fire-exposure temperatures.[195] Certain acrylonitrile copolymer fibres (which presumably have charring capabilities) enhance the efficiency of magnesium hydroxide in rubber and probably act as physical reinforcement of the barrier.[196] Polycarboxylic resins, perhaps aided by polysiloxanes, enhance the action of alumina trihydrate or magnesium hydroxide by forming barriers during fire exposure.[197]

In a few cases, simple hydrates can be used as flame retardants probably operating mainly by the heat sink/fuel dilution effect. Gypsum, calcium sulphate dihydrate, is a good example; because of its very low cost it has been used as part of the flame retardant system in carpet backing and in polyester resin.[198] Gypsum begins to lose its two moles of water at about 120 °C (too low for use in most thermoplastics). Various clays have flame-retardant action by endothermic water loss; indeed, this mode of flame retardancy was known and used in ancient times.

References

1 Little R W, *Flameproofing Textile Fabrics*, Reinhold, London, 1947.

2 Rosser W A, Wise H and Miller J, 'Mechanism of combustion inhibition by compounds containing halogen' in *Proc. 7th Symposium on Combustion*, Butterworth, London, 1959, 175.

3 Van Krevelen D W and Hoftyzer P J, *Properties of Polymers*, Elsevier Scientific Publishing Co., New York, 1976, pp 523–6; Van Krevelen D W, 'Flame Resistance of Polymeric Materials', *Polymer*, 1975, **16**(8), 615–20.

4 Minkoff G I and Tipper C F H, *Chemistry of Combustion Reactions*, Butterworth, London, 1962.

5 Petrella R V, 'Factors affecting the combustion of polystyrene and styrene' in *Flame Retardant Polymeric Materials*, M Lewin, S M Atlas and E M Pearce (editors) Vol. 2, Plenum Press, New York, 1978, 159–201.

6 Funt J and Magill J H, 'Estimation of the fire behavior of polymers,' *J. Fire Flam.*, 1975, **6**, 28.

7 Larsen E R, 'Fire Retardants (Halogenated)' in *Kirk-Othmer Encyclopedia of Chemical Technology*, 3rd edition, Vol. 10, 1980, 373–95.

8 Lewin M and Sello S B, 'Technology and test methods of flameproofing of cellulosics' in *Flame Retardant Polymeric Materials*, M Lewin, S M Atlas and E M Pearce (editors) Vol. 1, Plenum Press, New York, 1975, 19–125.

9 Lewin M, 'Flame Retarding of Fibers' in *Chemical Processing of Fibers and Fabrics*, M Lewin and S B Sello (editors), Vol. 2, Part B, Marcel Dekker New York, 1984, 1–141.

10 Larsen E R, *J. Fire Flamm./Fire Retard. Chem.*, 1974, **1**, 4–12.

11 Ewing C T, Beyler C and Carhart H W, 'Extinguishment of Class B flames by thermal and chemical action: principles underlying a comprehensive theory; prediction of flame extinguishing effectiveness', *J. Fire Protection Eng.*, 1994, **6**, 23–54.

12 Stuetz D E, DiEdwardo A H, Zitomer F and Barnes B P, 'Polymer combustion', *J. Polym. Sci., Polym. Chem. Ed.*, 1975, **13**, 585–621.

13 Basch A and Lewin M, 'Influence of fine structure on the pyrolysis of cellulose. I. vacuum pyrolysis', *J. Polym. Sci. Polym. Chem. Ed.*, 1973, **11**, 3071; 1974, **12**, 2063.

14 Lewin M and Basch A, 'Fire retardancy (cellulose)', in *Encyclopedia of Polym. Sci. and Technol. Supplement 2*, Wiley, New York, 1977, 340–62.

15 Kashiwagi T, *Chemical and Physical Processes in Combustion*, Fall Technical Meeting, Combustion Institute, Pittsburgh, PA, 1984, 1–53.

16 Lewin M and Basch A, 'Structure, pyrolysis and flammability of cellulose' in *Flame Retardant Polymeric Materials*, M Lewin, S M Atlas and E M Pearce (editors) Vol. 2, Plenum Press, New York, 1978, 1–41.

17 Weil E D, 'Phosphorus-based flame retardants' in *ibid.*, 103–33.

18 Lyons J W, *The Chemistry and Uses of Fire Retardants*, Wiley-Interscience New York, 1970, p 290.

19 Brauman S K, 'Phosphorus fire retardants in polymers. I. General mode of action', *J. Fire Retardant Chem.*, 1977, **4**, 18–37.

20 Basch A and Lewin M, 'Low add-on levels of chemicals on cotton and flame retardancy', *Textile Res. J.*, 1973, **43**, 693–4.

21 Back E L, 'Thermal auto-crosslinking in cellulose material', *Pulp and Paper Mag. Canada, Tech. Sec.*, 1967, **68**, T165–71.

22 Hendrix J E, Drake G L and Barker R H, 'Pyrolysis and combustion of cellulose. III. Mechanistic bases for the synergism involving organic phosphates and nitrogenous bases', *J. Appl. Polymer Sci.*, 1972, **16**, 257.

23 Economy J, 'Phenolic fibers' in *Flame Retardant Polymeric Materials*, M Lewin, S M Atlas and E M Pearce (editors) Vol. 2, Plenum Press, New York, 1978, 203–8.

24 Khanna Y P and Pearce E M, 'Synergism and flame retardancy' in *ibid.*, 43–62.

25 Sirdesai S J and Wilkie C A, 'RhCl(PPh$_3$)$_3$: A flame retardant for poly(methyl methacrylate)', *Polym. Preprints, Am. Chem. Soc. Polym. Chem. Ed.*, 1987, **28**, 149.

26 Roderig C, Basch A and Lewin M, 'Crosslinking and pyrolytic behavior of natural and man-made cellulosic fiber', *J. Polym. Sci., Polym. Chem. Ed.*, 1975, **15**, 1921–32.

27 Kashiwagi T, 'Polymer combustion and flammability – role of the condensed phase' in *Proc. 25th Symposium on Combustion*, The Combustion Inst., Pittsburgh, PA, 1994, 1423–37.

28 Pearce E M, 'Some polymer flammability structure relationships' in *Rec. Adv. in Polymeric materials*, M Lewin and G Kirshenbaum (editors) Vol. 1, Business Communications Co, Norwalk, USA, 1990, 36.

29 Pearce E M, Khanna Y P and Raucher D, 'Thermal analysis in polymer flammability' in *Thermal Characterization of Polymeric Materials*, E Turi (editor) Academic Press, New York, 1981, 793–847.

30 Lyon R, 'Mechanistic pyrolysis model for char-forming polymers' in *Rec. Adv. in Polymeric materials*, M Lewin (editor) BCC, Vol. 8, Norwalk, USA, 1997.

31 Lewin M, Basch A and Shaffer B, 'Pyrolysis of polymer blends', *Cellulose Chem. Technol.*, 1990, **24**, 477.

32 Endo M and Lewin M, 'Flame Retardancy of polypropylene by phosphorus-based additives' in *Advances in Flame Retardancy of Polymeric Materials* (*FR of Polym. Mat*), M Lewin (editor) Vol. 4, Business Communications Co, Norwalk, USA, 1993, 171.

33 Lewin M, 'Char and flame retardancy,' *Advances in Flame Retardancy of Polymeric Materals* (*FR of Polym. Mat*), 1995, **6**, 41.

34 Lewin M and Endo M, 'Intumescent systems for flame retarding of polypropylene' in *Fire and Polymers* II, G L Nelson (editor) ACS Symp. Series 599, 1995, 91.

35 Lewin M and Endo M, 'Intumescent systems for flame retardancy of polypropylene', PMSE ACS Preprints, 1994, **71**, 235.

36 Hastie J W, 'Molecular basis of flame inhibition', *J. Research NBS–A. Physics & Chem.*, 1973, **77A**(6), 733–54.

37 Hastie J W and McBee C L, 'Mechanistic studies of triphenylphosphine oxide-poly(ethylene terephthalate) and related flame retardant systems', *Natl. Bureau of Standards Report NBSIR 75–741*, Washington, DC, 1975.

38 Pitts J J, Scott P H and Powell D G, 'Thermal decomposition of antimony oxychloride and mode in flame retardancy', *J. Cell. Plast.*, 1970, **6**(1), 35–7.

39 Costa L, Goberti L, Paganetto P, Camino G and Sgarzi P, 'Thermal behavior of chlorine-antimony fire retardant systems', in *Proc. 3rd Meeting on FR Polymers*, Torino, 1989, 19.

40 Cleaver R F, 'Formulation and testing of fire-retardant GRP compounds', *Plastics and Polymers*, 1970, **38**(135), 198–205.

41 Gordon I, Duffy J J and Dachs N W, 'Fire retardant polymer composition with improved physical properties', US Patent 4000114, 1976.

42 Ilardo C S and Scharf D J, 'Flame retardant polymer composition', US Patent 4388429, 1983.

43 O'Brien D D, 'Flame retarded polyethylene wire insulation', US Patent 5358991, 1994.

44 Markezich R L and Mundhenke R F, 'Use of chlorine/bromine synergism to flame retard polymers', in *Recent Advances in Flame Retardancy of Polymeric Materials*, M Lewin (editor) Vol. **6**, Business Communications Co, Norwalk, USA, 1995, 177.

45 Ballistreri A, Roti S, Montaudo G, Papalardo S and Scamporrino E, 'Thermal decomposition of flame retardants. Chlorine–bromine antagonism in mixtures of halogenated polymers with Sb_2O_3', *Polymer*, 1979, **20**, 783–4.

46 Ballistreri A, Montaudo G, Puglisi C, Scamporrino E and Vitallini D, 'Intumescent flame retardants for polymers. I. The poly(acrylonitrile)–ammonium polyphosphate–hexabromocyclododecane system', *J. Appl. Polym. Sci.*, 1983, **28**, 1743.

47 Papa A J and Proops W R, 'Influence of structural effects of halogen and phosphorus polyol mixtures on flame retardancy of flexible polyurethane foam', *J. Appl. Polym. Sci.*, 1972, **16**, 236–73.

48 Green J, 'Phosphorus/bromine flame retardant synergy in polycarbonate blends' in *Recent Advances in Flame Retardancy of Polymeric Materials*, M Lewin (editor) Vol. 4, Business Communications Co, Norwalk, USA, 1993, 8.

49 Dombrowski R and Huggard M, 'Antimony free fire retardants based on halogens: phosphorus substitution of antimony' in *Proc. 4th Intern. Symp. Additives-95*, Clearwater Beach, Fl, 1996, 1.

50 Yang C-P and Lee T-M, 'Synthesis and properties of 4-hydroxy-3,5-dibromobenzyl phosphonates and their flame-retarding effects on ABS copolymer', *J. Polym. Sci., Polym. Chem. Ed.*, 1989, **27**, 2239–51.

51 Guo W, 'Flame-retardant modification of UV-curable resins with monomers containing bromine and phosphorus', *J. Polym. Sci.: Part A: Polym. Chem.*, 1992, **30**, 819–27.

52 Broadbent J R and Hirschler M M, 'Red phosphorus as a flame retardant for a thermoplastic nitrogen compound', *Eur. Polym. J.*, 1984, **20**, 1087–93.

53 Weil E D, 'Additivity, synergism and antagonism in flame retardancy' in *Flame Retardancy of Polymeric Materials*, W A Kuryla and A J Papa (editors) Vol. 3, Marcel Dekker, New York, 1975, 185–243.

54 Weil E D, 'Flame retardants (phosphorus compounds)' in *Kirk-Othmer Encyclopedia of Chemical Technology*, Vol. 10, 3rd edition, H F Mark et al (editors) John Wiley & Sons, New York, 1980, 396–419.

55 Weil E D, 'Phosphorus-based flame retardants' in *Handbook of Organophosphorus Chemistry*, R Engel (editor) Marcel Dekker, New York, 1992, 683–738.

56 Weil E D, 'Phosphorus-based flame retardants' in *Flame-Retardant Polymeric Materials*, M Lewin, S M Atlas and E M Pearce (editors) Vol. 2, Plenum Press, New York, 1978, 103–28.

57 Green J, 'Phosphorus-containing flame retardants', *Plastics Compounding*, 1984, **7**(7), 30, 32, 35, 37–8, 40.

58 Green J, 'Influence of coadditives in phosphorus-based flame retardant systems', *Plastics Compounding*, 1987, May–June, 57–64.

59 Green J, 'Mechanisms for flame retardancy and smoke suppression – a review', *J. Fire Sci*, 1996, **14**(6), 426–42.

60 Lewin M, 'Physical and chemical mechanisms of flame retarding of polymers', in *6th European Meeting on Fire Retardancy of Polymeric Materials*, Lille, Sept. 24–6, 1997.

61 Weil E D, 'Synergists, adjuvants and antagonists in flame retardant systems' in *Fire Retardancy of Polymeric Materials*, A Grand and C Wilkie (editors) Marcel Dekker, New York, 1999.

62 Granzow A, 'Flame retardation by phosphorus compounds', *Accounts Chem. Res.*, 1978, **11**(5), 177–83.

63 Brauman S K and Fishman N, 'Phosphorus flame retardance in polymers. III. Some aspects of combustion performance', *J. Fire Ret. Chem.*, 1977, **4**, 93–111.

64 Weil E D, 'Mechanisms of Phosphorus-Based Flame Retardants', paper given at Fire Retardant Chemicals Assoc., Orlando, Mar. 29–Apr. 1, 1992.

65 Weil E D, 'Flame Retardants – Phosphorus Compounds' in *Kirk-Othmer Encyclopedia of Chemical Technology*, 4th edition, Vol. 10, Wiley-Interscience, New York, 1994, 976–97.

66 Drake G L (Jr), 'Flame retardants for textiles' in *Kirk-Othmer Encyclopedia of Chemical Technology*, 3rd edition, Vol. 10, Wiley-Interscience, NY, 1980, 420ff.

67 Barker R H and Hendrix J E, 'Flame retardance of cotton', *Flame Retardancy of Polymeric Materials*, Vol. 5, W C Kuryla and A J Papa (editors) Marcel Dekker, New York, 1979, 1–65.

68 LeVan S, 'Chemistry of fire retardancy' in *The Chemistry of Solid Wood*, R Rowell (editor) American Chemical Society, Adv. Chem. Ser. 207, Washington, DC, 1984, 531–74.

69 Vandersall H J, 'Intumescent coating systems, their development and chemistry', *J. Fire Flamm.*, 1971, **2**, 97–140.

70 Camino G and Costa L, 'Mechanism of intumescence in fire retardant polymers', *Reviews in Inorganic Chemistry*, 1986, **8**(1–2), 69–100.

71 Camino G, Costa L and Trossarelli L, 'Study of the mechanism of intumescence in fire retardant polymers. I. Thermal degradation of ammonium polyphosphate–pentaerythritol mixtures', *Polym. Degrad. Stab.*, 1984, **6**, 243–52.

72 Camino G, Costa L and Trossarelli L, 'Study of the mechanism of intumescence in fire retardant polymers. II. Mechanism of action in polypropylene–ammonium polyphosphate–pentaerythritol mixtures', *Polym. Degrad. Stab.*, 1984, **7**, 25–31.

73 Camino G, Costa L and Trossarelli L, 'Study of the mechanism of intumescence in fire retardant polymers. III. Effect of urea on the ammonium polyphosphate–pentaerythritol system', *Polym. Degrad. Stab.*, 1984, **7**, 221–9.

74 Camino G, Costa L, Trossarelli L, Costanzi F and Landoni G, 'Study of the mechanism of intumescence in fire retardant polymers: IV. Evidence of ester formation in ammonium polyphosphate–pentaerythritol mixtures', *Polym. Degrad. Stab.*, 1984, **8**, 13–22.

75 Camino G, Costa L and Trossarelli L, 'Study of the mechanism of intumescence in fire retardant polymers. V. Mechanism of formation of gaseous products in the thermal degradation of ammonium polyphosphate', *Polym. Degrad. Stab.*, 1985, **12**, 203–12.

76 Camino G, Costa L, Trossarelli L, Costanzi F and Pagliari A, 'Study of the mechanism of intumescence in fire retardant polymers. VI. Mechanism of ester formation in ammonium polyphosphate-pentaerythritol mixtures', *Polym. Degrad. Stab.*, 1985, **12**, 213–28.

77 Camino G, Martinasso G and Costa L, 'Thermal degradation of pentaerythritol diphosphate, model compound for fire retardant intumescent systems. I. Overall thermal degradation', *Polym. Degrad. Stab.*, 1990, **27**, 285–96.

78 Bertelli G, Camino G, Marchetti E, Costa L and Locatelli R, 'Structural Studies on Chars from Fire Retardant Intumescent Systems', *Angew. Makromol. Chem.*, 1989, **169**, 137–42.

79 Weil E D, 'Melamine phosphate flame retardants', *Plastic Compounding*, 1994, May–June, 31–9.

80 Agunloye F F, Stephenson J E and Williams C M, 'Improved intumescent flame

retardant systems for polyolefins' in *Flame Retardants '94*, Interscience Communication London, Jan. 1994.

81 Willard J J and Wondra R E, 'Quantitative evaluation of flame retardant cotton finishes by the limiting-oxygen index (LOI) technique', *Text Res. J.*, 1970, **40**, 203–10.

82 Hendrix J E, Drake G L and Barker R H, 'Pyrolysis and combustion of cellulose. II. Thermal analysis of mixtures of methyl alpha-D-glucopyranoside and levoglucosan with model phosphate flame retardants', *J. Appl. Polym. Sci.*, 1972, **16**, 41.

83 Bakos D, Kosik M, Antos K, Karolyova M and Vyskocil L, 'The role of nitrogen in nitrogen–phosphorus synergism', *Fire Mat*, 1982, **6**(1), 10–12.

84 Holme I and Patel S R, 'A study of nitrogen–phosphorus synergism in the flame-retardant finishing of resin-treated polyester-cotton blends', *J. Soc. Dyers and Colourists*, 1980, **96**, 224–36.

85 Cullis C F, Hirschler M M and Tao Q M, 'Studies on the effects of phosphorus–nitrogen–bromine systems on the combustion of some thermoplastic polymers', *Eur. Polym. J.*, 1991, **27**(3), 281–9.

86 Sellman L-G, Ostman B A-L and Back E L, 'Methods of calculating the physical action of flame retardants', *Fire Mat*, 1976, **4**, 85–9.

87 Weil E D, Patel N, Huang C-H and Zhu W, 'Phosphorus-nitrogen synergism, antagonism and other interactions', *Proceedings of the 2nd Beijing International Symposium/Expedition on Flame Retardants*, Beijing, China, Oct. 11–15, 1993.

88 Weil E D, Patel N and Huang C-H, 'Phosphorus-nitrogen combinations – some interaction and mode-of-action considerations', in 4th Ann. Conf Recent advances in flame retardancy of polymeric materials, Business Communications Co, Norwalk, USA, May 18–20, 1993.

89 Weil E D and Patel N, 'Phospham – a stable phosphorus-rich flame retardant', *Fire Mat*, 1994, **18**, 1–7.

90 Zhu W and Weil E D, 'Intumescent flame-retardant system of phosphates and 5,5,5′,5′,5′,5′-hexamethyltris(1,3,2-dioxaphosphorinanemethan)amine 2,2′,2′-trioxide for polyolefins', *J. Appl. Polym. Sci.*, 1996, **62**, 2267–80.

91 Gnedin Ye V, Gitina R M, Shulyndin S V, Kartashov G N and Povikov S N, 'Investigation of phosphorus-containing foam-forming systems as combustion retardants for polypropylene', *Polym. Sci.*, 1991, **33**(3), 544–50.

92 Papa A J, 'Flame retarding polyurethanes' in *Flame Retardancy of Polymeric Materials*, Vol. 3, W C Kuryla and A J Papa (editors) Marcel Dekker, New York, 1975, 1–133 (see esp. 31ff).

93 Zabski L, Walczyk W and Jedlinski Z, 'Influence of phosphorus content and of increase of cyclic structures content in crosslinked cellular urethane copolymers upon their physical properties', *Chem. Zvesti* (Engl. trans.), 1976, **30**, 311–7.

94 Anderson J J, 'Retention of Flame Properties of Rigid Polyurethane Foams', *Ind. Eng. Chem. Prod. Res. Dev.*, 1963, **2**, 260–3.

95 Kresta J E and Frisch K C, 'Comparative study of the effect of phosphorus and chlorine in urethane and isocyanurate-urethane foams', *J. Cell. Plast.*, 1975, **11**, 68–75.

96 Piechota H, 'Some correlations between raw materials, formulation and flame retardant properties of rigid urethane foams', *J. Cell. Plast.*, 1965, **1**, 186–99.

97 Papa A J and Proops W R, 'Influence of structural effects of halogen and phosphorus polyol mixtures on flame retardancy of flexible polyurethane foams', *J. Appl. Polym. Sci.*, 1972, **16**, 2361–73.

98 Hall D R, Hirschler M M and Yavornitzky C M, 'Halogen-free flame retardant thermoplastic polyurethanes', *Fire Safety Science – Proceedings of the First International Symposium*, C E Grant and P J Pagni (editors) Hemisphere, Washington DC, 1986, 421–30.

99 Avondo G, Vovelle C and Delbourgo R, 'The role of phosphorus and bromine in flame retardancy', *Combust. Flame*, 1978, **31**, 7–16.

100 Suebsaeng T, Wilkie C A, Burger V T, Carter J and Brown C E, 'Solid products from thermal decomposition of poly(ethylene terephthalate); investigation by CP/MAS, C13 NMR and Fourier transform IR', *J. Polym. Sci.; Polym. Let. Ed.*, 1984, **22**, 625–34.

101 Brauman S K, 'Phosphorus fire retardance in polymers. IV. Poly(ethylene terephthalate)–ammonium polyphosphate, a model system', *J. Fire Ret. Chem.*, 1980, **7**, 61–8.

102 Gruntfest I J and Young E M, 'The action of flame proofing additives on PMMA', *ACS Div. Org. Coatings & Plastics Preprints*, 1962, **2**(2), 113–24.

103 Camino G, Grassie N and McNeill I C, 'Influence of the fire retardant, ammonium polyphosphate, on the thermal degradation of poly(methyl methacrylate)', *J. Polym. Sci., Polym. Chem. Ed.*, 1978, **16**, 95–106.

104 Wilkie C A, 'The Design of Flame Retardants' in *Fire & Polymers*, G L Nelson (editor) American Chemical Society, Symposium Series 425, Washington, DC, 1990, 178–88.

105 Brown C E, Wilkie C A, Smukalla J, Cody R B and Kinsinger J A, 'Inhibition by red phosphorus of unimolecular thermal chain-scission in poly(methyl methacrylate), investigation by NMR, FT-IR and laser desorption/Fourier Transform mass spectroscopy', *J. Polym. Sci., Part A, Polym. Chem.*, 1986, **24**, 1297–311.

106 Levchik G F, Levchik S V, Camino G and Weil E D, 'Fire retardant action of red phosphorus in nylon 6' in *6th European Meeting on Fire Retardancy of Polymeric Materials*, Lille, Sept. 24–6, 1997.

107 Martens M M and Kasowski R V (to E. I. Du Pont de Nemours) 'Fire Resistant Resin Compositions', US Pat. 5618865, 1997.

108 Levchik S V, Camino G, Costa L and Levchik G F, 'Mechanism of action of phosphorus-based flame retardants in nylon 6. I. ammonium polyphosphate', *Fire Mat*, 1995, **19**, 1–10.

109 Steutz D E, DiEdwardo A H, Zitomer F, Barnes B P, 'Polymer flammability. II.', *J. Sci., Polym. Chem. Ed.*, 1975, **13**, 585–621; *ibid.* 1980, **18**, 967–1009.

110 Martel B, 'Charring processes in thermoplastic polymers: effect of condensed phase oxidation on the formation of chars in pure polymers', *J. Appl. Polym. Sci.*, 1988, **35**, 1213–26.

111 Delobel R, Ouassou N, Le Bras M and Leroy J-M, 'Fire retardance of polypropylene: action of diammonium pyrophosphate-pentaerythritol intumescent mixture', *Polym. Degrad. Stab.*, 1989, **23**, 349–57.

112 Delobel R, Le Bras M, Ouassou N and Alistiqsa F, 'Thermal behaviors of ammonium polyphosphate-pentaerythritol and ammonium pyrophosphate-pentaerythritol intumescent additives in polypropylene formulations', *J. Fire Sci*, 1990, **8**, 85–108.

113 Delobel R, Le Bras M, Ouassou N and Descressain R, 'Fire retardance of polypropylene by diammonium pyrophosphate-pentaerythritol: spectroscopic characterization of the protective coatings', *Polym. Degrad. Stab.*, 1990, **30**, 41–56.

114 Bourbigot S, Le Bras M, Gengembre L, and Delobel R, 'XPS study of an intumescent coating. Application to the ammonium polyphosphate/pentaerythritol fire-retardant system', *Appl. Surface Sci.*, 1994, **81**, 299–307; *ibid.*, Part II, *Appl. Surface Sci.*, 1997, **120**, 15–29.

115 Nalepa C and Scharf D (to Hoechst Celanese Corp.) 'Two component intumescent flame retardant', US Pat. 5 204 392, 1993.

116 Sicken M and Wanzke W (to Hoechst AG) 'Flame-retardant plastics molding composition of improved stability', US Pat. 5 326 805, 1994.

117 Fontanelli R, Landoni G and Legnani G (to Montedison) 'Self extinguishing polymeric compositions', US Pat. 4 504 610, 1985.

118 Carnahan J, Haaf W, Nelson G, Lee G, Abolins V and Shank P, 'Investigations into the mechanism for phosphorus flame retardancy in engineering plastics', *Proc. 4th Intl. Conf. Fire Safety*, Product Safety Corp., San Francisco, 1979.

119 Weil E D, Zhu W and Patel N, 'A systems approach to flame retardancy and comments on modes of action', *Polym. Degrad. Stab*, 1996, **54**, 125–36.

120 Halpern Y and Niswander R H (to Borg-Warner Corp.) 'Pentaerythritol Phosphate', US Pat. 4 454 064, 1984.

121 Termine E J and Taylor K G, 'A new intumescent flame retardant additive for thermoplastic and thermoset applications', 'Conference Additive Approaches to Polymer Modification, SPE RETEC', Society of the Plastics Industry, Brookfied, Connecticut, Sep. 24–6, 1989, paper no. R89–160.

122 Davis J and Huggard M, 'The technology of halogen-free flame retardant additives for polymeric systems', *6th Ann. BCC Conference on Flame Retardants*, Stamford, Business Communications Co, Norwalk, USA, 1995.

123 Huggard M T and White P R (to Albright & Wilson Americas, Ltd.) 'Flame retardant thermoplastic resin composition with intumescent flame retardant', US Pat. 5 137 937, 1992.

124 Marchal A, Delobel R, Le Bras M and Leroy J-M, 'Effect of intumescence on polymer degradation', *Polym. Degrad. Stab*, 1994, **44**, 263–72.

125 Lewin M, 'Physical and chemical mechanisms of flame retardancy of polymers' in *Fire Retardancy of Polymers: The Use of Intumescence*, M Le Bras, G Camino, S Bourbigot and R Delobel (editors) Royal Society of Chemistry, Cambridge, UK, 1998, 3–34.

126 Bourbigot S, Le Bras M, and Delobel R, 'Carbonization mechanisms resulting from intumescence association with the ammonium polyphosphate–pentaerythritol fire retardant system,' *Carbon*, 1993, **31**(8), 1219–30; Part II, *ibid.*, 1995, **33**(3), 283–94.

127 Takeda K, Amemiya F, Kinoshita M and Takayama S, 'Flame retardancy and rearrangement reaction of polyphenylene–ether/polystyrene alloy', *J. Appl. Polym. Sci.*, 1997, **64**(6), 1175–83.

128 Lewin M, Isaacs P, Sello S B and Stevens C, 'Flame-resistant cellulose esters', *Text. Res. J.*, 1974, **47**, 700–7.

129 Lewin M, Isaacs P, Sello S B and Stevens C, 'Flame retardant modification of cellulose', *Textilveredlung*, 1973, **8**, 158–61.

130 Ermolenko I N, Lyubliner I P and Gulko N V, 'Chemically Modified Carbon Fibers', VCH Publishers, New York, 1990, 202–3.

131 Oh S G and Rodriguez N M, '*In situ* electron microscopy studies of the inhibition of graphite oxidation by phosphorus', *J. Mater. Res.*, 1993, **8**(11), 2879–88.

132 McKee D W, Spiro C L and lamby E J, 'The inhibition of graphite oxidation by phosphorus additives', *Carbon*, 1984, **22**(3), 285–90.

133 Gibov K M, Shapovalova L N and Zhubanov B A, 'Movement of destruction products through the carbonized layer upon combustion of polymers', *Fire Mat*, 1986, **10**, 133–5.

134 Brauman S, 'Phosphorus fire retardance in polymers. I. General mode of action', *J. Fire Ret. Chem.*, 1977, **4**, 18–37.

135 Brauman S, 'Phosphorus fire retardance in polymers. II. Retardant-polymer substrate interactions', *J. Fire Ret. Chem.*, 1977, **4**, 38–58.

136 Bertelli G, Camino G, Marchetti E, Costa L and Locatelli R, 'Structural studies on chars from fire retardant intumescent systems', *Angew. Makromol. Chemie*, 1989, **169**, 137–42.

137 Mori H, 'Teijin's flame retardant Tetoran Extar' in *Proceedings of the 1975 Symposium on Textile Flammability*, LeBlanc Research Corp., East Greenwich, RI, 1975, 124–54.

138 Jann A K (to Dow) 'Flame-resistant thermoplastic alkenylaromatic polymers', Ger. Pat. 1495419 (July 16, 1970); *Chem. Abst.*, **73**, 67229.

139 Gouinlock E V, Porter J and Hindersinn R, 'Mechanism of the fire-retardance of dripping thermoplastic compositions', *J. Fire and Flammability*, 1971, **2**, 206–18.

140 Prindle J C, Nalepa C J and Kumar G (to Albemarle Corp.) 'Fire retardant styrene polymer compositions', Eur. Pat. 806451-A1, Nov. 12, 1997.

141 Swihart T J and Campbell P E, 'How silicones affect flame retardancy', *Text. Chem. Color.*, 1974, **6**, 109–12.

142 Mach H-R, 'How can the low flammability of a polyester fibre type be retained in textile printing?' *Melliand Textilberichte* (English), 1990, (1), E29–33.

143 Ravey M, Keidar I, Weil E D and Pearce E M, 'Flexible polyurethane foam. II. Fire retardation by tris(1,3-dichloro-2-propyl) phosphate, Part B. Examination of the condensed phase (the pyrolysis zone)', *J. Appl. Polym. Sci.*, 1998, **68**, 231–54.

144 Serenkova I A and Shlyapnikov Yu A, 'Phosphorus containing flame retardants as high temperature antioxidants', Proc. Intl. Symposium on Flame Retardants, Nov. 1989, Intl. Academic Publ. (Pergamon), Beijing, 1989, 156–61.

145 Tkac A, 'Radical processes in polymer burning and its retardation', *J. Polym. Sci. A, Polym. Chem. Ed.*, 1981, **19**, 1495–508.

146 Citovichy P, Kosik M and Spilda I, 'Thermoanalytical investigation of grafted polypropylene with fixed flame retardant compounds', *Thermochem. Acta*, 1985, **93**, 171–74.

147 Silberberg J and Weil E D (to Stauffer Chem. Co.) 'Inorganic filler material and polymer composition containing the same', US Pat. 4251436, 1981.

148 Walter S, 'Filled polypropylene can be flexible – and flame retardant, too', *Plast. Eng.*, 1981, **37**(6), 24–7.

149 Rosser W A, Jr, Inami S H and Wise H, 'Quenching of premixed flames by volatile inhibitors', *Combust. Flame*, 1966, **10**, 287.

150 McHale E T, 'Survey of vapor phase chemical agents for combustion suppression', *Fire Research Abstracts & Reviews*, 1969, **11**(2), 90.

151 Hastie J W and Bonnell D W, 'Molecular chemistry of inhibited combustion systems', Natl. Bureau of Standards Report NBSIR 80-2169, Washington, DC, 1980.

152 Hastie J W, 'Molecular basis of flame inhibition', *Comb Flame*, 1973, **21**, 178, 401.

153 Hastie J W and McBee C L, 'Mechanistic studies of triphenylphosphine oxide-poly(ethylene terephthalate) and related flame retardant systems', *Natl. Bureau of Standards Report NBSIR 75-741*, Washington, DC, 1975.

154 Weil E D, Jung A K, Aaronson A M and Leitner G C, 'Recent basic and applied research on phosphorus flame retardants' in *Proc. 3rd Eur. Conf. on Flamm. and Fire Ret.*, V M Bhatnagar (editor) Technomic Publ., Westport, CT, 1979.

155 Kashiwagi T, Gilman J W and Nyden M R, 'New Flame Retardant Additives' in *6th European Meeting on Fire Retardancy of Polymeric Materials*, Lille, Sept. 24–6, 1997.

156 Basch A, Nachumowitz S, Hasenfrath S and Lewin M, 'The chemistry of THPC-urea polymers and relationship to flame retardancy in wool and wool–polyester blends. II. Relative flame retardant efficiency', *J. Polym. Sci., Polym. Chem. Ed.*, 1979, **17**, 39.

157 Basch A, Zwilichowski B, Hirschman B and Lewin M, 'The chemistry of THPC-urea polymers and relationship to flame retardancy in wool and wool–polyester blends. I. Chemistry of THPC–urea polymers', *J. Polym. Sci., Polym. Chem. Ed.*, 1979, **17**, 27–38.

158 Bright D A, Dashevsky S, Moy P Y and Tu K-M, 'Aromatic oligomeric phosphates: effect of structure on resin properties', paper at SPE ANTEC, Atlanta, April 26–30, 1998.

159 Boscoletto A B, Checchin M, Milan L, Pannochia P, Tavan M, Camino G and Luda M P, 'Combustion and fire retardance of poly(2,6-dimethyl-1,4-phenylene ether) – high impact polystyrene blends. II. Chemical aspects', *J. Appl. Polym. Sci.*, 1998, **67**, 2231–40.

160 Larsen E R, 'Halogenated fire extinguishants: flame suppression by a physical mechanism?' in *Halogenated Fire Suppressants*, R G Gann (editor) ACS Symposium Series 16, American Chemical Society, Washington DC, 1975.

161 Larsen E R and Ludwig R B, 'On the mechanism of halogen's flame suppressing properties', *J. Fire Flamm.*, 1979, **10**, 69–77.

162 Ravey M, Keidar I, Weil E D and Pearce E M, 'Flexible polyurethane foam. II. Fire retardation by tris(1,3-dichloro-2-propyl) phosphate, Part A. Examination of the vapor phase (the flame)', *J. Appl. Polym. Sci.*, 1998, **68**, 217–28.

163 Weil E D, 'Additivity, synergism and antagonism in flame retardancy' in *Flame Retardancy of Polymeric Materials*, Vol. 3, W C Kuryla and A J Papa (editors) Marcel Dekker, New York, 1975, 186–243.

164 Broadbent J R A and Hirschler M M, 'Red phosphorus as a flame retardant for a thermoplastic nitrogen-containing polymer', *Eur. Polym. J.*, 1984, **20**(11), 1087–93.

165 Cheng S and Li J, 'Mechanism of flame retardant action of tris(2,3-dichloropropyl) phosphate on epoxy resin', *Proceedings of the International Symposium on Flame Retardants*, Beijing, China, Nov. 15, 1989.

166 Morgan A W, 'Formulation and testing of flame-retardant systems for plasticized poly(vinyl chloride)' in *Advances in Fire Retardants*, Vol. 1, V M Bhatnagar (editor) Technomic Publishing Co., Westport, CT, 1972.

167 Moy P, 'FR characteristics of phosphate ester plasticizers with antimony oxide', *Plastics Engineering*, 1997, Nov., 61–3.

168 Bonsignore P V and Manhart J H, 'Alumina trihydrate as a flame retardant and smoke suppressive filler in reinforced polyester plastics' in *Proc. 29th Ann. Conf. Reinf. Plast. Compos. Inst. SPE*, 1984, 23C, 1–8.

169 Scharf D, Nalepa R, Heflin R and Wusu T, 'Studies on flame retardant intumescent char, Part I.' in *Proc. Int. Conf. on Fire Safety*, 1990, **15**, 306–15.

170 Bourbigot S, Le Bras M, Delobel R and Tremillon J-M, 'Synergistic effect of zeolite in an intumescence process', *J. Chem. Soc., Faraday Trans.*, 1996, **92**(18), 3435–44.

171 Granzow A and Savides C, 'Flame retardancy of polypropylene and impact polystyrene: phosphonium bromide/ammonium polyphosphate system', *J. Appl. Polym. Sci.*, 1980, **25**, 2195–204.

172 Savides C, Granzow A and Cannelongo J F, 'Phosphine-based flame retardants for polypropylene', *J. Appl. Polym. Sci.*, 1979, **23**, 2639–52.

173 Weil E D, 'Phosphorus-containing polymers' in *Encyclopedia of Polymer Science and Engineering*, Vol. 11, J Kroschwitz (editor) Wiley, New York, 1988, 96–126.

174 Weil E D, 'Recent advances in phosphorus-containing polymers for flame retardant applications' in *Proc. Int. Conf. on Fire Safety*, 1987, **12**, 210–18.

175 Deshpande A B, Pearce E M, Yoon H S and Liepins R, 'Some structure–property relationships in polymer flammability: studies on poly(ethylene terephthalate)', *J. Appl. Polym. Sci., Appl. Polym. Symp.*, 1977, **31**, 257–68.

176 Stackman R, 'Flammability of phosphorus-containing aromatic polyesters: A comparison of additives and comonomer flame retardants' in *Modification of Polymers*, C H Carraher and M. Tsuda (editors) ACS Symp. Ser. 121, Washington, DC, 1980, 425–34.

177 Stackman R, 'Phosphorus based additives for flame retardant polyester – 1. low molecular weight additives', *Ind. Eng. Chem. Prod. Res. Dev.*, 1982, **21**, 328–31.

178 Stackman R, 'Phosphorus based additives for flame retardant polyester – 2. polymeric phosphorus esters', *Ind. Eng. Chem. Prod. Res. Dev.*, 1982, **21**, 332–6.

179 Wyld O, 1735, 'Preventing paper, linen, canvas, etc. from flaming or retaining fire etc.,' British Pat. 551.

180 Cullis C F and Hirschler M M, *The Combustion of Organic Polymers*, Clarendon Press, Oxford, 1981, 256–61.

181 Shen K K and Ferm D J, 'Boron compounds as fire retardants' in *Flame Retardants – 101: Basic Dynamics*, Fire Retardant Chemicals Association, March 24–7, 1996, 137–46.

182 Kuckro G, 'Flame Retardant Composition', US Pat. 5 710 202, 1998.

183 Shen K, 1999, 'Modes of action of firebrake ZB as a fire retardant', article in preparation.

184 Shen K, 'Zinc borate as a flame retardant in halogen-free wire and cable systems', *Plastics Compounding*, 1988, **11**(7), 26–8, 30–2, 34.

185 Bourbigot S, LeBras M, Leeuwendal R and Schubert D, 'Zinc borate/metal hydroxide formulations in designing FR-EVA materials' in *9th Ann. Conf.*

Recent advances in flame retardancy of polymeric materials, Business Communications Co, Norwalk, USA, 1998.

186 McKee D W, 'Borate treatment of carbon fibers and carbon–carbon composites for improved oxidation resistance', *Carbon*, 1986, **24**(6), 737–41.

187 Jourdain J, Le Bras G and Combourien J, 'Kinetic studies on reactions of hydrogen, oxygen and hydroxyl radicals with hydrogen bromide, and chlorinated and brominated compounds of boron and phosphorus; application to flame retarding', *Colloq. Int. Berthelot-Vieille-Mallard-LeChatelier [Actes]*, 1st, 1981, (1), 296–300; *Chem. Abs.* **98**, 128 732.

188 Yang Y, Shi X and Zhao R, 'Behavior of zinc borate in flame retardation', *Engineering Chemistry (Huagang Yejin)*, 1995, **16**(4), 358–61; English version in preparation for *J. Fire Sciences*, 1999.

189 Walter M D and Wajer M T, 'Magnesium hydroxide: a viable alternative in flame retardants,' 1996, in *Functional Fillers Conference*, Oct. 28–30, 1996, Amsterdam.

190 Rothon R N and Hornsby P R, 'Flame Retardant Effects of Magnesium Hydroxide', *Polym. Degrad. Stab*, 1996, **54**(2–3), 383–5.

191 Martin F J and Price K R, 'Flammability of epoxy resins', *J. Appl. Polym. Sci.*, 1968, **12**, 143–58.

192 Hshieh F-Y, 'Shielding effects of silica-ash layer on the combustion of silicones and their possible applications on the fire retardancy of organic polymers', *Fire Mat*, 1998, **22**(2), 69–76.

193 Weil E D, 'Meeting FR goals using polymer additive systems', 1995 in *Improved Fire- and Smoke-Resistant Materials for Commercial Aircraft Interiors, a Proceedings*, Publ. NMAB-477-2, National Academy Press, Washington, DC, 1995, 129–49.

194 Innes J and Innes A, 'Technology Changes in Metal Hydrate Flame Retardants', 1998, in *Wire & Cable Focus Conf.*, Greensboro, NC, Sep. 15, 1998.

195 Weil E D, Lewin M and Lin H S, 'Enhanced flame retardancy of polypropylene with magnesium hydroxide, melamine and novolac', *J. Fire Sci.*, 1998, **16**, 383–404.

196 Miyata S and Imahashi T (to Kyowa Chemical Industry) 'Fire retardant resin composition', US Pat. 5 094 781, May 1992.

197 Smith P J and Mortimer J (to Mobile Corp.) 'Low toxicity fire retardant thermoplastic material', US Pat. 5 017 637, 1991 and 5 218 027, 1993.

198 Adams R W, 'Gypsum as a flame retardant filler', *Plastics Engineering*, 1988, **44**(3), 59–61.

199 'Mechanism of Flame Inhibition. II: A New Principle of Flame Suppression,' *Fire Flamm/Fire Retard Chem.*, 1975, Supplement 2, 5–20.

200 'On the Mechanism of Halogen's Flame-Suppressing Properties', *J. Fire Sci*, 1979, **10**, 69–77.

201 Lewin M and Endo M, 'Flame retardancy of polypropylene by intumescent systems: synergism, char and cone calorimetry,' *Advances in Flame Retardancy of Polymeric Materals* M Lewin (editor), Vol. 5, 1994, 56.

202 Kandola B K, Horrocks A R, Price D and Coleman G V, 'Flame retardant treatments of cellulose and their influence on the mechanism of cellulose pyrolysis' J. Macromoi. Sci, Rev. Macromoi. Chem. Phys., 1996, C36, 721–94.

3

Toxicity of fire retardants in relation to life safety and environmental hazards

DAVID PURSER

Fire Research Station, Building Research Establishment,
Watford, UK

3.1 Introduction

The use of Fire retardant (FR) treatments in applications such as building materials, furnishings and consumer durables has increased enormously during the second half of the twentieth century. This has been driven partly by increasing concerns and regulation regarding the ignitability and post ignition fire behaviour of materials and products, and partly as a result of the use of FR technology to widen the application of materials with an otherwise poor fire performance. The result is a large and increasing tonnage of FR substances in materials and products throughout their life cycle during manufacture, use and disposal. As pedigree chemicals, FR additives have been subject to the same routine toxicity evaluations as other chemical substances, but little attention has been directed to the toxicity of the thermal decomposition products of FR additives and FR treated materials when they are heated, involved in fires or incinerated for waste disposal.

Toxicity issues with regard to fire retardants have been an area of increasing concern in recent years. One issue currently receiving considerable attention is that of possible environmental contamination from small amounts of highly toxic combustion products released during accidental fires and during waste incineration, especially chlorinated and brominated dioxins and dibenzofurans from halogenated FR systems. These concerns have led to considerable restrictions being put on the use of particular FR systems by specifiers, and to pressure (particularly within Europe) for regulatory restrictions, often in the absence of an effective cost/benefit risk assessment. Another issue is concern that the yields of toxic products (particularly acid gases and carbon monoxide) and their rate of evolution during fires might be greater for FR-treated materials than those from untreated materials, and that this might result in an increased toxic hazard. Further concerns relate to processing and recycling in terms of workplace and environmental hazards as well as technical difficulties in recycling some materials.

By far the majority of injuries and deaths arising from fires, as well as the environmental hazards from combustion processes result from exposure to toxic products, so that the toxicity of combustion products is a major concern.[1] However, when evaluating any particular FR additive or system, it is important to consider the toxicity cost/benefit relationships of fire retarded materials not just in terms of the toxicity (toxic potency, i.e. dose required to produce a given toxic effect) of fire retardants and their combustion products but mainly in the context of the toxic hazards and toxic risks arising from their use.

Once a fire has started, the toxic hazard to a building occupant (or to the environment generally) depends mainly upon the rate of decomposition of the materials in the fire, the yields of toxic products evolved and to some extent the final size of the fire. Toxic risk depends upon the probabilities of occurrence of different fire scenarios and the extent of the hazards should such fires occur. In terms of the toxic risk from fires, the cost/benefit relationship for fire retardant use therefore depends upon how many fires and their subsequent hazards are prevented by the use of the FR-material and the extent of the toxic hazards from the remaining fires occurring that involve FR-materials. The toxicity of fire retarded materials therefore becomes an issue when the toxic hazards (both direct and environmental) from fires involving fire-retarded materials becomes significantly greater than those from fires involving equivalent non-FR materials. The toxicity of FR materials becomes a more serious issue when the toxic risk arising from the use of a fire retardant becomes greater than that when the fire retardant is not used.

An example of an issue involving a risk assessment problem is provided by the debate concerning the use of brominated fire retardants in printed circuit boards and casings of television sets. In this case the cost/benefit problem relates to the potential benefits in terms of fire prevention resulting from the use of the retardants compared to the potential disbenefits of possible environmental hazards resulting from the release of toxic products at different stages of the product life cycle. In recent years there has been a concern, particularly in Europe, that polybrominated ether fire retardants used in such situations might evolve toxic dibenzodioxins and dibenzofurans when heated and that this might present a health hazard.[6,7] This has led to a reduction in the use of such FR-treatments and there is now some evidence for an increase in the occurrence of fires in television sets in Europe.[8] A television fire can present a serious toxic hazard to building occupants, even if the fire does not spread to other room contents items, as is often the case. If the fire spreads and destroys part of the building then a potential environmental hazard arises from the contamination of air and water by the total toxic effluent. Due to the mix of materials involved in most large fires and the mixed elemental composition of these materials, it is likely that such fires will release considerable quantities of toxic products, including the very chlorinated and brominated dioxins and dibenzofurans that cessation of use of the brominated fire retardants was designed to prevent. On the other hand, it may be that substitute materials with a better fire performance might be considered for television sets, or less environmentally hazardous FR-systems, so that the overall risks might be improved. In order to make informed judgements of toxic risks associated with different FR-systems and the risks of using non-FR-materials, it is important that valid investigations are made of the toxic product yields and toxic potencies from the materials involved and their alternatives. These should include realistic large- and small-scale experimental investigations of toxic substance yields and toxic hazards in full-scale fires involving a range of realistic fire scenarios. These need to be combined with comprehensive surveys of end-use risk scenarios.[4,5]

3.2   Toxic combustion products from fires: general

During a fire the toxic products most hazardous to building occupants are organic irritants (e.g. acrolein, crotonaldehyde, formaldehyde, phenol, styrene), inorganic irritants (e.g. hydrogen halides, nitrogen oxides, sulphur oxides and phosphates) and asphyxiant gases (CO, HCN and CO_2). Organic and inorganic irritants are important mainly because, in combination with smoke, they impede escape attempts due to painful effects on the eyes and respiratory tract. Asphyxiant gases are important mainly because they cause incapacitation and death in fires.[2] Fires inside buildings (or transport systems) are likely to be fatal to any occupant remaining in the vicinity of the fire due primarily to the effects of asphyxiant gases (especially CO and HCN) and heat. However, the main determinant of survival in a fire is whether or not the occupants are incapacitated before they are able to escape. This depends primarily upon the effects of smoke, irritants, asphyxiants and heat on escape capability.[2]

Outside or following a fire the most important products are substances causing potential toxic environmental contamination such as acid residues, certain metals and exotic toxic products such as polyaromatic hydrocarbons, isocyanate derivatives, dioxins and dibenzofurans.[3]

3.2.1 FR-systems and toxicity

FR-systems are generally designed to provide a particular level of resistance to ignition or flame spread, usually defined in terms of standard small- (or sometimes large-) scale tests. Their main benefit is usually therefore to prevent or delay fires growing from small ignition sources. Since any fire potentially presents a serious toxic hazard to occupants, FR-systems generally present some benefits in terms of toxic risk for ignition sources up to those for which they are designed to provide protection. When fires do occur in FR-systems there may still be some reduction in the rate of development of toxic hazard if the FR-system delays flame spread and fire growth, as is often the case. The yields of toxic products in fires involving FR-systems (mass of toxic products per mass of materials involved in the fire) may be greater or less than those from non-FR equivalents. Some FR-systems are designed to act in the solid phase by promoting char formation or the formation of other barriers to decomposition (e.g. borates, silicon systems, phosphate systems, inert fillers, certain aromatic polymers or co-polymers). These tend to prevent or inhibit the decomposition of the parent material and therefore reduce the yields of toxic products released into the fire atmosphere, thereby reducing its toxic potency. Since the rate of decomposition is also reduced, these systems potentially provide a double benefit by reducing both the rate of decomposition and the toxic product yields. Unfortunately, although there may be benefits from these processes, some of these FR treatments, particularly some phosphate systems, may release products acting in the gas phase which may themselves be toxic or lead to an increased formation of other toxic fire products.

FR-systems designed to act in the gas phase tend to do so by inhibiting gas-phase reactions either by dilution or though the inhibition of free-radical flame reactions. The addition of alumina trihydrate to materials provides an effective and toxicologically benign mechanism for achieving fire retardancy. This material reduces the mass of combustible material as a filler, and on decomposition releases water, which absorbs heat and dilutes fire gases, inhibiting the flame. The overall effect is generally a reduction in both decomposition rate and toxic product yields. The commonly used halide and metal halide FR-systems, sometimes also involving volatile phosphates or nitrogen compounds, act primarily by inhibition of flame reactions. Although such systems may be beneficial by reducing toxic risk they usually provide increased yields of toxic products in fires. Whether or not this leads to an increased toxic hazard in any particular fire depends upon whether there is a decrease or otherwise in the fire growth rate compared with that in a non-FR equivalent, and whether this offsets the disbenefit of increased toxic product yields.

Toxic product yields may be increased partly due to increased yields of asphyxiant gases (particularly carbon monoxide and hydrogen cyanide) and of organic irritants (resulting from impaired combustion efficiency) and partly due to the release of irritant or asphyxiant decomposition products from the fire retardants themselves (hydrogen and metal halides, phosphates, sulphates, nitrogen oxides or hydrogen cyanide). Whether or not there is a net decrease or increase in toxic hazard will depend upon the particular FR-system/material combinations in use and the particular fire scenario. Where FR-treatments result in small ignition sources failing to produce propagating fires or slowing propagation, then there may be a reduction in toxic hazard, particularly in combination with efficient fire detection. Where non-flaming decomposition occurs in a confined space or where an ignition source is large enough to overcome the FR-system, then the toxic hazard may be similar to, or somewhat worse than, that resulting from the use of a non-FR system.

3.2.2 Environmental toxicity issues

This problem of the brominated fire retardants raises another issue. The toxicity of FR-treatments may be a problem not just in fires but also during the entire normal life cycle of the product and in its disposal. Of particular concern are situations where the FR-compounds themselves are significantly toxic and where they may be released from the product during use or following disposal. Also of concern are situations where methods of disposal involving combustion such as incineration or disposal on open fires may lead to the release of environmentally-threatening toxic products. With regard to the polybrominated ethers the greatest yield of brominated dioxins and dibenzofurans is likely to occur during non-flaming thermal decomposition or combustion in smouldering or flaming fires. An investigation of three domestic fires involving television cases containing brominated fire retardants showed brominated dioxin surface deposits at concentrations of up to 14.9 ppm.[9] It has also been suggested that some release might occur from televisions at temperatures attained during normal use.[10] More recently, work in Sweden has shown levels of polybrominated ethers in human breast milk to have increased more than 50 times over the last 25 years and doubled since 1992, while employees working with brominated flame retardants have been shown to have 50 times higher blood levels than normal. Concerns have been raised about possible neurotoxic effects from brominated fire retardants and Sweden is reported to be considering a ban on brominated fire retardants.[11]

When such toxicity issues arise it is often possible to develop alternative FR-treatments, and this is one issue driving research in the FR industry. A number of brominated fire retardants have been developed with the intention of avoiding any potential dioxin release problem.[12] Another example arose some years ago in the United Kingdom when an issue was identified with regard to the flammability of childrens' sleepwear. An increasing incidence of burns and deaths was identified, through the fire statistics of fires resulting from children wearing loose sleepwear (particularly girls' nightdresses), standing close to domestic fires. Legislation was introduced requiring FR-treatment of such garments or the use of less flammable materials such as 100% polyester or nylon. A particular fire retardant (TRIS) was often used. However, it was subsequently found that this could leach out of the garment when wet (such as from urine). The substance was found to be a potential carcinogen[13] and there was concern that dermal absorption in children might lead to a potential problem. The use of this substance was therefore discontinued and a number of alternative approaches have been developed.[14]

In general, toxicity and toxic hazard issues relating to direct exposure to fire retardants depend upon the toxicity profile of each substance, its bioavailability, bioaccumulation and biotransformation. The direct toxicity of most common fire retardants appears to be low, although full data are not always readily available on the wider range of compounds in use.[15] The main concerns arising currently appear to be related to a greater degree of bioavailability and bioaccumulation than might have been expected. This results in exposure of the whole population, generally to low levels, so that subtle toxic effects resulting from chronic exposure become an issue. The consideration of toxicity in relation to the use of fire retardants therefore raises a number of complex issues. Their resolution sometimes depends upon detailed considerations of individual cases, but it is often difficult to estimate the overall risk or benefits from a particular course of action.

With regard to environmental contamination, the greatest concerns have been expressed in relation to halide and halide metal systems, particularly with respect to the potential release of chlorinated and brominated dibenzodioxins and dibenzofurans during fires and during incineration.[4] In order to make a realistic assessment of the toxicity cost/benefits of using such FR-systems, it is important to consider the contribution to environmental contamination from these substances during combustion processes in relation to environmental contamination from other sources, which are often very much greater.[3] In terms of accidental fires, it is important to consider the contribution to environmental contamination (and other hazards) from fires involving these FR-systems with the contribution from other halogenated materials in the same fires, and from other accidental non-FR fires which might have been prevented, or reduced in extent, had halogenated FR-systems been used.

A further complication arises when an intention to control the fire performance of a particular product or material is translated into a test regime. Such controls are often based upon very simplistic small-scale tests for particular properties such as ignitability, flame spread or toxic potency. The use of such tests in combination with pass-fail or indexation criteria may present a simple set of performance targets for industrial producers to meet and for specifiers and regulators to use. However, it is often difficult to relate performance in such tests to full-scale performance in the end-use situation. Certainly, in the toxicity area many small-scale tests in current use for specification may produce misleading or even counterproductive results in terms of end-use performance or hazard.[16]

3.2.3 Evidence of beneficial effects of fire retardants

The benefits of fire retardant usage may be estimated in retrospect by an examination of fire statistics. In the United Kingdom many fire deaths and injuries, the majority resulting from the effects of toxic smoke, have been the outcome of fires involving upholstered furniture. Table 3.1 shows data for the early post war period, the peak period for fire deaths during the late 1980s and the current situation. Fire deaths rose from around 500 per year during the late 1950s and peaked during the early 1970s, occurring at around 950 per year up to the late 1980s. In 1988 the United Kingdom upholstered furniture fire regulations were introduced requiring cigarette and match ignition resistance, resulting in a cessation of the use of non-combustion modified polyurethane foams[17] in new furniture. Furniture sold since 1988 typically contains brominated/antimony trioxide or chlorinated phosphate treatments for textiles and chlorinated phosphate/melamine treatments for foam fillings. Since 1988 there has been a gradual decline in UK fire deaths, which have occurred at around 750 per year since 1994, with a reduction in deaths resulting from furniture as the item first ignited. There is therefore some evidence that the use of fire retardant systems in UK upholstered furniture has helped to reduce fire deaths, although there has been no corresponding impact on fire injuries, particularly toxic smoke injuries, which have increased steadily throughout this period and up to the present. However, fire and smoke deaths and injuries are still well in excess of those occurring during the late 1950s before modern synthetic materials and FR-treatments were widely used in furniture.

Televisions provide another example of the benefits of the application of fire retardant products and improved design to a consumer product.

Table 3.1 Average total fire deaths and injuries and deaths and injuries caused by toxic smoke in the United Kingdom

	1955–1960	1985–1990	1995–1998
Total deaths	500	950	750
Total injuries	3000[a]	11 000[a]	12 000[a]
Toxic smoke deaths	100	600	450
Toxic smoke injuries	100	3000	6000

[a] Excluding check up.
Copyright HER.

Television fires were identified as a particular problem in the United States and Europe in the late 1960s. Fire performance was improved in the early 1970s as a result of voluntary standards introduced in the United States and a European Community Directive (73/23/EEC) in Europe. These involved a variety of FR-treatments of television casings and printed circuit boards. This resulted in an estimated reduction of 73% in residential fires due to televisions in the United States between 1983 and 1991, and a reduction of 79% in reported television fires in the United Kingdom.[15] This pre-dates the recent problems with European televisions related to reduced levels of FR-treatments.

In this chapter the main aim is to examine the parameters determining the toxic potency of thermal decomposition and combustion products from fire retarded materials in relation to acute toxic hazards in fires and environmental hazards. In particular, an examination is made of the evolution of toxic products in fires and their capacity to impair escape attempts or cause incapacitation or death during a fire. Also considered are effects immediately after a fire and potential long term effects on health.

3.3   Toxic effects of smoke products

3.3.1 Life-threatening effects of exposure during fires

The concentrations of toxic products and their effects change rapidly during a fire exposure. Survival depends upon the relationship between the timing of the growing toxic hazard in the fire and the time required for occupants to escape.[2,18] Some toxic products are important because they may impede or delay escape during the early stages of a fire, while others are more important as causes of incapacitation, preventing escape, and as causes of death.

With regard to a toxic hazard assessment the major considerations are:

1   The time when partially incapacitating effects are likely to occur which might delay escape.
2   The time when incapacitating effects are likely to occur which might prevent escape, compared with the time required for escape.
3   Whether exposure is likely to result in permanent injury or death.

Incapacitating effects include:

(a)   Impaired vision resulting from the optical opacity of smoke and from the painful effects of irritant smoke products and heat on the eyes. This depends upon the concentrations of smoke particulates and the concentrations of acid gases and organic irritants (hydrogen halides, oxides of nitrogen, sulphur and phosphorus, acrolein, formaldehyde and other organic substances).

(b) Respiratory tract pain and breathing difficulties or even respiratory tract injury resulting from the inhalation of irritant smoke which may be very hot. In extreme cases this can lead to collapse within a few minutes from asphyxia due to laryngeal spasm and/or bronchoconstriction. Lung inflammation may also occur, usually after some hours, which can also lead to varying degrees of respiratory distress.

(c) Asphyxiation from the inhalation of toxic gases resulting in confusion and loss of consciousness. This depends upon the concentrations of asphyxiant gases (CO, CO_2, HCN and low O_2 as well as the respiratory effects of irritants).

(d) Pain to exposed skin and the upper respiratory tract followed by burns or hyperthermia, due to the effects of heat, preventing escape and leading to collapse.

All of these effects can lead to permanent injury, and all except (a) can be fatal if the degree of exposure is sufficient. During the early stages exposure to irritant smoke is likely to be the most important factor in impeding and slowing escape. The effects of smoke obscuration and sensory irritation are immediate and the severity depends upon the concentration present. After the initial effects of smoke exposure, asphyxiation becomes a more serious hazard. This is partly because concentrations of asphyxiant gases usually build up to high concentrations some minutes after smoke becomes a problem, and partly because the toxic effects depend upon a dose inhaled over a period of time by the occupants. Also, asphyxiant gases have minor effects at low doses but cause severe intoxication with collapse and unconsciousness once a sufficient dose has been inhaled.

Up to a certain level of severity, the hazards listed in the list (a) to (d) cause a partial incapacitation by reducing the efficiency and speed of escape. These effects lie on a continuum from little or no effect at low levels to relatively severe incapacitation at high levels, with a variable response from different individuals. It is important to make some estimate of effects that are likely to delay escape, which may result in fewer occupants being able to escape during the short time before conditions become so bad that escape is no longer possible. Most important in this context is exposure to optically dense and irritant smoke, which tends to be the first hazard confronting fire victims. For more severe exposures a point may be reached where incapacitation is predicted to be sufficiently bad as to prevent escape. For some forms of incapacitation, such as the point where asphyxiation leads to a rapid change from near normality through a brief period of intoxication, to loss of consciousness, this point is relatively easy to define. For other effects an end point is less easily defined; for example the point where smoke becomes so irritating that pain and breathing difficulties lead to the cessation of effective escape attempts, or the point where pain and burns prevent movement. Nevertheless, it is considered important to attempt some estimate of the point where conditions become so severe in terms of these hazards that effective escape attempts are likely to cease, and where occupants are likely to suffer drastic incapacitation or injuries.

3.3.2 Toxic environmental hazards threatening long-term health

The most important toxic gases threatening survival in fires tend to be the least important environmental hazards to health. The main environmental health hazards are smoke particles containing adsorbed carcinogens such as polyaromatic hydrocarbons, dibenzodioxins and dibenzofurans. Other carcinogens released in fires include benzene and acrylonitrile. Ultrafine smoke particles are currently viewed with concern as possible causes of premature deaths from cardiovascular and respiratory diseases.

3.4   Methods available and hazards to be assessed

3.4.1 Main factors determining hazards to life

For toxic fire effluent, hazards to life can be expressed in terms of three major parameters: the first two listed below relate to the fire itself and determine the concentrations of toxic gases in the fire effluent. The third relates to the toxic effects. They are as follows:

1   The fire growth curve in terms of the mass-loss rate of the fuel (kg/s) and the volume into which it is dispersed (kg/m^3).
2   The yield of toxic products and smoke in the fire (for example, kg of CO per kg of material burned).
3   The toxic potency of the products (the concentration or exposure dose needed to cause toxic effects).

The hazard from smoke obscuration also depends upon the mass loss rate of the fuel, the smoke yield and the degree of obscuration required to affect escape capability (expressed in terms of optical density per metre [OD/m] or similar units). The hazard from heat depends upon the heat release rate of the fire and the subsequent energy balance of the effluent. The hazard from convected heat is expressed in terms of the air temperature at the exposed skin of the occupant and that from radiant heat in terms of the radiant flux to the exposed skin (kW/m^2).

There are a number of ways in which data on these parameters can be obtained directly, or by which they may be covered by a particular test strategy, with varying degrees of confidence depending upon the types of test used.

The most comprehensive method is to carry out full-scale tests of the situation under investigation, or large-scale tests providing a partial simulation. In this case the mass loss rate of the fuel, the volume of dispersal of the effluent, the concentrations of toxic gases, the smoke obscuration, effluent temperature and radiant flux can all be measured directly. Otherwise, data may be obtained from a variety of sources, usually involving the use of mathematical models of fire growth, with small-scale test results as input data. The third set of data has been obtained historically by exposing animals to the mixed test effluents from burning materials, or by studying the toxic effects of individual fire gases in animals and humans. Research in this area has shown that in many cases the major toxic effects are caused by a small number of well-known toxic gases, so that toxic potency can now be predicted to a considerable extent based upon existing knowledge if the concentrations of CO, CO_2, HCN, O_2 and some relevant irritant gases are measured.[2,19] If these data are available it should be possible to predict the major toxic effects such as time to incapacitation or death for humans (where data relate to full-scale tests or fire simulations) or rodents (where analytical data are obtained from small-scale toxicity tests).[2,20] However, without animal exposures it is not possible to make a full assessment of irritancy, nor to detect unusual acute toxic effects and more subtle long-term effects such as those from halogenated dioxins and dibenzofurans.[2,16]

3.4.2 Practical methods for toxic hazard assessment

There are essentially two ways in which toxic and other life threat hazards in fire can be assessed:

1   From large-scale fire data including the concentration/time profiles of the major toxic gases, smoke obscuration, temperature and radiant heat flux and existing knowledge of the toxic and physiological effects in humans of exposure to these effluent components. These data may be obtained directly from large-scale fire tests, from design fire data or from fire data estimated using mathematical models.
2   From a battery of small-scale tests and mathematical models, or simple large-scale tests. The essential components for a toxic hazard analysis are:
    (a)   the toxic potency data for materials (lethal mass loss exposure dose [gm^{-3} min] obtained from small-scale tests using animal exposures or analytical methods;
    (b)   the mass loss/concentration curve for the fire.

In principle it is possible to treat smoke obscuration in the same way, by measuring smoke yield from a material in a small-scale test and relating it to the mass loss/concentration curve for the fire. It is also possible to predict effluent temperature and radiant flux in a fire from small-scale test data.

3.4.3 Basic mechanisms of flame retardancy and implications for toxicity and toxic hazard

3.4.3.1 *Factors determining the toxic potency of materials under fire conditions*

Fire effluent consists of a mixture of gases, liquid droplets and solid particles representing the thermal decomposition and combustion products from fires. The chemical and physical composition of fire effluent depend principally upon:

1 The elemental composition of the material decomposed.
2 The organic composition of the material decomposed.
3 The thermal decomposition conditions in the fire.[2,21,22]

For materials 'normally' involved in fires (i.e. materials used in building construction, furnishings and other contents, and indeed almost all combustible materials) the main elements are carbon, hydrogen and oxygen, so that the bulk of all combustion products consists of products formed from these elements. The next most important elements present in common materials are halogens (mostly chlorine) and nitrogen. Chlorine comprises up to 50% (approximately) of the mass of polyvinyl chloride and is present at smaller percentages in many other materials, particularly fire retarded materials. Nitrogen constitutes approximately 5 and 40% of the mass of some common materials (such as polyamides, polyurethanes, urea formaldehyde resins and polyacrylics). Other elements are usually present at a few per cent, and are often used to modify the material properties or combustion performance (antimony, halogenated phosphorus compounds, sulphur, aluminium salts).

3.4.3.2 *Products of combustion*

Inorganic products

In fires, almost all inorganic anions are released as acid gases at high yields (sometimes approaching 100%). Thus HCl, HBr, HF, SO_2, and P_2O_5, are usually present at varying levels in the effluent from most mixed fuel fires, often partly derived from substances used as fire retardant treatments. For nitrogen present in the materials the fate is more complex and depends to a large extent on the decomposition conditions.[18,21,22] The main products are N_2, HCN and NO_x. Generally, a significant proportion of the nitrogen from well ventilated fires is released as NO_x, while for more vitiated, ventilation-controlled fires (i.e. most fires in buildings after the early stages) a significant percentage of the nitrogen is released as HCN. Other elements, including anions and metals may be found in varying amounts in soot deposits, depending upon the materials involved in the fire. However, depending upon the decomposition conditions, a considerable fraction of phosphorus and metal salts tends to remain in the char and residues rather than entering the fire plume. Deposition of heavy metals tends to be greater near the fire than for lighter materials.

Carbon compounds

The major products of fires are the carbon compounds, and the yields of different compounds depend mainly upon the thermal decomposition conditions in the fire, and to some extent upon the organic chemical composition of the materials burned. They also depend upon the effect of fire retardants in modifying the combustion process. In terms of the fire chemistry, the basic scenarios and the hazard development, fires can be classified into three basic types:

1 Smouldering/non-flaming fires.
2 Early, well ventilated fires.
3 Ventilation-controlled flaming fires:
    (a) small, vitiated flaming fires
    (b) post-flashover, ventilation-controlled fires.

The general conditions in these fire types are shown in Table 3.2. This fire classification has been developed as an ISO fire classification for the revised version of the current technical report on fire models.[23] It is published in the current British Standard.[24]

When materials are decomposed the first step is pyrolysis, by which the material is broken down by heat into a range of organic fragments. These consist primarily of an aliphatic series of hydrocarbons from methane upwards, with aromatic compounds formed by ring cyclisation, or by the thermal decomposition of substances containing aromatic moieties (e.g. styrene monomer formed from the decomposition of polystyrene). If the molecule contains oxygen and if (as in most cases) decomposition occurs in air, then these products are partially oxidised to produce carbonyl compounds such as acrolein and formaldehyde and other species such as phenol. In addition to these acutely toxic compounds, the organic mixture usually contains systemically toxic and carcinogenic substances such as benzene and polyaromatic hydrocarbons. Other substances such as acrylonitrile, toluene diisocyanate, dioxins and dibenzofurans may also be present. Some of these

Table 3.2 Revised classification of fire types

Fire stage or type	Temperature (°C)		Oxygen to fire (%)	Fire effluents	
	Fire	Hot layer		Oxygen from fire (%)	CO₂/C (v/v)
1. NON-FLAMING					
a Self-sustaining	450–600	RT[a]	21	>20	1–5
b Oxidative pyrolysis from externally applied radiation	300–600	<50	21	>20	1–5
c Non-oxidative pyrolysis from externally applied radiation	300–600	<50	0	0	<5
2. WELL VENTILATED FLAMING The fire size is small in relation to the size of the compartment, the flames are below the base of the hot layer and fire size is fuel controlled	>700	RT to 500	>15	5–21	>20[b]
3. LESS WELL VENTILATED FLAMING The fire size may be large in relation to the size of the compartment, the flames are partly above the base of the hot layer and fire size is ventilation controlled					
a Small vitiated fires in closed compartments	>700	RT to 500	<15	0–12	2–20
b Post-flashover fires in large or open compartments	>700	500 to 1000	>15	0–12	2–20

[a] RT = Room Temperature. During smouldering or the early stages of small well ventilated or vitiated flaming fires the hot layer temperature may be barely elevated over room temperature.
[b] May be lower if the burning materials contain fire retardants. In order to determine whether flaming decomposition conditions in a particular apparatus fall into category 2 or category 3 it is necessary to use a non-fire retarded reference material capable of efficient combustion.

toxic compounds such as benzene and pyridine can be present at concentrations in the 10–20 mg/m^3 concentration range, approximately equal to the occupational exposure limit (OEL); others such as isocyanates have been detected in the smoke plume. These organic fragments provide the volatile fuel for combustion, yielding carbon oxides and water. Based upon these influences the conditions in the three major fire types are as follows:

3.4.3.3 Types of fire

Smouldering/non-flaming fires

These involve slow thermal decomposition without flames, so that a serious hazard inside a building requires several hours to develop. The decomposition may be induced by heat supplied externally or may be self-sustaining. Slow thermal decomposition results in oxidative non-flaming conditions. The products are very rich in organic compounds (approximately 50% or the mass decomposed), which are usually highly irritant to the respiratory tract. Inorganic acids provide a further source of irritants. Another major toxic product is carbon monoxide. Such fires are usually small and develop slowly. They can present a serious toxic hazard over a period of an hour or more due to the slow build-up of carbon monoxide and organic irritants. Inorganic acid gases may also be released slowly if the appropriate anions are present in the material. In general, the toxicity of fire retarded materials would not be expected to differ greatly from non-fire retarded equivalents under these conditions. The presence of flame retardants may have a considerable influence on the probability that a transition to flaming combustion will occur, which may have profound effects on subsequent fire hazards.

Well ventilated flaming fires

These fires occur when there is plenty of air available so that the ratio of fuel to air is low. Under these conditions combustion is most efficient, so that for most non-fire retarded materials, the main products are carbon dioxide, water and heat and the yields of smoke and toxic products tend to be low initially. The toxic potency and toxic hazards from simple CHO-containing polymers are therefore small to start with, but the fire is likely to grow quickly, producing considerable quantities of heat and carbon dioxide while consuming oxygen. The behaviour of fire-retarded materials during this phase of a fire depends to some extent on the nature of the fire, the materials being burned and the mechanism of flame retardancy. Fire retardants tend to act either mainly in the gas phase, by inhibiting flame reactions, or in the solid phase, by inhibiting combustion. Once the degree of ignition resistance afforded by the FR-treatment bas been overcome, then the fire retarded material will decompose in the flaming mode.

As the fire develops a more hazardous mixture of products may be formed. Carbon monoxide and carbon dioxide can be significant toxic products, and many inorganic products may be released as acid gases. Some materials (particularly if treated with fire retardants) are unable to burn efficiently, producing high yields of CO and organic products, particularly those treated with fire retardants acting in the gas phase. The early stages of most fires in buildings and fires outside tend to fit into this category. Such fires inside buildings are usually too small to present a significant environmental hazard outside the building, but may grow into post-flashover ventilation controlled fires (see below).

Ventilation controlled flaming fires

These fires occur when the air supply is restricted in comparison with the fuel available for combustion. Most fires in buildings become ventilation controlled after the early stages. They may consist of pre-flashover fires in enclosed spaces or large, post-flashover fires where all surfaces are ignited in high temperature (often as high as $1000\,^{\circ}C$) conflagrations in very large or ventilated spaces. Ventilation controlled fires, both pre- and post-flashover are the main threat to building occupants and the main threat to the environment beyond the building of origin. The restricted ventilation results in high yields of carbon monoxide, carbon dioxide, hydrogen cyanide, organic products, smoke and inorganic acid gases. Ventilation controlled fires therefore tend to be a worst case for toxicity, since they produce large amounts of effluent containing high yields of toxic products. These high yields of toxic products are obtained from both normal and fire retarded materials, so that the increased yields from some fire retarded materials compared to untreated materials seen under well ventilated conditions are less evident under ventilation controlled conditions.

As stated, smoke and toxic products are important in fires: firstly, in impeding escape attempts, secondly, in causing incapacitation and thirdly, as causes of death. One measure of toxic potency with respect to FR and non-FR materials is the exposure dose of combustion products released under different fire conditions required to cause death. Lethal exposure doses are expressed in terms of the LCt_{50}, the exposure dose required to kill 50% of exposed rats, which is considered to provide an approximate indication of the likely lethal exposure dose to humans. This is estimated from small scale toxic potency tests on materials using either direct animal exposures or using measurements of the yields of toxic products in small scale tests and calculating the lethal toxic potency by Fractional Effective Dose methods.[2,20]

A difficulty with small-scale tests is the extent to which the decomposition conditions in the test, and hence the yields of toxic products, reflect those in particular types or stages of full scale fires. Table 3.2 has been produced as a guide to the generic decomposition conditions of different fire types. Small-scale test methods have often been developed without consideration of the extent to which they address any particular fire type but by an examination of the conditions under which they operate and the CO_2/CO yields in the effluent it is possible to classify at least some test methods. The results can then be used as an indication of the ranges of lethal toxic potencies for different classes of common materials (including fire retarded materials) under different decomposition conditions. Different test methods have been reviewed[2,22] and ISO guidance on the selection and use of small-scale fire tests has been provided.[23] Unfortunately, by far the majority of data available are for non-flaming oxidative pyrolysis from externally applied radiation (fire type 1b), while some data are available for well ventilated flaming (fire type 2). Few data are available for the most important fire conditions (fire types 3a and b). Based upon a review of usable published data.[22] Table 3.3 has been developed as an indication of the ranges of lethal toxic potencies for different classes of common materials.

As the table shows, most materials are least toxic (highest values) under well ventilated flaming conditions since the yields of smoke and toxic products are at a minimum under these conditions. Toxic potency is approximately doubled under non-flaming conditions, due to the high yields of organic irritants and carbon monoxide. Under vitiated or post-flashover

Table 3.3 Approximate lethal exposure doses ($LCt_{50} g m^{-3} min$), and lethal concentrations ($LC_{50} g m^{-3}$) for common materials under different fire conditions[a]

Material	Non-flaming		Early flaming		Post-flashover	
	LCt_{50}	LC_{50}	LCt_{50}	LC_{50}	LCt_{50}	LC_{50}
Cellulosics	730	24	3120	104	750	25
CHO. polymers	500	17	1200	40	530	18
PVC	500	17	300	10	200	7
Wool/nylon (low N_2)	500	17	920	31	70	2
Flexible polyurethane	680	23	1390	46	200	7
Rigid polyurethane	63	2	100	3	54	2[b]
Modacrylic/PAN	160	5	140	5	45	1.5[b]

[a] LC_{50}s are for a 30-minute exposure time with a 14 day observation period.
[b] estimates based upon limited available data.
(adapted from Purser[2,22]).
Copyright HER.

combustion conditions toxic potency is also approximately double that under well ventilated conditions for most materials, mainly due to high yields of CO (or HCN) and to a lesser extent organic irritants. Nitrogen-containing materials show the highest toxic potencies due to the evolution of HCN, particularly under vitiated combustion conditions. PVC shows less variation with different combustion conditions. This is because PVC produces HCl at similar yields under all combustion conditions, and high yields of CO under both well ventilated and vitiated flaming conditions.

Since toxic hazard depends mainly upon the rate of decomposition of materials, basically the heat release rate (HRR), and the toxic potencies of the products released, it is of interest to compare the range of decomposition rates shown by common materials with the range of toxic potencies. Stevens and Mann[15] report that HRR comparisons of product performance suggest a range of 100:1 between the best and worst performing products, with fire retardants producing a reduction of a factor of 10 or more. This compares with a toxic potency range of around 200:1 between the best and worst performing materials in Table 3.3.

3.5   Effects of specific fire-retardant systems

3.5.1 Inorganic fillers

3.5.1.1 Toxic potency of combustion products in fires

Inert fillers such as clays and a variety of salts improve the fire performance of materials and their toxic potency by reducing the organic content of the material. Some also show active fire retardant properties by promoting char formation or releasing inert gases which dilute the combustible gases produced. Alumina trihydrate is an example which both reduces organic content and releases water on heating. This both absorbs heat and dilutes flammable gases. Table 3.4 shows the yields of major toxic products and calculated toxic potency data for a low smoke and fume cable material containing alumina trihydrate compared with a low density polyethylene cable material under a range of fire conditions using the Purser tube furnace.[25] The yields are expressed in mass charge terms (yield per gram of material placed in the furnace). Organic carbon indicates the approximate mass yield of carbon in the form of organic products, an indication of the comparative yields of organic irritants and other potentially toxic organic compounds. For this method the yields of major toxic products are measured under appropriate decomposition conditions and the toxic potency (predicted LC_{50} concentration) is calculated in mass charge and mass loss terms according to a toxic potency model.[20,25]

Table 3.4 Toxic product yields (mass charge) and toxic potency of 100% organic cable material (low density polyethylene) compared with a low smoke and fume cable material containing aluminium hydroxide (LSF)

Material and decomposition condition		CO g/g	CO_2 g/g	Organic Carbon g/g	Smoke OD/g/m^2	LC_{50} g/m^3 mass charge	LC_{50} g/m^3 mass loss
LDPE:							
Non-flaming	350 °C	0.16	0.03	0.69	0.59	17	16
Well ventilated	650 °C	0.00	2.58	0.14	0.17	70	70
Small vitiated	650 °C	0.06	0.48	0.70	0.10	28	28
Post flashover	825 °C	0.07	0.48	0.68	0.31	24	24
LSF:							
Non-flaming	350 °C	0.04	0.09	0.17	0.17	60	25
Well ventilated	650 °C	0.00	0.98	0.04	0.07	168	100
Small vitiated	650 °C	0.03	0.33	0.15	0.07	73	35
Post flashover	825 °C	0.05	0.38	0.20	0.08	50	31

Adapted from Purser et al 1994.[25] Copyright HER.

When yields of toxic products are expressed in terms of mass charge (yield per gram or material heated) rather than mass loss (yield per gram of material decomposed) the low smoke and fume (LSF) cable material has a lower yield of CO, CO_2, organic products and smoke under all fire conditions and the toxic potency performance is considerably better (i.e. higher LC_{50} concentrations) than the low density polyethylene, also over a range of fire conditions. When expressed in mass loss terms the yields of toxic products are still generally lower and the LC_{50} concentrations higher than for the LDPE. This is at least partly due to the fact that part of the mass loss of LSF is in the form or water from the alumina trihydrate.

Table 3.5 shows the effect of borax–boric acid topical FR treatment on cotton twill. The treatment formed a glass-like substance with heavy char formation, resulting in a reduced yield of carbon monoxide, smoke and organic irritants under non-flaming decomposition conditions. This resulted in a decreased toxic potency at both 400 °C and 700 °C. It also inhibited ignition to the extent that flaming decomposition was not obtained.[26]

Depending upon the particular application and formulation inert fillers can be highly effective in reducing ignitability, burning rate and toxic product yields. For example in the CBUF project[27] foam mattresses fully impregnated with alumina trihydrate were found to be highly ignition resistant. Such treatments therefore reduce both toxic hazard and toxic risk, although there may be difficulties with other desired non-fire performance characteristics (such as tensile strength or comfort).

Table 3.5 Toxic product yields (mass charge) and toxic potency of cotton twill untreated and treated with borax–boric acid under non-flaming and flaming conditions at 400 °C and 700 °C at a mass charge concentration of 20 g/m^3

Material and decomposition condition		CO g/g	CO$_2$ g/g	Acrolein g/g × 1000	Fomaldehyde g/g × 1000	Smoke OD/m	LC$_{50}$ g/m^3 mass charge	LC$_{50}$ g/m^3 mass loss
Cotton twill:								
Non-flaming	400 °C	0.21	0.75	2.10	24.66	0.16	23	21
Non-flaming	700 °C	0.33	0.70	5.83	17.74	0.06	14	14
Well ventilated	700 °C	0.04	1.61	0.00	14.40	0.00	57	55
Cotton twill: borax/boric acid								
Non-flaming	400 °C	0.11	0.36	0.47	13.87	0.01	51	30
Non-flaming	700 °C	0.24	1.06	0.82	6.38	0.00	25	22

Adapted from: T. Wright PhD Thesis 1997.[26] Copyright HER.

3.5.2 Halogen acid vapour-phase fire retardants

Halogen acids act as fire retardants primarily by inhibiting flaming in the vapour phase through free-radical scavenging mechanisms. They also dilute the organic content of polymers, reducing the heat of combustion, and dilute the organic content of the gas phase. These mechanisms tend to reduce toxic hazard and risk by rendering halogenated materials less easily ignited and slower burning than non-halogenated equivalents, but the reduction of combustion efficiency and addition of irritant acid gases tend to increase toxic potency especially under flaming conditions.

3.5.2.1   Lethal toxic potencies of halogenated materials – small scale test data

The reduction in combustion efficiency, in terms of increased carbon monoxide yield, is proportional to the chlorine content of the material.[28,76] This is illustrated by heavily chlorinated materials such as PVC, which contain up to 50% chlorine by mass. As Table 3.3 shows, under non-flaming conditions the lethal toxic potency (in terms of the LCt$_{50}$) of PVC (both plasticised and non-plasticised) tends to be similar to most other materials at around 500 g m^{-3} min, but under well ventilated flaming decomposition conditions the LCt$_{50}$ is around 300 g m^{-3} min, compared with LCt$_{50}$ in excess of 1000 g m^{-3}/min for non-halogenated (and non-nitrogen containing) plastics and cellulosics (NB the lower the LCt$_{50}$ the higher the toxic potency). Under vitiated flaming conditions such as in enclosed fires and post-

Table 3.6 Toxic product yields and toxic potencies (mass loss) of different PVC types under a range of fire conditions using DIN and Purser tube furnaces

Material and decomposition condition		CO g/g	CO_2 g/g	HCl g/g	Organic carbon g/g	Smoke $OD/g/m^2$	LC_{50} g/m^3 mass charge	LC_{50} g/m^3 mass loss
PVC (non-plasticised)[a]								
Non-flaming	3W	0.01		0.56				8
Well ventilated	6W	0.17		0.55				7
PVC (plasticised)[b]								
Non-flaming	380°C	0.02	0.00	0.36			20	15
Flaming	650°C	0.31	2.39	0.53			8	7
PVC (plastic + $CaCO_3$)[c]								
Non-flaming	350°C	0.01	0.09	0.32	0.44	0.76	28	12
Well ventilated	650°C	0.06	1.27	0.16	0.12	0.29	34	22
Small vitiated	650°C	0.08	0.74	0.17	0.33	0.30	29	16
Post flashover	825°C	0.17	0.83	0.41	0.29	0.34	18	9

Data from Hartzell et al,[29] Purser et al.[2,3,25] Copyright HER.

flashover fires the difference is less, since all materials burn inefficiently, but is still evident. Table 3.6 shows examples of CO and HCl concentrations obtained using DIN and Purser tube furnace methods, with calculated LC_{50} concentrations. The first is for a non-plasticised PVC and the second for a plasticised PVC. This illustrates that it is possible to reduce the yield of hydrogen chloride to some extent by the use of fillers which are either inert or reactive. An example is calcium carbonate, which was used in a plasticised cable material decomposed under a range of fire conditions in the Purser furnace. This reduced the PVC content of the material, and under both well ventilated and small vitiated flaming conditions reduced the yield of HCl. Under post-flashover conditions the calcium carbonate decomposed before reacting with the HCl so that the HCl yield was increased. As the table shows the toxic potency under well ventilated flaming conditions is driven by the high yield of CO, which is only slightly less than under vitiated flaming conditions, and also by the high yield of HCl. These two toxic gases also maintain a high toxic potency under vitiated conditions. The toxic potency is somewhat less than the average figure shown in Table 3.3, presumably due to the calcium carbonate filler. Special formulations of PVC designed to reduce HCl emissions have reduced toxic potency levels to LCt_{50} exposure doses of above $1000\,g\,m^{-3}$ min under both non-flaming and well ventilated flaming decomposition conditions.

Another point illustrated in Table 3.6 is that the use of inert fillers results in significant differences between mass loss and mass charge LC_{50} concentrations. Thus when considering the use of PVC or other materials in terms of the mass used for a particular application, the lower toxicity produced by the filled materials may present an advantage.

The data presented in Table 3.6 were obtained from flow-through methods, where the fire gases have a limited residence time in the apparatus. Other methods maintain a static atmosphere in a chamber for 30 minutes. In these methods the HCl concentration decreases rapidly as the gas is absorbed on chamber walls and reacts with metal surfaces, so that the resultant toxic potency estimates are low. This is also likely to happen to some extent in full-scale fires, although in such fires the critical exposure time for building occupants is usually only a few minutes, so that losses may be small during the critical exposure period.

PVC is an example of a material with inherent fire retardant properties. The fluorocarbons constitute another class. They have excellent fire resistant properties, decompose at high temperatures and it is difficult but not impossible to ignite them. The toxic potency of fluoropolymers is very complex and has been reviewed by Purser.[30,31] Most work has been done on PTFE but other per-fluorinated polymers appear to behave similarly to PTFE. For per-fluorinated polymers the toxic potency under flaming conditions is around $200\,\mathrm{g\,m^{-3}\,min}$, which is not dissimilar to that of PVC. Under non-flaming decomposition conditions the toxic potency depends very much upon the exact decomposition conditions, varying over a very wide range from $0.5\text{–}87\,\mathrm{g\,m^{-3}\,min}$, depending upon the extent to which extreme toxic potency particulates are formed.

The materials described are inherently fire retarded materials with high halogen contents. Halogens, particularly bromine and chlorine, are used in FR compounds added to a variety of materials or components in composites to modify ignition and combustion behaviour. These elements may be present as low mass percentages of materials of composites (for example, brominated antimony trioxide fire retardants in backing layers of upholstered furniture covering materials or polybrominated ether fire retardants used in television sets) and are therefore unlikely to make a major direct contribution to toxic potency during combustion of such materials. However, their use may result in significant effects on toxic potency by modifying combustion efficiency and increasing the yields of smoke and asphyxiant gases.

Table 3.7 shows an example of a bromine-antimony FR back coated cotton print, in which the HBr yield was 1% of the mass of material decomposing and contributed 1.4% to the overall lethal toxic potency. When compared with untreated cotton print decomposed under non-flaming conditions at 400°C the calculated LC_{50} concentration of the

Table 3.7 Toxic potency of cotton print with bromine–antimony trioxide FR back coating: non-flaming conditions at 400 °C and 700 °C at a mass charge concentration of 20 g/m³ (DIN tube furnace)

Material and decomposition condition		CO g/g	HBr g/g	Formaldehyde g/g × 1000	CO₂ g/g	Smoke OD/m	LC_{50} g/m⁻³ mass charge	LC_{50} g/m³ mass loss
Untreated cotton								
Non-flaming	400 °C	0.28	0	7.1	0.64	4.9	18.6	17.8
Flaming	700 °C	0.04	0	0.2	1.77	0	50.9	50.1
FR cotton Br–Sb:								
Non-flaming	400 °C	0.21	0.01	4.6	0.77	5.0	29.1	27.4
Non-flaming	700 °C	0.43	0.01	2.0	0.78	1.4	14.2	13.9

Adapted from T. Wright PhD Thesis 1997. Copyright HER.

treated materials is actually slightly higher (less toxic) than the untreated cotton. At 700 °C the untreated cotton flamed while the treated cotton failed to flame. Although it is therefore not possible to make a comparison at the higher temperature under the same decomposition conditions it is interesting to note that the toxic potency of the treated material was worse by a factor of 3.6 than the untreated material both in mass charge and mass loss terms. This was mainly due to the factor of 10 difference between CO yields under flaming and non-flaming conditions at 700 °C. When untreated cotton was decomposed under non-flaming conditions at 700 °C the CO yield was similar to that obtained from the FR-treated material, which illustrates the importance of testing materials under defined decomposition conditions. An important consideration in this case is that despite the higher toxic potency the material still resisted ignition.

Another example (Table 3.8) is a thermoplastic polyurethane untreated and then treated with a decabromophenyl oxide and antimony trioxide tested under flaming decomposition conditions using the Purser furnace method. In this case flaming occurred in both materials, but for the FR-treated material flaming was intermittent, with high yields of CO, HCN and irritants released during the non-flaming periods and during the transition from non-flaming to flaming behaviour. The result was a 20-fold higher irritancy for the FR-treated material and a 5-fold higher mass loss (8-fold higher mass charge) toxic potency compared with the untreated material.

Halogens are also used in combination with phosphates. These systems also have a low halogen content and are considered in the section on solid phase fire retardants.

Table 3.8 Toxic potency of thermoplastic polyurethane untreated and treated with decabromophenyl oxide and antimony trioxide at 600 °C at a mass charge concentration of 8 g/m^3 under flaming decomposition conditions (Purser tube furnace)

Material and decomposition condition		CO g/g	CO$_2$ g/g	HCN mg/g	Smoke OD/m	Irritancy RD$_{50}$ g/m^{-3}	LC$_{50}$ g/m^3 mass charge	LC$_{50}$ g/m^3 mass loss
Thermoplastic polyurethane: Steady flaming	600 °C	0.11	6.40	3.32	0.07	4	64	30
Thermoplastic polyurethane + Br–Sb: unsteady flaming	600 °C	0.77	1.73	13.65	1.23	0.2	8.2	6.6

Adapted from Purser.[5] Copyright HER.

3.5.2.2 Effects of halogenated compounds in full-scale fires

The results from small-scale combustion toxicity tests on halogenated materials are useful in that they illustrate the effects of halogens on smoke and toxic product yields and on lethal toxic potency under defined thermal decomposition conditions. In particular, they tend to show the rather negative aspects of increased yields of smoke, irritants and carbon monoxide. Where toxic hazards to building occupants of full-scale fires are concerned the data from small-scale tests provide only part of the information needed to determine toxic hazard. Of course, one major benefit of using an FR-system should be that few propagating fires will result following exposure to a variety of ignition sources. However, when flaming (or in some cases smouldering) ignition occurs and growing fires result, the main cause of severe incapacitation and death is exposure to asphyxiant gases, particularly carbon monoxide and hydrogen cyanide. When comparing halogenated or any other FR- or non-FR-treatments a major consideration is therefore the time from ignition to occupant asphyxiation. Also important in this context are any effects likely to delay or prevent escape, which will increase the probability that occupants will remain in the building long enough to be incapacitated by asphyxiants, or in some cases by irritants or by heat. It is therefore important to estimate the time from ignition to that when the concentration of irritant smoke is likely to reduce escape efficiency. For halogenated systems the contribution made to the irritancy of the fire atmosphere by acid gases, which would not normally be present otherwise, is of obvious concern. In terms of escape impairment, much lower concentrations have more drastic consequences than those leading to death. For example, it is considered that for a fairly typical fire exposure of up to 5 minutes' duration, an exposure to smoke containing around 200 ppm HCl is likely to impair escape attempts, while around 900 ppm is likely to cause incapacitation and 1.5% HCl is considered to constitute a lethal exposure over 5 minutes.[2,77]

For these reasons, fire scenarios in which irritant smoke containing high concentrations of acid gases is released at an early stage, followed by large amounts of CO and other asphyxiants, are of particular concern. In practice, the developing toxic hazard in any full-scale fire depends upon interactions between properties of materials, product design, the fire enclosure and ventilation. One important parameter, especially during the early stages of fires, is the extent to which halogenated FR-treatments delay flame spread and fire involvement of the item first ignited, and subsequent flame spread to other items and across surfaces, compared to that in non-FR-treated systems. Another important parameter is the yield of toxic products from the materials decomposed. Depending upon the performance of the particular system, the rate of contamination of the fire enclosure during the early stages may be slower or faster than for the equivalent non-FR system. From an early stage of the fire scenario, the shape and size of the fire enclosure, and the disposition of any vents will have a considerable influence on the rate of hazard developments. For example, if the fire enclosure has a large volume and/or open vents such as doors or windows, and the FR-treatment results in a slow-growing, but smokey fire, then the smoke layer may descend slowly from the ceiling and it may be some time before occupants become exposed. However, if a small fire occurs in a small unvented enclosure, such as a domestic room, then even a relatively small amount of irritant smoke can cause rapid contamination. When fires involving non-FR material occur in larger enclosures, then although the yields of irritant smoke and asphyxiants may be lower initially than for FR-treated systems, this may soon be overtaken by the more rapid rate of fire growth and effluent production. Not only can the rate of toxic effluent production soon overtake that of the FR-system, but as the fire scenario changes the yields of asphyxiants can increase to match or overtake those from the fire-retarded system. This is because the more rapid fire growth in a non-FR-system tends to use up available oxygen more rapidly, leading to vitiated combustion conditions which produce increased CO and HCN yields. The ways in which these interacting factors can influence the development of toxic hazard in different scenarios can be illustrated by some examples of full-scale fire tests and reports of fire incidents.

Irritant smoke has been reported as a significant problem during enclosed smouldering or flaming fires involving primarily articles with a high halogen content. In such situations exposed subjects have reported severe eye and respiratory tract pain causing difficulties in escape when smoke and other toxic gases were at relatively minor levels (personal communication, Kent, Essex and Bedforshire Fire and Rescue services). In at least one case exposures resulted in fatalities (New York Telephone Exchange fire).[32]

Another situation where there may be a problem is in large, rapidly growing fires where survival depends on occupants escaping within a few minutes. An example occurs in post-crash aircraft cabin fires. Figure 3.1 shows the concentrations of acidic gases (HCl and HF), CO and smoke in the cabin of a passenger aircraft fuselage during a post-crash fire test.[33] The combined concentrations of HF and HCl exceed 200 ppm by 0.5 minutes after ignition, and exceed 3000 ppm by 3 minutes. By 4 minutes the exposure to CO (and other asphyxiant gases) would still be only half that required for incapacitation and the smoke optical density is just reaching a significant level. Over the period up to 4 minutes the main toxic threat to the occupants is therefore the acid gases. It is likely that these would cause some degree of impairment of escape efficiency from 0.5 minutes and may cause very severe effects from 3 minutes.[34] In another similar full-scale fire test on a furnished passenger aircraft fuselage, an average concentration of 1027 ppm HCl and 1229 ppm HBr were measured over a period between 1 and 4 minutes, which is likely to have had a similar severe effect on any passengers attempting to escape. These concentrations were measured at

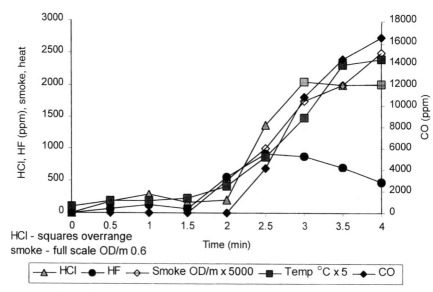

3.1 Toxic gases heat and smoke during the early stages of an aircraft cabin fire. Data from Sarkos et al 1982[33]

the breathing zone of a cabin occupant some distance from the fire.[18] Since acid gases tend to decay in fire atmospheres it is likely that the concentrations would have been higher nearer the fire. It is therefore possible for halogenated materials, which covered a very large surface area of the cabin and which were also possibly present in seating and other components, to present a significant toxic hazard in some situations.

In other situations, the presence of halogenated FR-systems may be beneficial in terms of overall hazard development and fire risk. With domestic furniture in the United Kingdom, ignition-resistance of acrylic fabrics is improved by latex back-coatings containing brominated fire retardants. Although this may reduce the incidence of propagating fires from small ignition sources, it does not necessarily improve burning behaviour once the initial ignition resistance is overcome, and in some cases may increase the amount of smoke and toxic gases produced by combustion.[27] Figures 3.2 and 3.3 summarise the results of a series of full-scale fire tests involving a range of furniture in which the item first ignited was a typical design of domestic armchair in two test rigs.[34] The ignition source was a No. 7 wood crib, (roughly equivalent to two sheets of newspaper) which was used to overcome the ignition resistance of all materials used. Test 11 was carried out in an enclosed apartment rig consisting of a fire room, corridor and target room. Tests 16, 17, 18, 21 and 23 were carried out in a closed two-storey domestic house. The tests were designed to investigate the development of hazards from furniture fires in typical domestic enclosures. For all these fires the rigs were enclosed (all external doors and windows shut) apart from various small openings (such as air bricks) used to vary background ventilation. Varying levels of fire room door openings were used. The house fires were set in the downstairs lounge with the doorway to the hall and stairs either open or closed, and one upstairs bedroom door open. The total open volume of the two rigs was approximately $100\,m^3$. Smoke, heat and toxic gases were measured at a number of locations throughout the rigs. The results are expressed in terms of time to loss of tenability for smoke, sensory irritancy, asphyxiants and heat for an occupant of the fire room or in the bedroom remote from the fire. Tenability criteria were assessed according to the method of Purser.[2,18]

The tests were not primarily intended for the comparison of different FR systems but do provide some illustrations of performance during typical fire scenarios in typical domestic spaces. For all except test 16 the fire room door was open. The fires in the house were very similar to those in the single storey apartment. The upper three fires (11, 21, and 23) in Fig. 3.2 and 3.3 show rapid loss of tenability due to irritant smoke and asphyxiants (particularly hydrogen cyanide), at between approximately 2–3 minutes after ignition in the fire room and after approximately 3–4.5 minutes in the remote bedroom. These all had a common feature of non-fire-retarded acrylic covers over

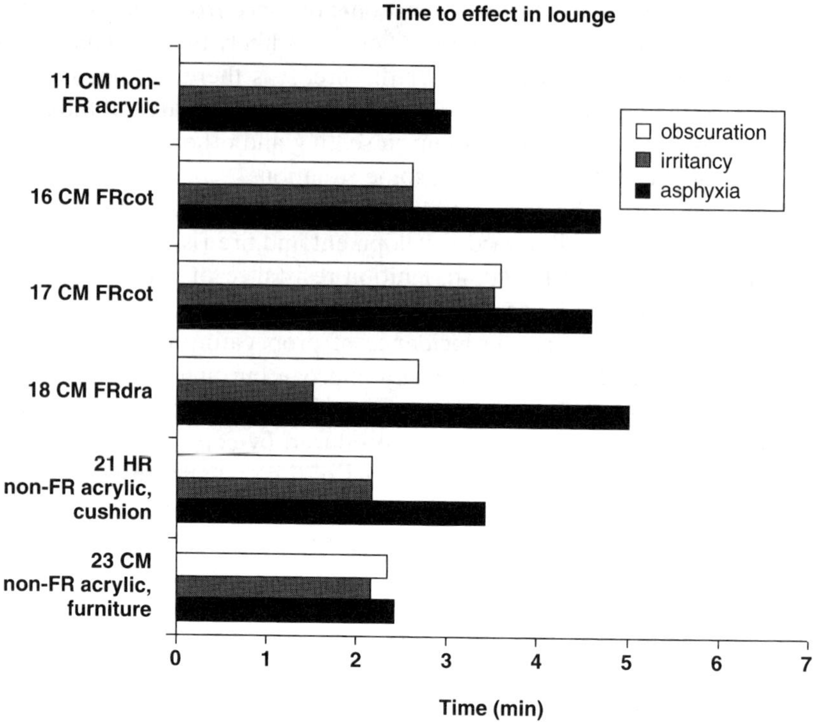
Time to effect in lounge
11 CM non-FR acrylic
16 CM FRcot
17 CM FRcot
18 CM FRdra
21 HR non-FR acrylic, cushion
23 CM non-FR acrylic, furniture
obscuration
irritancy
asphyxia
0
1
2
3
4
5
6
7
Time (min)

3.2 Comparison of time to effect in burn room (lounge) for armchair with non-FR covers (11,21,23) and FR covers (16,17,18). Fires were conducted in an enclosed apartment (11) or house (16–23) with the fire room door open (except for 16)

11 CM non-FR acrylic (armchair, non fire retarded acrylic covers, combustion modified foam – apartment rig)
16 CM FRcot (as for 17 but fire room door closed)
17 CM FRcot (FR back coated cotton covers, combustion modified foam – house)
18 CM FRdra (FR back coated acrylic covers, combustion modified foam – house)
21 HR non-FRacr, cush (non-FR acrylic covers, high resilience modified foam, foam scatter cushions – house)
23 CM non-FRacr, furn (non-FR acrylic covers, combustion modified foam, fully furnished room – house)

polyurethane foam. The fires grew until the oxygen concentration entering the fire decreased to around 14–16% at which point the fires self-extinguished, leaving the house or apartment filled with a uniform atmosphere containing a dense irritant smoke and lethal concentrations of asphyxiant gases. These three fires provide examples of the rapid fire growth and hazard development associated with old-style furniture. The lower three fires shown in the figures (tests 16, 17 and 18) used armchairs which all had FR-treated covers over combustion modified foam. For tests 16 and 17 the covers

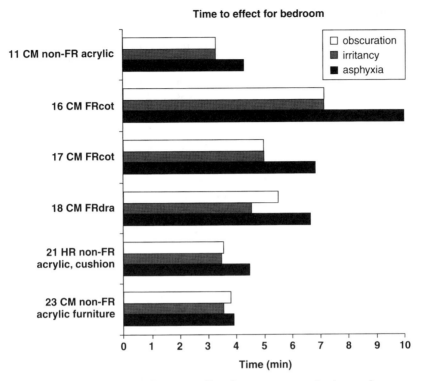

3.3 Comparison of time to effect in open remote bedroom for armchair with non-FR covers (11,21,23) and FR covers (16, 17,18). Fires were conducted in an enclosed apartment (11) or house (16–23) with the fire room door open (except for 16)

11 CM non-FR acrylic (armchair, non-fire retarded acrylic covers, combustion modified foam – apartment rig)

16 CM FRcot (as for 17 but fire room door closed)

17 CM FRcot (FR back coated cotton covers, combustion modified foam – house)

18 CM FRdra (FR back coated acrylic covers, combustion modified foam – house)

21 HR non-FRacr, cush (non-FR acrylic covers, high resilience modified foam, foam scatter cushions – house)

23 CM non-FRacr, furn (non-FR acrylic covers, combustion modified foam, fully furnished room – house)

were cotton with an FR-treatment containing bromine (4.4%) and some chlorine (0.9%). For test 18 the covers were back-coated acrylic (Dralon, Bayer) and also contained bromine (2.4%) and chlorine (3.6%). These provide examples where the FR-treatment to the covers was a definite advantage by slowing the rate of development and size of the fire, so that predicted times to asphyxia were improved by approximately two minutes compared to the chairs with untreated covers. For the chair covered with the back-coated acrylic there were some problems with smoke and irritants during the early stages of the fire.

However, in terms of hazard development, two other features of the system involving interactions between the fire and the building are important. One is the position of the fire room door and the other is detection. For Test 16 the fire room door was closed. This meant that during the early stages the fire effluent filled the enclosed lounge rapidly, so that smoke and irritants became a problem after 2.5 minutes. However, the hallway and open upstairs bedrooms were protected and took 7 minutes to become smoke logged. Because the chair had FR-treated covers, the time to asphyxia was relatively long at almost 5 minutes, even though the room was enclosed, while the tenability time for asphyxiation exceeded 10 minutes in the upstairs bedroom. The importance of time-to-detection is that the hazard to occupants depends upon the time available for escape, which in turn depends upon the time from detection to loss of tenability. Time-to-detection in these experiments depended upon the type of smoke detector (ionization or optical) and its position (fire room or hallway). In general, the time available for escape was very short (approximately 1 minute) for the armchairs without FR-treated covers and longer (approximately 2 minutes) for the armchairs with FR-treated covers.

Another aspect of the interaction between the burning materials and the enclosure is illustrated in Fig. 3.4 which shows the relationship between plume carbon monoxide concentrations and plume oxygen concentrations for armchair fires in a $100 \, \text{m}^3$ volume enclosed apartment and house. The squares show fires in the apartment rig and the triangles house fires. The solid triangles show chairs with FR-treated covers. Based upon the mechanism of fire retardancy of halogenated fire retardants and the results of small-scale toxicity test data, it is to be expected that the yield of CO under well ventilated flaming conditions, such as those at the beginning of the chair fires, would be higher for the chairs with FR-treated covers than for those with untreated covers. During the later stages of the fires, when conditions were oxygen vitiated, the differences might be somewhat less. In practice, the CO yield was also influenced by the size of the fires. Figure 3.4 shows that the armchairs with untreated covers (shown by the open squares and triangles) produced larger fires, using up more of the available oxygen in the fire plume than did the armchairs with FR-treated covers (shown by the solid triangles). This increased vitiation tended to increase the yield of CO in the fire plume, so that in practice there was a higher CO concentration in the plumes from the untreated chairs.

In summary, halogenated FR-treatments are likely to reduce hazard for scenarios where they prevent the ignition of a propagating fire or where they slow the rate of fire development and fire size. Halogenated systems may increase some aspects of toxic hazard when small or smouldering fires

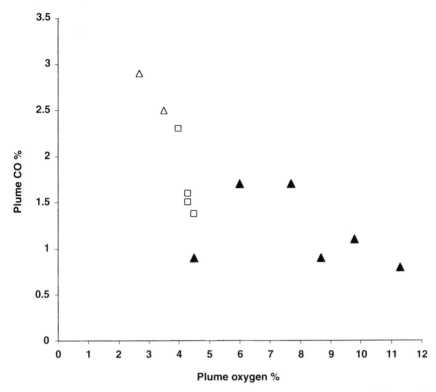

3.4 Relationship between plume carbon monoxide concentrations and plume oxygen concentrations for armchair fires in 100 m³ enclosed apartment (squares) and houses (triangles). Open symbols show armchairs with non-FR covers, closed symbols show armchairs with FR-covers

produce significant quantities of irritant smoke during the early stages or where large fires cause rapid releases of acid gases from items or surfaces containing halogens. It is also evident that the yields of toxic gases from materials in small-scale tests provide only part of the data needed for a toxic hazard assessment.

3.5.2.3 Environmental problems caused by combustion of halogenated materials

Life-threatening exposure to toxic fire effluent fortunately affects only a relatively small proportion of the population annually, but exposure to low levels of environmental contamination from combustion products through air, food and water affects the whole population daily. The extent to which these environmental pollutants, particularly those from halogenated materials, present hazards to long-term health has become an issue of considerable public concern and the subject of fierce political debate, especially in Europe. The subject is scientifically complex and the issues are difficult to resolve. Good objective data are also scarce, so that within this review it is possible only to present a very general view of the issues involved.

The main area of current concern centres on halogenated dioxins and dibenzofurans. These are likely to be evolved to some extent when any organic material containing halogens is combusted. Particular concerns have been raised with regard to polyvinylchloride (PVC), polychlorinated biphenyls (PCB), polychlorinated phenols and brominated ether fire retardants, but other halogenated fire retardants also lead to some releases as well as other materials, such as wood (especially wood treated with pentachlorophenol) and straw stubble.[35]

There are two particular issues:

1   To what extent does the production, use and disposal of organohalogen compounds, particularly PVC, lead to general levels of environmental contamination by dioxins, dibenzofurans and phthalates which constitute a risk to public health?
2   To what extent does the exposure of people to combustion products from PVC and other halogenated materials during and after fires in buildings constitute a risk to their long-term health?

In order to answer these questions it is important to establish two other points:

3   To what extent do the levels of general environmental contamination and levels of contamination during and after building fires from dioxins and other toxic products result from the combustion of PVC and other halogenated materials?
4   To what extent do these levels of contamination present hazards to health?

Dioxins, dibenzofurans and their effects on health

Dioxins are polychlorinated dibenzo-*p*-dioxins (PCDDs), furans are polychlorinated dibenzo-*p*-furans (PCDFs). These PCDDs and PCDFs are halogenated aromatic compounds substituted in several positions by one or more chlorine atoms. They are formed as trace by-products in processes involving chlorine and organic compounds. There are many possible types that can be formed including some 75 different chlorinated dioxins and 135 chlorinated furans. The toxic potencies of the different substances vary over a wide range, and 17 are of concern as the most toxic. In order to simplify this complex picture the toxicity of a mixture of dioxins and furans is commonly expressed as a 'toxic equivalent' (TEQ). This is obtained by multiplying the concentration of individual dioxin and furan species ('congeners') by a suitable toxic equivalent factor (TEF) and summing the results to obtain an overall toxic potency. TEF values are calculated relative to the most toxic congener 2,3,7,8-TCDD, which is assigned a TEF of 1, others congeners may have TEFs as low as 0.01.[35] This is important when considering reported levels of contamination by dioxins and furans, since the total mass of compounds may be much greater than the toxic equivalent mass.

Daily intakes from different sources

Dioxins are fat soluble so levels are quoted for body fat or in the fat content of the blood. A mean background level of 57 pg TEQ/g fat (1.4 ng total PCDD/Fs) has been reported in human fat tissue in the Welsh population.[36]

There are three routes of intake for dioxins, from food, from breathing air and from skin contact. By far the largest source for the general population is in food, especially meat, fish and dairy products. There has been a considerable decrease in total daily intake from food in the United Kingdom from peak levels occurring during the 1980s as illustrated by the following figures:

1982   240 pg   TEQ/person per day.
1988   125 pg   TEQ/person per day.
1992   69 pg   TEQ/person per day.[37]

According to the Ministry of Agriculture, Fisheries and Food (MAFF), the decrease in intake has been caused mainly by changes in dietary habits and a fall in the average fat content of many foodstuffs, rather than a decline in environmental PCDD/F levels.[37] However, as will be discussed in a later section, there is strong evidence for a considerable decline in levels of dioxin contamination of land and food since the mid 1980s. Both water and air[46,35,39] are very minor sources of dioxin intake: the figures for air are given as follows:

Four UK sites: 6.8 pg/m³ for 17 dioxin and furan congeners
Hamburg: 0.02 pg 2,3,7,8 TCDD/m³ and total of 0.1 pg TEQ/m³.[7,8,9]

Assuming 20 m³ air breathed each day and 100% absorption, this represents a total possible intake of:

UK: 140 pg/person per day of these congeners ≡ 1.4–14 pg TEQ
           depending upon the cogeners present
Hamburg: total 2.1 pg TEQ/person per day.[35]

Cigarette smoke contains dioxins and furans with a TEQ value of 1.81 ng TEQ/m³ sufficient to deliver 4.3 pg TCDD/kg bw per day, equivalent to

Table 3.9 Average daily intake of dioxins and furans from all sources for a 70 kg adult (bw = body weight)

Source	pg TEQ /kg bw per day	pg TEQ/person per day
Food	1	69
Air	0.03	2.1
Consumer products	0.01	0.7
Water	very little	
Total	1.04	71.8

Adapted from Committee on Toxicology report.[42] Copyright HER.

approximately 0.08 pg TEQ/kg bw per day or 5.8 pg TEQ/person/day smoking 20 cigarettes.[38] Due to their poor aqueous solubility only minute quantities would be absorbed following prolonged skin contact.[35] For example, a baby wearing 6 nappies/day made from chlorine-bleached paper is estimated to absorb less than 0.001 pg/kg bw/day. Therefore firefighters, fire investigators or fire victims should absorb very little through dermal contact with soot contaminated with dioxins.

On the basis of the figures quoted in the previous sections the average daily intake of dioxins and furans as TEQ for a 70 kg adult is estimated as shown in Table 3.9. The data in Table 3.9 show that with the possible exception of smokers and occupationally exposed groups, by far the greatest source of human dioxin and furan intake is through food. This therefore raises the questions of the hazards associated with such intake and in the context of this chapter the extent to which food content arises as a result of combustion processes.

Toxic effects of dioxins and recommended maximum
daily intake levels

Information on the toxic effects of dioxins and furans is derived from human epidemiology studies and detailed animal experimental studies.[35,39] Human studies include health effects and background levels in different human populations (including industries using compounds known to be contaminated with dioxins and furans) and accidental exposures (such as the Seveso incident resulting in high level exposures derived from an explosion in a trichlorophenol plant).[40] By taking results from studies in a range of animal species together with the human data, it is possible to make estimates of likely effects in humans. Safety factors of one or more orders of magnitude are used to set recommended maximum daily intake levels for guidance on the significance of environmental exposure levels. As more information becomes available these recommended levels are reviewed and updated.

The best indicator of exposure to high levels of PCDD/Fs in humans is the skin disease chloracne, associated with transient effects on liver enzymes. This is associated with very high levels of dioxins in the body and major incidents have not been found for more than a decade.[35,40] Apart from this there are concerns about three major toxic effects found in animal studies: immunotoxicity, reproductive toxicity and carcino-genicity. So far it has not been possible to determine whether PCDD/Fs are immunotoxic to humans and studies have produced conflicting findings.[41] Hormones affecting reproduction may be involved, with decreased levels of testosterone reported in male workers manufacturing 2,4,5-trichlorophenol.[39] Reproductive toxic effects on fertility and offspring body weight have been reported in Rhesus monkeys. Animal studies indicate that dioxins are among the most potent carcinogens known, but epidemiological evidence linking dioxins with cancer in the general human population is lacking, although there is evidence for elevated cancer risk in workers using phenoxy herbicide preparations which contain dioxins as contaminants.[39]

Exposure limits and guidelines

Dioxins and furans produce chloracne at high doses, may harm human metabolism, development and reproduction and may also constitute a human cancer hazard.[35,39,41,42,43] These adverse effects may occur at levels less than ten times above the current average body burdens,[39] but humans may not experience adverse health effects at current body burdens of dioxins and furans.[44] The UK committee on toxicology (COT)[35,39] recommendations were to reduce human exposures, identify major sources and reduce environmental inputs to reduce levels in food and human tissues.[35]

The currently recommended exposure limits and guidelines for dioxins are expressed as ADI (acceptable daily intake) or TDI (tolerable daily intake). The United Kingdom Committee on Toxicity (1995) has accepted the World Health Organisation recommendations[43] which were based upon the lowest observed adverse effect in animals, based upon carcinogenic, hepatotoxic, immunotoxic and reproductive toxicity endpoints.[35] Based upon a no effect level of 1000 pg/kg bodyweight, a factor of 100 was applied to take into account toxicokinetic differences and uncertainties regarding reproductive toxicity, giving a recommended value of 10 pg/kg body weight. (WHO 1991).[43] More recent work has shown adverse effects on animals at lower levels corresponding to human daily intake in the range 14–73 pg/kg bw per day. To arrive at a figure for tolerable daily intake on uncertainty factor of 10 was applied by WHO, giving a new range of 1–4 pg/kg bw per day. This is approximately the current total daily intake in the UK of approximately 1 pg TEQ/kg bw per day for a 70 kg person. The American

Table 3.10 Exposure limits and guidelines for dioxins and furans

Agency	Exposure limit/kg bodyweight/day	Exposure limit/ person/day (assuming 70 kg)
UK – COT	10 pg TEQ/kg/day TDI	700 pg TEQ/day TDI
WHO	1990: 10 pg TEQ/kg/day ADI now reduced to:	700 pg TEQ/day ADI
	1–4 pg TEQ/kg/day TDI*	70 pg TEQ/day TDI
US – EPA	6.4 fg TCDD/kg/day TDI	4.5 pg TCDD/kg/day TDI

Chemosphere 40, 1095 (May 2000).
Adapted from several sources. Copyright HER.

EPA is currently recommending a lower level based upon assumptions regarding possible carcinogenicity.[45] These limits and guidelines are summarized in Table 3.10. Based upon the data from Table 3.9, the average UK citizen is absorbing approximately the maximum acceptable daily intake recommended by the WHO and UK COT, but considerably more than the EPA recommended levels.

The contribution of PVC and other halogenated compounds to environmental dioxins and furans

An important consideration with respect to the use of PVC and halogenated fire retardants is the extent to which they contribute to background levels of environmental dioxin and furan contamination. Authoritative reviews of sources of dioxins and furans entering the environment include Eduljee (1988),[46] DOE Pollution Paper No. 27(1989),[35] including the advice of the COT for the Department of the Environment, and a recent review by Alcock and Jones.[47] These show that there are many sources of dioxin input into the environment, some resulting from man-made organochlorine compounds, and some from natural processes involving the combustion of organic materials with inorganic chlorine salts (for example: coal fires, forest fires, stubble burning). Up to approximately the late 1980s, probably the three main sources were dioxins as contaminants in polychlorinated biphenyls (PCBs) used in transformers, penta-chlorinated phenols (used as wood preservatives) and municipal waste incineration (MSW). Other sources were industrial and hospital incinerators, steel processing, coal fires, leaded petrol combustion in motor vehicles, certain herbicides and occasional accidents such as Seveso.[40] Two more personal sources to the individual were hexachlorophene (a common antiseptic once also used in toothpaste) and cigarette smoke. For the majority of these chlorinated chemicals small amounts of dioxins often occurred as contaminants from the manufacturing process, or could be formed when the material was burned.

Of these sources, PCBs were banned in the 1970s and PCP (pentachlorophenol) manufacture ceased in the 1980s. PVC and other halogenated FR-materials could contribute only to municipal and other waste incineration. This depends upon the dioxin output from incinerators and its relationship to the amount and nature (organic or inorganic) of the chlorine load in the incinerator fuel. In old style incinerators, even when high temperatures were used, dioxins could not only be released from the fuel, but could be formed secondarily from hydrogen chloride and organic residues in the flue, which was often the main route of formation. There was felt to be a poor relationship between the chlorine content of the fuel, in particular the amount of PVC and related compounds present, and the dioxin content of the flue gases, although this view has been challenged in recent work.[48] Modern incinerators are designed to remove dioxins from the flue gases, or prevent their secondary formation, and since the modifications were made the stack emissions from incinerators are greatly reduced by factors of between 100 and 1000.[49] This compares with a factor of 10 reduction achieved by removing PVC in a situation where it constituted 50% of chlorinated waste.[49]

Trends in dioxin sources and levels of
environmental contamination

Time series examination of environmental dioxin levels in a variety of places and materials reveals a reasonably consistent pattern.[47] These are illustrated in Fig. 3.5. Studies of lake bed sediments in remote locations in the United States (solid line marked with diamond) and in park grass samples (solid line) collected since 1860 in north London show steady levels of contamination up to around 1950. This is followed by a very large increase over the period from approximately 1955 to 1980, then followed by large and rapid decline during the late 1980s and early 1990s. This has been accompanied by decreases in levels of air contamination (measured in Germany), decreases in contamination of sewage sludge, decreases in contamination levels in wild life such as sea birds and decreases in contamination in livestock tissue and especially large decreases in food fats (including cows' milk) shown as square symbols in Figure 3.5, and in human milk.

This generally improving picture runs counter to the data for PVC production and use, which started in the later 1950s. The broken line in Figure 3.5 shows the greatly increasing world production since 1950. Consumption in Europe has risen from low levels in the early 1960s to 600 000 tonnes in

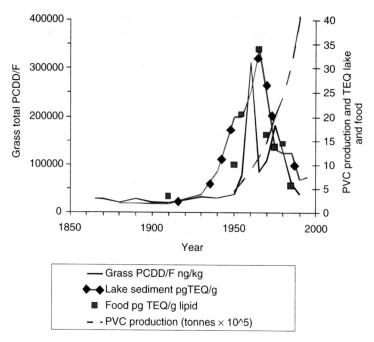

3.5 Trends in environmental dioxins compared with annual PVC production

the early 1990s, and continues to rise, while PCDD/F concentrations in a wide range of environmental media are on the decline. This enormously increasing tonnage of PVC and materials containing chlorinated and brominated fire retardants in use does present an enormously increasing disposal and recycling problem. Just one example of this kind of problem is presented by the huge increase in the amount of electrical cable in building plenum spaces. New cable, much of which consists of PVC or other halogenated materials, is continually being introduced into both new and existing buildings, and in the latter case old cable is seldom removed. It is estimated that in buildings in the United States, the total amount of plenum cable present (expressed in terms of millions of metres) increased from about 500 to about 1300 between 1991 and 1996. This represents an average annual growth rate of 46% and the total is predicted to reach nearly 2700 million metres during 2000 compared with 108 million metres in 1983 (a 25 fold increase). All of this material (and that derived from halogenated polymers from other sources) will eventually have to be disposed of or recycled and it is possible that as the previous major sources of dioxin contamination are eliminated, the amount produced by the combustion of PVC and other halogenated materials in landfill, accidental fires and modern incinerators will come more into prominence. However, current data show emissions from such sources to be low.[47,49]

The general picture with regard to the environmental hazards from PVC and halogenated fire retardants is therefore not a simple one. Certainly as far as dioxin release is concerned there seems to be little evidence that PVC, chlorinated or brominated fire retardants have been a major source up to recent times. However, the release and combustion of increasingly large tonnages of PVC and other halogenated materials do need consideration.

Long-term health risks of dioxins and furans

Although general environmental contamination by dioxins and furans arises from a number of sources, a possible concern is that significant local environmental contamination may arise from particular fire incidents in buildings and that this might represent a health hazard. Health hazards could arise to occupants of buildings during fires, to the population of areas surrounding fires, to members of the emergency services attending fires or to workers involved in post-fire investigation and decontamination. The health hazards to people from exposure to dioxins arising from the combustion of halogenated materials can only be considered realistically in relation to the health hazards presented by other combustion products from those materials in building fires and the health hazards from other materials burning in typical fires. It is also important to consider the nature, circumstances and frequency of exposures, as well as to compare the intake of dioxins and other toxic substances from fires with the likely levels of intake from other sources.

The most important possible routes by which building occupants or the population around building fires are likely to be exposed to dioxins and furans is by inhalation of smoke particulates during a fire, aerosolised soot afterwards and oral ingestion arising from contamination of food from contact with soot deposits.

There are a number of potential sources of dioxins in most building fires, but it is likely that PVC is the major chlorine source in most buildings, and it has been noted that dioxin yields from decomposition of copper covered PVC wiring are high.[49,50] The yields of dioxins and furans depend very much on the decomposition conditions in the fire. Incinerator and other studies have shown[49,50] that the yields can be much greater when a furnace is first started and relatively cold (by a factor of 5)[49] than when it is hot, and that 300–400 °C is the optimum formation temperature. It is therefore to be expected that dioxin yields will vary considerably depending upon the type and size of fire and its ventilation. Vitiated fires of the kind found in buildings, particularly during stages before serious structural breaching, are likely to produce high dioxin yields, while yields may be lower in very large conflagrations. Some data are available from a few accidental fires and some small-scale experiments.[49,50,78]

There is evidence from an EPA study[78] that yields of dioxins and furans from small, inefficiently burning fires are much greater than those obtained from incinerators. In this study, samples of household waste were burned in a 55 gallon oil drum, a common method of waste disposal. In one experiment involving waste with a 4.5% PVC content the total yield of PCDDs and PCDFs was 493 ng/g waste decomposed. This compares with a figure of 0.0035 ng/g waste decomposed in a modern municipal waste incinerator, a factor of more than 100 000 times greater yield. On this basis the total dioxin and furan output of a household using this method to dispose of its waste could be approximately equivalent to the entire output of a Municipal Waste Incinerator. Assuming a factor of ×0.1 for the TEQ cogeners present, this could represent 49.3 ng TEQ/g waste decomposed.

Two major considerations in terms of human exposure are the concentrations in the effluent plume and the concentrations in soot deposits. It is very difficult to obtain good quantitative data in such situations due to difficulties in sample collection and difficulties in knowing the mass of PVC, halogenated fire retardants and other fuels involved.

Soot samples from a number of accidental fires studied involving PVC and other materials seem to have dioxin contamination in a range from approximately 5–400 ng TEQ/g soot.[49,50,51] A high value of 390 ng/g was obtained in some soot samples taken from a PVC warehouse fire in Canada, while other parts of the same fire had levels of 0.3 ng/g.[49] The levels reported for the Dusseldorf airport fire of 42.6 ng TEQ/g might be considered fairly typical,[51] the German authorities citing levels in fires of up to 200 ng TEQ/g. The Dusseldorf dioxins are now considered not to have been derived mainly from PVC. Taking the EPA household waste fire data, which are likely to be similar to the yield in a small domestic fire, it is possible to estimate likely soot dioxin concentrations from the dioxin and smoke particulate yields in several different experiments over a range of approximately 700–7000 ng TEQ/g soot which are one to two orders of magnitude greater than those recovered from fire incidents. It is estimated that this could represent a concentration of around 20 ng TEQ/m^3 in a dilute smoke plume (OD/m 0.01) in the vicinity of such an oil barrel fire. This compares to a figure of 5–40 ng TEQ/m^3 measured in the effluent plume from old-fashioned incinerators in Sweden, which if diluted by a factor of 100 would give a maximum of 0.4 ng TEQ/m^3 in a diluted plume.

Based on these figures it is possible to estimate possible hazards to a person working in the vicinity of a burning building during a fire and exposed to a dilute smoke plume, or inside a building after a fire and exposed to soot (see Table 3.11). Assuming a person was exposed for a

Table 3.11 Possible total dioxin and furan intake (pg TEQ) for a person near or inside a building during or after a fire

Dioxin concentration of smoke particles and soot ngTEQ/g	Intake	Dose received pg TEQ	Fraction of maximum acceptable daily intake 700 pgTEQ/day
Old Swedish incinerator plume (assuming 1/100 plume dilution factor)	0.05–0.40 ngTEQ/m^3 diluted smoke inhalation for 1 hour (1 m^3)	50–400	0.7–5.7
EPA household waste study 700–7000[a]	Inhalation 100 m visibility smoke dilution 2–20 ng TEQ/m^3 for 1 hour (1 m^3)	2000–20 000	29–290
EPA household waste study 700–7000[a]	Dust inhalation 1 mg/m^3 for 5 hours (5 m^3)	3 300–33 000	47–470
German fire residue maximum 200	Dust inhalation 1 mg/m^3 for 5 hours (5 m^3)	1000	14
EPA household waste study 700–7000[a]	Oral intake 0.1 g soot	70 000–700 000	1000–10 000
German fire residue maximum 200	Oral intake 0.1 g soot	20 000	286

[a] assuming TEQ = total dioxin and furan content × 0.1.

1 hour period to the dilute smoke plume and inhaled 1 m^3 of air, then the inhaled total dose of dioxins and furans could total 400 pg TEQ for a 1/100 diluted incinerator plume or between 1980 and 19 800 pg TEQ for a similar dilution of the household waste fire case (dilution to give a visibility of 100 m through the smoke). This can be compared to the maximum tolerable daily intake of 70 pg TEQ/person (from Table 3.10).

Another potential source of inhalation exposure is the inhalation of soot dust while working in a damaged area after a fire without respiratory protection. Assuming a level of 1 mg/m^3, this might result in the inhalation of up to 5 mg soot during a 5 hour period, representing 3304–33 043 pg TEQ for the household waste fire case. Using the German figure of 200 ng TEQ/g gives a dust intake figure of 1000 pg TEQ.

The other main route of ingestion is likely to be that of oral ingestion through contamination of food by contact with dirty hands or clothing. If 100 mg of soot (a very small pinch) was ingested, this would represent a dose of 66 100–661 000 pg TEQ for the household waste case or 20 000 pg TEQ for the German data.

The data are summarized in Table 3.11. They indicate that a diluted smoke plume from a fire is likely to contain considerably more dioxins and furans than the diluted smoke plume from even an old style incinerator, and that being enveloped in smoke from a fire dilute enough to provide 100 m visibility could result in 300 times the daily maximum acceptable intake. The inhalation of soot dust after a fire could provide a similar or somewhat greater hazard depending upon the dust concentration and duration of exposure. However oral intake with food is likely to be the most hazardous for someone working in a fire-contaminated area. This could provide up to 10 000 times the maximum acceptable daily intake, but the hazard could be minimised by simple hygiene precautions. Dust and smoke inhalation could also be minimised by the use of simple respiratory protection such as a disposable dust mask.

An important consideration with regard to dioxin intake is that it is a long term accumulation and removal problem. Exposure during any one day or even over a week is less important than the overall cumulative exposure over a long period extending to years. The health significance would therefore depend on how often a person was exposed to such levels. It is also important to remember that the recommended maximum exposure levels include a safety factor of around an order of magnitude or more. On this basis it would seem that there is a potential small but significant risk to health for a person in contact with dioxin contaminated soot from fires on a regular basis. It must be remembered that soot and smoke also contain other carcinogenic and systemically toxic compounds such as polyaromatic hydrocarbons (PAHs), so that inhalation, oral ingestion and prolonged dermal contact with smoke and soot should be avoided as far as possible. The hazard from dioxins and furans due to contact with soot and combustion process was recognised in the Department of the Environment pollution paper No. 27.[35] Section 8.5 page 41 states:

Occupational exposure may also occur for persons involved with combustion processes. These may include the operation of all types of combustion plant and incinerators, including the handling of ash; the burning of chlorinated materials e.g. PCP-treated timber and chlorinated plastics; or open fires; and the smelting of plastics-coated scrap metal. Similarly entry into areas contaminated by soot as a result of fire engulfment of PCP-containing materials may give rise to exposure to PCDDs and PCDFs.

3.5.3   Phosphorus-based fire retardants

3.5.3.1   General

Organic and inorganic phosphorus compounds act as fire retardants mainly by promoting char formation, although volatile phosphorus compounds also have some vapour phase free-radical inhibiting properties. Phosphorus is often used in combination with other free-radical inhibiting of char forming fire retardants such as halogens and melamine. Phosphorus is used in many forms, including elemental red phosphorus, as an inorganic compound (such as ammonium polyphosphate) or in organic form (such as phosphate esters).[52,53] Phosphorus-containing fire retardants may be non-reactive (finishes on fabrics, surface coatings, fillers in resins), or they may be reactive, combining with the polymeric structure during processing.[52,53,54] These substances act in a number of different ways to impede combustion and the wide variety of chemical structures and reactions involving phosphorus compounds can lead to a wide variety of phosphorus-containing products. In the vapour phase a variety of potentially toxic phosphorus-containing products may be formed.

Potential toxicity issues relating to the use of these compounds as fire retardants include:

1   The direct toxicity of organophosphorus (OP) fire retardants if released from materials.
2   The direct toxicity of OP compounds and inorganic phosphorus compounds formed in the vapour phase during combustion.
3   The effects of OP fire retardants on the yields of other toxic products during fires and the toxic effects of halogen and nitrogen compounds from OP fire retardants released during combustion.

It has been proposed that considerable toxicological problems may exist in the decomposition products of some flame retardants,[54] and at least one example exists in the form of TMPP (trimethylol propane phosphate), a highly neurotoxic product formed in the vapour phase during fires involving certain materials. Apart from this there is very little information on the chemistry and toxicology of the decomposition products from the majority of phosphorus-containing fire retardants.

3.5.3.2   Direct toxicity of OP fire retardants

As a class OP compounds are often neurotoxic agents.[55] These include some of the most toxic compounds known. Neurotoxic effects have acute and chronic phases. The main acute effect is that some OPs act as anti-cholinesterases. Anticholinesterases inhibit the action of cholinesterase and thereby potentiate the effects of acetycholine. This can result in a range of effects including paralysis and death. Over a longer period, even after a single exposure in some cases, neurotoxic OPs can cause nerve degeneration (neuropathies).[55] It is therefore very important to ensure that OP fire retardants and their combustion products do not show anticholinesterase or neurotoxic properties, particularly if they can be shown to leach out of treated materials and become absorbed systemically. Fortunately, OP fire retardants currently in use have not been reported to show anti-cholinesterase activity, but at least one has been shown to produce longer term neurotoxicity (triorthocresyl phosphate [TOCP],[56,57,58] and another has been shown to be a carcinogen (tris-[2,3-dibromopropyl] phosphate [TRIS]).[59] A problem is that routine toxicity investigations are unlikely to reveal delayed neurotoxic effects, because the young rodents used in these tests are usually unaffected, while delayed neurotoxicity does occur in humans, chickens and cats, so that special test procedures are available for investigating suspect compounds. A single dose of TOCP has been shown to cause ataxia with a distal axonopathy in hens over a 21-day period.[58]

3.5.3.3 The direct toxicity of phosphorus-containing compounds during combustion

During fires a certain amount of phosphorus is released as phosphorus pentoxide, which becomes hydrolysed to phosphoric acid: This can contribute to the sensory and lung irritant effect of smoke. An acute inhalation toxicity study of phosphorus pentoxide[61] gave a 1 hour LC_{50} in rats of $1.217\,g/m^3$ (206 ppm at $20\,°C$). Death was caused by lung congestion, haemorrhage and oedema with extensive necrosis and inflammation of the larynx and trachea. Due to the small amounts of phosphorus usually present in materials and the low yield into the vapour phase, acidic phosphorus compounds may make only a minor contribution to overall irritancy of fire effluent. Similarly, small amounts of the highly toxic gas phosphine (PH_3), have been detected in fire effluents from phosphorus-containing materials.[53] Phosphine is a potent lung oedemogen (lung irritant) with a 1 hour LC_{50} in rats of 44 ppm.[60]

In addition to these inorganic irritants a major main concern is that highly toxic OP compounds may be formed in the vapour phase, arising from the decomposition of the fire retardant itself, or through combination with other fire products. Virtually no work has been done to examine the chemical forms or yields of phosphorus compounds in thermal decomposition products, or their toxicity for the majority of fire retardant compounds, especially with regard to long term toxic effects. L'homme et al[61] examined the effects of pyrolytic (in helium) and oxidative thermal decomposition on trialkyl phosphates (trimethyl [TMP] and triethyl [TEP] and also a triaryl phosphate (triphenyl [TPP]). As with OP insecticides,[62] so the trialkyl phosphates were found to be thermally unstable, with scission of the C–O bond at 200–300 °C. This yielded phosphorus pentoxide with various aliphatic scission and condensation products, mainly methane and ethane under pyrolytic conditions and CO_2 with traces of aldehydes under oxidative conditions. TPP was more thermally stable, decomposing only above 600 °C, with scission of both the P–O and C–O bond. The authors state that all phosphorus was recovered as phosphoric acid (resulting from hydrolysis of phosphorus pentoxide), with small amounts of red phosphorus being formed under pyrolytic conditions. TPP is volatile, acting in the vapour phase as well as the solid phase as a fire retardant. It is therefore possible that some TPP may exist in fire effluent. Although the authors did not analyse specifically for organic phosphorus compounds, which might have been present in small amounts, the work established that phosphate esters are in general easily destroyed by heat to release inorganic phosphorus oxides and acid. The main expected toxic hazard would therefore be from the irritant effects of inhaled phosphorus pentoxide, adding to the general irritant effects of the smoke.

In these studies the OP compounds were decomposed alone. When fire retardants are added to materials the inorganic phosphate released may combine with other substances such as alcohols in the solid or vapour phase to form new phosphate esters, which may survive in the cooling smoke or char. An example of such a mechanism in the solid phase occurs during char formation in intumescent coatings containing ammonium polyphosphate and pentaerythritol.[53] On heating, ammonia and water are evolved with the formation at 250 °C of a bicyclic phosphate, followed by char formation, and the fate of the bicyclic compound is unknown. A possible concern is that this compound might be neurotoxic as is the caged bicyclic phosphate ester trimethylol propane phosphate (TMPP),[67] or even that this might lead to the formation of TMPP in the vapour phase (for a review of TMPP see Purser.[65] However, when Wyman et al (1987)[63] exposed rats to the thermal decomposition products of lubricants containing pentaerythritol and tricresyl phosphate, no signs of neurotoxicity were seen. Nevertheless, the formation of the above ester in the solid phase, and of TMPP in the vapour phase, clearly demonstrates that OP esters can be formed during the thermal decomposition of materials treated with phosphorus-based fire retardants. Other toxicity results that are difficult to explain in term of normal toxic products have been obtained during tests on certain FR fabrics.[64,65]

3.5.3.4   The effects of OP fire retardants during fires

Another source of toxic compounds in fires results from the effect of phosphorus-based fire retardants on the yields of other toxic gases released from materials. A further possibility is the formation of toxic phosphorus products due to combination with combustion products from treated materials in the gas phase. Since fire retardants tend to reduce combustion efficiency they can lead to increased yields of CO, organic products and smoke in fires, but phosphate-based fire retardants acting mainly in the solid phase should reduce the yields of toxic effluents by encouraging char formation. Nevertheless, phosphorus-based fire retardants do release some phosphorus-containing products into the vapour phase. Two systems in which phosphorus-based fire retardants are commonly used are in flexible polyurethane foams (FPU) in furniture and in furnishing and other fabrics. Studies of fire-retarded and non-fire-retarded polyurethane foams have been carried out under both flaming and non-flaming conditions using the NIST (former National Bureau of Standards) combustion toxicity method. This method, in which samples of material are decomposed in a cup furnace, has proved a reasonably good model for the conditions during early, well ventilated flaming fires. In one study, two FPUs were examined, one containing a chlorinated phosphate so that it was cigarette and flame ignition resistant[66] (Table 3.12). Under non-flaming conditions at 357–400 °C the LC_{50} of the standard foam was $34 \, g/m^3$ mass loss (i.e. when 34 g of foam were decomposed into each cubic metre of air) compared with $23 \, g/m^3$ for the FR foam. Deaths occurred after exposure, apart from one rat exposed to the FR foam, indicating that the main agents responsible were most likely to have been lung irritants or other toxic species rather than asphyxiant gases. Under flaming conditions at 450 °C no deaths occurred at concentrations of up to $40 \, g/m^3$ mass loss for the standard foam, while the LC_{50} of the FR foam was $25 \, g/m^3$ mass loss. The increased toxicity was partly due to a threefold increase in HCN yield and a doubled CO yield, which caused deaths during exposure, but because the majority of deaths occurred after exposure it is likely that the main cause of death was lung irritation from isocyanate-derived compounds and other pyrolysis products escaping the flame zone, or from

Table 3.12 Toxic product yields and toxic potency of flexible polyurethane foam (FPU) untreated and treated with a chlorinated phosphonate fire retardant under non-flaming and flaming decomposition conditions (NBS [NIST] cup furnace)

Material and decomposition condition		CO g/kg	CO_2 g/kg	HCN g/kg	CO_2/CO v/v	CO/HCN v/v	LC_{50} g/m³
FPU:							
Non-flaming	400 °C	46	73	0.35	1/1	71/1	34
Flaming	450 °C	23	1579	1.7	43/1	15/1	>40
FPU + Cl–P:							
Non-flaming	375 °C	35	35	0.2	1/1	123/1	23
Flaming	450 °C	45	1533	5.24	20/1	10/1	25

From: Purser[5]. Data from Braun et al.[66]

some other factor related to the OP fire retardant, as under non-flaming conditions.

These results show that under non-flaming and particularly under flaming conditions the toxic potency of the FR-material was greater than that of the untreated material, due a reduced combustion efficiency and increased yields of CO, HCN and other toxic products. While it is possible that there may have been a contribution to the toxicity from OPs, in general the results obtained from acute experiments on flexible polyurethane foams do not suggest that OPs form the major toxic atmosphere components, although anticholinesterase activity and delayed neurotoxicity have not been tested.

3.5.3.5 *Effects of phosphorus-based FR treatments on the toxicity of combustion products from fabrics*

Other small-scale studies have examined the toxic potencies of treated and untreated cotton and polyester fabrics using such methods as that of the DIN tube furnace among others.[64,66] In one series of studies FR-polyester materials were found to be approximately twice as toxic as non-FR polyester materials. When cotton/polyester fabrics were tested some of which were treated with tetrakis(hydroxymethyl)phosphoniumhydroxide (THPOH) the retardant caused a 3–9 fold increase in toxic potency. In another series of studies Kallonen et al[64] tested a range of FR and non-FR fabrics using the DIN furnace at 500 °C and 800 °C. The phosphorus-containing fire retardants used were tetrakis(hydroxymethyl)phosphonium chloride and urea concentrate (Proban®) and N-methylol-dimethyl-3-phosphonopropionamide (Pyrovatex®). The majority of rats survived immediate exposure to cotton under both non-flaming and flaming conditions, but under non-flaming conditions there was a 42% mortality over the next 14 days, most probably due to pulmonary irritation. The FR-cotton failed to flame and the yields of CO and HCN were higher than for the untreated cotton, which for Pyrovatex/cotton may explain the few deaths occurring during exposure. For Proban/cotton the pattern of toxicity is less obvious. During decomposition at 500 °C four of the rats died with only 1% carboxyhaemoglobin despite exposure to a CO concentration of 3100 ppm, indicating that they died early during the exposure from some unknown toxic effect. A similar effect occurred at 700 °C. These experiments illustrate that the presence of phosphorus-based fire retardants can increase the yields and toxic potency of normal toxic products by altering combustion behaviour, as did the treatment of foams. They also illustrate the possibility of other toxic effects of unknown cause which were only revealed by animal exposures to the combustion products.

3.5.3.6 *Formation of high toxic potency neurotoxic caged biphosphorus esters*

In 1975 Petajan et al[67] reported the formation of a neurotoxic OP product in the combustion products from a non-commercial rigid PU-foam treated with a phosphorus-containing fire retardant. The substance was trimethylol propane phosphate (TMPP) (4-ethyl-1-phospha-2,6,7-trioxabicyclo (2.2.2) octane-1-oxide). It had an extreme toxic potency and rapid action causing *grand mal* epileptic seizures and death. The product was formed by the reaction of a propoxylated trimethylolpropane polyol component of the foam with the phosphate fire retardant (0,0-diethyl N,N-bis(2-hydroxymethyl) aminomethylphosphonate. Thermal decomposition of the foam released the propoxylated trimethylol propane polyol adduct, which decomposed to form trimethylol propane. This combined with reactive phosphorus species to form principally TMPP.

Woolley and Fardell[68] studied the yields of TMPP during thermal decomposition of various types of flexible and rigid PU foams. Based on the toxic potency of TMPP and other major toxic products (CO and HCN) and the yields at which they were released, TMPP might make a major contribution to the overall toxic potency of these foams under non-flaming conditions, but not under flaming conditions. Evidence for this was obtained by Wright and Adams[69] and further discussion of the issue is given in Purser.[65] The problem has largely been eliminated by avoiding combinations of trimethylol propane polyols with phosphorus fire retardants, although some specialised lubricants and hydraulic fluids have been shown to produce TMPP under certain conditions.[70]

3.5.3.7 *Toxic hazards from phosphorus treated materials and products*

The work described in the sections on the combustion of phosphorus-containing compounds indicates that some phosphorus-based FR-treatments can increase the toxic potency of the thermal decomposition and combustion products from materials. This results particularly when the phosphorus is released into the vapour phase, sometimes accompanied by halogen or nitrogen containing components. This is not to say that the toxic hazard during a large-scale fire would necessarily be increased by the use of such FR-treatments.

These treatments are intended to increase ignition resistance (thereby reducing ignition risk) and to reduce the rate of flame spread and fire growth. An example is provided by the FR-treated PU foams for which the small scale results already discussed have shown increased toxic potencies from the FR-treated materials.[5,66] Armchairs were fabricated from these foams with Haitian cotton covers and burned in an enclosed room-corridor-room apartment rig.[66] From the results obtained it is possible to estimate the development of the toxic hazard in the rig[5] and time to loss of tenability for any occupant. The results following flaming ignition of the arm chairs are illustrated in Fig. 3.6 for a potential occupant of the fire room. The results show the CO time-concentrations curves (which provide an indication both of the rate of fire growth and the concentration of a major toxic species) and the accumulating FED for incapacitation. The rate of fire growth is greater in the non-FR-treated chair and therefore the rate of accumulation of toxic effluent in the rig. Loss of tenability for asphyxiant gases is predicted when FED > 1, and this occurs after 8 minutes for the non-FR-treated chair and after 11 minutes for the FR-treated chair. However, when the smouldering ignition of the chairs was obtained by using cigarettes, the rate of accumulation of CO in the rig was more rapid for the FR-treated chair, so that incapacitation is predicted after 54 minutes but after 63 minutes for the non-FR chair. This finding is countered to some extent by the fact that while the non-FR chair eventually

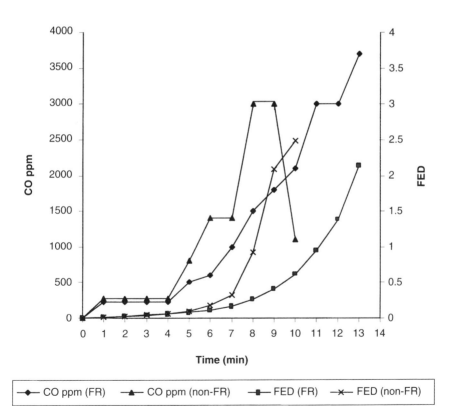

3.6 Comparison of tenability time after flaming ignition of FR and non-FR foam chairs in fire room

went from smouldering to flaming spontaneously, the FR chair was deliberately ignited to study post-smouldering ignition. Overall, it could be argued that the FR-treated chairs would present an improved fire risk, with an improved fire hazard performance under flaming ignition conditions. These benefits would more than offset the small potential increased hazard from smouldering conditions, particularly if the hazard is considered in conjunction with the use of efficient detection.

3.5.4 Melamine and melamine/chlorinated phosphate systems

In this section the performance of mixed fire retardant treatments for upholstered furniture is considered. These include the use of chlorinated and brominated FR-systems for fire retardant cotton covers in conjunction with melamine and chlorinated phosphate treatments for polyurethane foams.

Melamine is used alone or in combination with other FR-additives in a variety of materials. As a fire retardant it exhibits a wide range of mechanisms all of which contribute to its effectiveness. These include endothermic reactions, inert gas dilution due to ammonia and nitrogen formation, free radical scavenging and the promotion of char formation.[70] When used with phosphates it helps to retain phosphorus in the solid, further promoting char formation. A central aspect of these properties is the high nitrogen content of melamine and this is also its greatest potential disadvantage, because organic nitrogen tends to form hydrogen cyanide in fires, particularly under vitiated combustion conditions, and this presents a serious asphyxiant toxic hazard.[2,18] Under well-ventilated combustion conditions nitrogen-containing materials produce less HCN but more oxides of nitrogen, which are dangerous lung irritants.[2] HCN is particularly dangerous because it causes rapid incapacitation at low concentrations and is therefore considered an important factor in preventing escape from fires, so that victims are more likely to be trapped and die from a combination of toxic gases including carbon monoxide.[2,18] Cyanide production during thermal decomposition of melamine has been shown by Morikawa.[72] Other work has shown that thermal decomposition of melamine-treated flexible polyurethane foams yielded six times more cyanide than non-melamine foams.[73] Cyanide production was particularly increased when char was decomposed.[74]

Melamine reduces ignitability and burning rate once an item is ignited. It is therefore useful in reducing fire risk in terms of both the probability of ignition and the rate of fire growth. However, the extent to which it reduces full-scale fire hazard will depend upon the extent to which increased cyanide yield is offset by reduced rates and extent of burning. This is particularly important in the context of the United Kingdom, where all polyurethane foams used in upholstered furniture since 1988 are combustion modified (combustion modified – CM or combustion modified high resilience – CMHR foams), containing various formulations of melamine and chlorinated phosphates. The effect of this and changes in furniture fabrics on the UK fire statistics has already been described. Full-scale fire tests can also be used to investigate the effects of such combined FR-treatments on the development of toxic hazard when a sufficiently large ignition source is used to overcome ignition resistance and produce a propagating flaming fire. In this context the main consideration is the rate of development of toxic hazard compared with the time to detection and the time required for occupants to escape.

The results of a programme of such tests conducted in a typical design of apartment (tests CDT10–13) and UK house (tests CDT14–23) are summarised in Figs. 3.2 and 3.3. Figure 3.7 shows an example of one of the house fire tests. The lower part shows the concentrations of the key toxic gases and smoke during the fire. The upper part shows the increase in Fractional Effective Dose or Fractional Irritant Concentration with time for each hazardous fire component. An endpoint is considered to be achieved when each parameter crosses unity on the y-axis. Thus for this example the irritancy criterion is breached at around 1.5 minutes and the smoke density criterion at around 2.5 minutes. These indicate increasingly unacceptable effects on escape efficiency. Incapacitation is predicted at just under 5 minutes, primarily due to the asphyxiant effects of hydrogen cyanide. Incapacitation due to heat is predicted at 6 minutes and, had there been no cyanide, incapacitation due mainly to carbon monoxide is predicted at 7 minutes. The chair had back-coated acrylic covers with a 3.6% chlorine and 2.4% bromine content. A CM foam was used containing melamine (total nitrogen content 11%) and a chlorinated phosphate (chlorine 2.2%, phosphorus 0.8%). The results show a problem with the early evolution of a dense irritant smoke, but a reasonably slow rate of burning giving 5 minutes to loss of tenability. The deleterious effects of HCN release from the covers and foam are illustrated by the reduction in time to incapacitation by 2 minutes compared with the effects of CO.

The importance of the performance of the system of different components is illustrated in Figs. 3.2 and 3.3 by the results for test CDT10 and CDT21 (HR foam non-FR acrylic covers) and tests CTD11 and CDT23 (CM foam non-FR acrylic covers). The combustion modified foam appears to convey no advantage in terms of time to incapacitation in the absence of FR-covers. The combination of either FR-back-coated acrylic (CDT18 and CDT22) or FR-cotton (containing 4.4% bromine and 0.9% chlorine)

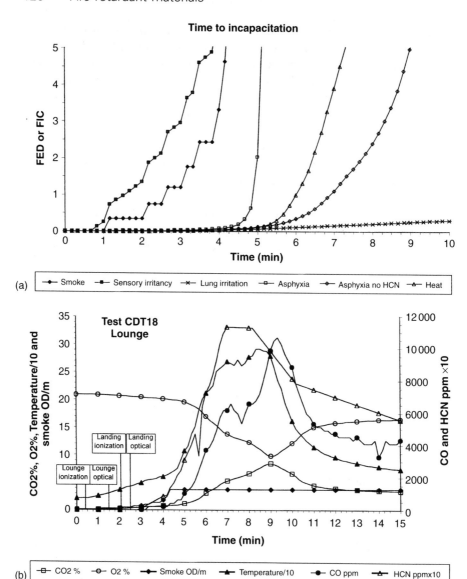

(b) Toxic gases, smoke and heat during single armchair fire in open ground floor lounge of a two-storey house. Triggering times for ionisation and optical smoke detectors in lounge and upstairs hall are also shown.

(CDT16,17,19 and 20) with the CM foam is more effective, producing times of predicted incapacitation of between 4 and 5 minutes in the fire room. In all cases time to incapacitation was dominated by the effects of hydrogen cyanide.

Another series of large-scale studies carried out at NIST[75] compared the hazard from FR- and non-FR-treated materials and products in a variety of configurations. The products and materials included: television cabinet housings (high impact polystyrene ± decabromodiphenyl oxide (12% by mass) and antimony oxide (4%), business machine housings of polydephenylene oxide and other fractions ± a triaryl phosphate ester to give 1% phosphorus by mass, upholstered chairs PU foam with chlorinated phosphate, organic brominated retardant and 35% alumina trihydrate (4.75% Br, 2.6% Cl, 0.32% P and 10% Al) with nylon covers, cable array ethylene vinyl acetate copolymer with clay as insulation ± antimony oxide (18.9 parts) and covers of chlorosulphonated polyethylene, laminated circuit boards of polyester resin also a brominated fire retardant 10%, antimony oxide 3% and hydrated alumina 30%. These products were used in a range of experiments. The overall findings were that the FR-treated materials provided a greater than 15 fold increase in escape time for room occupants than for a non-FR room. The production of combustion products was as follows: the amount of material consumed in the fire was less than half that from non-FR tests; the FR test released 75% less heat than the non-FR tests and 66% less toxic gases with no significant difference in smoke production. The use of FR-products was therefore found to reduce the overall fire hazard in these tests.

3.6   Conclusions

The majority of fire injuries and deaths result from exposure to toxic products. The extent of such exposure (toxic risk) depends upon the probability that fires will occur and the extent of the subsequent toxic fire hazard. Fire retardant systems are used in a wide range of materials and products to control fire risk and fire hazard. They achieve this by improving ignition resistance and in some cases by improving the rate of fire growth once defined levels of ignition resistance are overcome. This strategy is successful providing that the use of fire retardants does not increase the toxic hazards in fires by increasing the yields of toxic products to an extent which negates any benefits arising from improved ignition and reaction to fire properties. It is also important that other toxicity issues (direct toxicity and environmental issues) do not present unacceptable problems.

In general, research and testing has been concentrated on ignition and reaction to fire properties rather than the evaluation of toxicity and toxic hazard properties of fire retardant systems. Some systems improve both fire performance and toxic potency properties of materials and products, others improve fire performance but tend to have negative effects on toxicity. Whether or not an increased toxic potency leads to an increase in toxic hazard depends upon the interaction between toxicity, the burning rate and the overall fire scenario.

Toxicity issues exist in relation to fire retardants with respect to the direct toxicity of fire retardant compounds, the toxicity of combustion products from fire retardants, and the effects of fire retardants on the yields and toxicity of other combustion products from treated materials with respect to both direct exposure and environmental contamination. These issues need to be considered in the overall context of the impact of fire retardant design and use on fire risks and hazards.

Halogenated fire retardant systems tend to present potential toxicity problems during fires due to increased yields of asphyxiants, smoke and irritants as well as the direct release of irritant acid gases. There is also the potential problem of environmental contamination from the release of halogenated dioxins and dibenzofurans. These problems also apply to combined halogen-phosphate systems. Phosphate systems have the advantage that they do not produce environmentally persistent toxic combustion products, and depending upon the system, may reduce both the yields of toxic products and the fire growth rate during fires. Apart from a few special cases there is little evidence for the release of exotic toxic organophosphorus products during fires, although few studies have been performed. Some organophosphorus fire retardants are directly neurotoxic or carcinogenic.

In practice when the toxic hazards from fires involving fire retardants are compared with those from non-fire retarded equivalents the results depend upon the particular system involved and the fire scenario. Where the use of a particular system leads to a significant decrease in fire incidence and/or a decrease in the rate of fire growth in fires that do occur, the benefits are likely to outweigh considerably any disbenefits resulting from increased toxic product yields. Systems that reduce fire risk, fire growth and toxic product yields are likely to be particularly beneficial.

When the incidence of environmental contamination due to dioxins and dibenzofurans from all sources is examined then there has been a considerable decrease since the mid 1980s, despite the considerable increased use of PVC and halogenated fire retardants. At present, there is little evidence that the use of halogenated FR-systems is likely to lead to a significant general environmental contamination problem from accidental fires.

References

1 *United Kingdom Office Fire Statistics*, Her Majesty's Stationery Office, London, Published Annually.

2 Purser D A, 'Toxicity assessment of combustion products', in *SFPE Handbook of Fire Protection Engineering*, P J DiNenno (ed) Quincy, MA. National Fire Protection Association, Section 2 85–146, 1995.

3 Purser D A, *Acute Environmental Hazards in the Vicinity of Large Fires*, Society of Chemical Industry Fire Chemistry Discussion Group. University of Salford, UK, 1996.

4 Purser D A, *The Performance of Fire Retarded Materials in Relation to Toxicity, Toxic Hazard and Toxic Risk*, Society of Chemical Industry Fire Chemistry Discssion Group. University of Lancaster, UK, 1998.

5 Purser D A, 'The development of toxic hazard in fires from polyurethane foams and the effects of fire retardants', in *Flame Retardants '90*, The British Plastics Federation. Elsevier, London, 206–21, 1990.

6 Buser H-R, 'Polybrominated dibenzofurans and dibenzo-p-dioxins: thermal reaction products of polybrominated diphenyl ether flame retardants', *Environ. Sci. Technol.*, **20** 404–8, 1986.

7 Thoma H, Hauschulz G, Knorr E and Hutzinger O, 'Polybrominated dibenzofurans (PBDF) and dibenzodioxins (PBDD) from the pyrolysis of neat brominated diphenylethers, biphenyls and plastics mixtures of these compounds', *Chemosphere*, **16** 277–285, 1987.

8 Sperring K, *Current State of Regulations Relating to Brominated Fire Retardants*, Society of Chemical Industry Fire Chemistry Discussion Group. University of Lancaster, UK, 1998.

9 Zelinski V, Lorenz W and Bahadir M, 'Brominated fire retardants and resulting PBDD/F in accidental fire residues from private residences', *Chemosphere*, **27** 1519–28, 1993.

10 *Stern Magazine*, 1990.

11 'Sweden to ban brominated flame retardants', Report by Swedish National Chemicals Inspectorate (Kemi) and associated press reports, ENDS Daily – 15/03/99.

12 Troitzsh J H, 'New trends in the use of halogenated flame retardants in Germany', in *International Progress in Fire Safety, Proceedings of the Fire Retardant Chemicals Association meeting*, New Orleans, 141–50, 1987.

13 Tesoro G C, *J. Fire Ret. Chem.*, **6** 239, 1979.

14 Horrocks A R, 'Developments in flame-retarding polyester/cotton blends', *J. Soc. Dyers Chem.*, **105** 346–9, 1989.

15 Stevens G C and Mann A H, *Risks and Benefits in the Use of Flame Retardants in Consumer Products*, University of Surrey, Guildford. Department of Trade and Industry ref. URN 98/1026, 1999.

16 Toxicity Testing of Fire Effluent: *Guidance to Regulators and Specifiers on the Assessment of Toxic Hazard in Fires in Buildings and Transport*. ISO/IEC TR 9122–6, 1993.

17 HMG, Statutory Instruments (i) 1988, No. 1324, *Consumer Protection, The Furniture and Furnishing (Fire) (Safety) Regulations 1988*; (ii) 1989, No. 2358, *Consumer Protection; Public Health, England and Wales; Public Health Scotland, The Furniture and Furnishing (Fire) (Safety) (Amendment) Regulations*, 1989.

18 Purser D A, 'Behavioural impairment in smoke environments', *Toxicology*, **115** 25–40, 1996.

19 *Toxicity Testing of Fire Effluent: Prediction of Toxic Effects of Fire Effluent*. ISO/IEC TR 9122–5, Geneva 1993.

20 *Determination of Lethal Toxic Potency of Fire Effluents*. ISO/IEC 13344, Geneva 1995.

21 Woolley W D and Fardell P J, 'The prediction of combustion products', *Fire Research*, **1** 11–21, 1977.

22 Purser D A, 'The harmonization of toxic potency data for materials obtained from small and large scale fire tests and their use in calculations for the prediction of toxic hazard in fire', *Proceedings of First International Fire and Materials Conference*, Washington DC USA. 1992. Interscience Communications, London, 179–200, 1992.

23 *Toxicity Testing of Fire Effluent: the Fire Model*. ISO/IEC TR 9122–4, 1993.

24 *British standard code of practice for assessment of hazard to life and health from fire: Part 2: Guidance on methods for the quantification of hazards to life and health and estimation of time to incapacitation and death in fires*. BS 7899 Part 2 1999.

25 Purser D A, Fardell P J, Rowley J, Vollam S and Bridgeman B, 'An improved tube furnace method for the generation and measurement of toxic combustion products under a wide range of fire conditions', in *Proceedings of the 6th International Conference Flame Retardants '94*, London. Interscience Communications, 1994.

26 Wright T, *Environmentally friendlier flame retardant systems*. PhD Dissertation. Leeds University, Leeds, 1997.

27 *Fire Safety of Upholstered Furniture – the Final Report on the CBUF Research Programme. European Commission Measurement and Testing Report* EUR 16477 EN, ed. B Sunsdtröm Interscience Communication, London.

28 Schnipper A, Smith-Hansen L and Thomasen E S, 'Reduced combustion efficiency of chlorinated compounds, resulting in higher yields of carbon monoxide', *Fire Mat*, **19** 61–4, 1995.

29 Hartzell G E, Grand A F and Switzer W G, 'Modelling of toxicological effects of fire gases: VI. Further studies on the toxicity of smoke containing hydrogen chloride', in *Advances in Combustion Toxicology*, Volume 2, ed. G E Hartzell, Technomic, Lancaster PA, 285–308, 1989.

30 Purser D A, 'Recent developments in understanding the toxicity of PTFE thermal decomposition products', *Fire Mat*, **16** 67–75, 1992.

31 Purser D A, Fardell P J and Scott G E, 'Fire Safety of PTFE-based materials used in buildings,' *Building Research Establishment Report 274*, Garston, UK, 1994.

32 New York Telephone Exchange fire.

33 Sarkos C P, Hill R G and Howell W D, 'The development of a full-scale wide body test article to study the behaviour of interior materials during a postcrash fire,' *AGARD Lecture Series No. 123 Aircraft Fire Safety*. 6:1–6:21, 1982.

34 Purser D A, 'Modelling time to incapacitation and death from toxic and physical hazards in aircraft fires', in *Conference Proceedings No. 467. Aircraft Fire Safety. NATO-AGARD*, Sintra, Portugal, 41-1–41-13, 1989.

35 DOE *Report of an Interdepartmental Working Group on Polychlorinated Dibenzo-para-dioxins (PCDDs) and Polychlorinated dibenzofurans (PCDFs)* (Pollution Paper No. 27). Department of the Environment, London, HMSO, 1989.

36 Duarte-Davidson R, et al. 'The relative contribution of individual PCBs, PCDDs and PCDFs to toxic equivalent values derived for bulked human adipose tissue samples from Wales, United Kingdom,' *Arch. Environ. Contam. Toxicol.*, **24** 100–27, 1993.

37 MAFF. *Dioxins in food-UK Dietary Intake (Food Surveillance Paper No. 71).* Ministry of Agriculture, Fisheries and Food, London, UK. HMSO, 1995.

38 Muto H and Takizawa Y, 'Dioxins in cigarette smoke', *Arch. Environ. Hlth.*, **44** 171–4, 1989.

39 Humfrey C, Taylor M and Amaning K, *IEH Report on Health Effect of Waste Combustion Products, Report R7*. Medical Research Council, Institute for Environmental Health, Leicester, 1997.

40 Tschirley F, 'Dioxin', *Scientific American*, **254** 29–35, 1986.

41 Agency for toxic substances and disease registry, US Department of Health and Human Services. 'Dioxin toxicity,' *Am. Fam. Physic.*, **47** 855–61, 1993.

42 *COT Statement by the Committee on Toxicity of Chemicals in Food, Consumer Products and the Environment on the EPA Draft Health Assessment Document for 2,3,7,8-tetrachloro-p-dioxin and Related Compounds.* (Available from the Department of Health, Skipton House, 80 London Road, London), London, 1995.

43 WHO (191) *Consultation on Tolerable Daily Intake from Food of PCCDs and PCDFs*. Bilthoven, Netherlands, World Health Organisation Regional Office for Europe, 1991.

44 Safe S, 'Polychlorinated dibenzofurans: environmental impact, toxicology, and risk assessment', *Toxic Substances J.*, **11** 177–222, 1991.

45 ENVIRON. *An Expert Panel Review of the US Environmental Protection Agency's Draft Risk Characterisation of the Potential Health Risks of 2,3,7,8-tetrachloro-p-dioxin (TCDD) and Related Compounds.* (Available from the Science Advisory Board of the US Environmental Protection Agency, Washington DC.) Washington, 1995.

46 Eduljee G H, 'Dioxins in the environment', *Chemistry in Britain*, December 1988.

47 Alcock R E and Jones K C, 'Dioxins in the environment: A review of trend data', *Environmental Science and Technology*. **30** 3133–43, 1996.

48 Thomas V, *Toxicological and Environmental Chemistry*, **50**(1) 1995.

49 Maklund S, *Dioxin Emissions and Environmental Emissions. A study of Polychlorinated Dibenzodioxins and Dibenzofurans in Combustion Processes*. University of Umea, Umea, 1990.

50 Carroll W, 'Is PVC in house fires the great unknown source of dioxin?' *Fire and Materials* **20** 161–6, 1996.

51 Chlorophiles Web Site, www.ping.be/~ping5859/index.html

52 Pierce E M and Liepins R, 'Flame Retardants', *Environ. Hlth. Perspect.*, **11** 59–69, 1975.

53 Stevenson J E and Guest R, 'New developments in organic flame retardants.' *Proceedings of International Progress in Fire Safety*, Fire Retardant Chemicals Association, New Orleans 141–150, 1987.

54 Liepins R and Pierce E M, 'Chemistry and toxicity of flame retardants for plastics,' *Environ. Hlth. Perspect.*, **17** 55–63, 1976.

55 Ballantyne B and Marrs T, 'Overview of the biological and clinical aspects of organophosphates and carbamates,' in *Clinical and Experimental Toxicology of*

Organophosphates and Carbamates, B Ballantyne and T Marrs (eds) Butter-
worth-Heinemann Oxford, 3–14, 1992.

56 Aring C D, 'The systemic nervous affinity of triorthocresyl phosphate (Jamaica
    ginger palsy)', *Brain*, **65** 34–47, 1942.

57 Abou-Donia M B, 'Organophosphorus ester-induced delayed neurotoxicity',
    Ann. Rev. Pharmacol. Toxicol., **21** 511–48, 1981.

58 PrenticeD E and Roberts N L, 'Acute delayed neurotoxicity in hens dosed with
    tri-orthor-cresyl phosphate (TOCP): correlation between clinical ataxia and neu-
    ropathologic findings', *Neurotoxicology*, **4** 271–6, 1983.

59 Dybing E, Omischikinski J G, Soderlund E J, et al. 'Mutagenicity and organ
    damage of 1,2-di-bromo-3-chloropropane (DBCP) and tris (2,3-dibromopropyl)
    phosphate (TRIS-BP): role of metabolic activation,' in *Reviews in Biochemical
    Toxicology*, Vol. 10, 139–86, Elsevier Amsterdam, 1989.

60 Ballantyne B, 'Acute inhalation toxicity of phosphorus pentoxide smoke,' *Toxi-
    cologist*, **1** 140, 1981.

61 L'homme V, Bruneau C and Soyer N, et al. 'Thermal behaviour of some organic
    phosphates', *Ind. Eng. Chem. Prod. Res. Dev.*, **23** 98–102, 1984.

62 Smith W M and Ledbetter J O, 'Hazards from fires involving organophosphorus
    insecticides', *Am. Ind. Hyg. Assoc. J.*, **32** 468–74

63 Wyman J F, Porvaznik M and Serve P, et al. 'High temperature decomposition
    of military specification L-23699 synthetic aircraft lubricants', *J. Fire Sci.*, **5**
    162–77, 1987.

64 Kallonen T, von Wright A and Tikkanen L, et al. 'The toxicity of fire effluents
    from textiles and upholstery materials', *J. Fire Sci.*, **3** 145–60, 1985.

65 Purser D A, 'Combustion toxicology of anticholinesterases', in *Clinical and
    Experimental Toxicology of Organophosphates and Carbamates*. 386–95. B Bal-
    lantyne and T C Marrs (eds) Butterworth-Heinemann Oxford, 1992.

66 Braun E, Levin B C and Paabo M, et al. *Fire Toxicity Scaling*. US Department
    of Commerce National Institute of Standards and Technology Report No.
    NBSIR 87-3510, 1987.

67 Petajan J H, Voorhees K J and Packham S C, et al. 'Extreme toxicity from com-
    bustion products of a fire-retardant polyurethane foam,' *Science*, **187** 742–44,
    1975.

68 Woolley W D and Fardell P J, *Formation of a Highly Toxic Organophosphorus
    Product (TMPP) During Decomposition of Certain Polyurethane Foams under
    Laboratory Conditions*, Building Research Establishment Ltd. Fire Research
    Station. Fire Research Note No. 1060, Garston, Watford, UK, 1976.

69 Wright P L and Adams C H, 'Toxicity of combustion products from burning poly-
    mers: development and evaluation of methods', *Environ. Hlth. Perspect.*, **17**
    75–83, 1976.

70 Kalman D A, Voorhees K J and Osborne D, et al. 'Production of bicyclophos-
    phate neurotoxic agent during pyrolysis of synthetic lubricant oil', *J. Fire Sci.*, **3**
    322–9, 1985.

71 Weil E D and Choudhary V C, 'Flame-retarding plastics and elastomers with
    melamine', *J. Fire Sci.*, **13** 104–26, 1995.

72 Morikawa T, 'Evolution of hydrogen cyanide during combustion and pyrolysis',
    J. Combust. Toxicol., **3** 315–30, 1978.

73 Levin B C, 'New research avenues in toxicology: 7-gas N-gas model, toxicant sup-
    pressants, and genetic toxicology', *Toxicology*, **115** 89–106, 1996.

74 Levin B C, Paabo M, Fultz M L and Bailey C S, 'Generation of HCN from flexible polyurethane foam decomposed under different combustion conditions', *Fire Mat.*, **9** 125–34, 1985.

75 Babrauskas V, Harris R H, Gann R G, Levin B C, Lee B T, Peacock R D, Paabo M, Twilley W, Yoklavich M F and Clark H M, *Fire Hazard Comparison of Fire-retarded and Non-fire-retarded Products. NIST Special Publication 749.* US Department of Commerce, National Institute of Standards and Technology. 1–85, 1988.

76 Hietaniemi J, Kallonen R and Mikkola E, Burning characteristics of selected substances: production of heat, smoke and chemical species. *Fire Mat*, **23** 171–85, 1999.

77 *Code of Practice for Assessment of Hazard to Life and Health from Fire. Part 2 Guidance on Methods for the Quantification of Hazards to Life and Health and Estimation of Time to Incapacitation and Death in Fires.* British Standard 7899-2, 1999.

78 *Evaluation of Emissions from the Burning of Household Waste in Barrels.* Volume 1. Technical Report. United States Environmental Protection Agency Report EPA-600/R-97-134a, 1–69, 1997.

79 Van Leeuwen F X, Feely M, Shrenk D, Larsen J C, Farland W and Younes M, *Dioxins: WHO's tolerable daily intake (ITDI) revisited.* Chemosphere 40, 1095–101.

4
Textiles

A RICHARD HORROCKS

Faculty of Technology, Bolton Institute, Bolton, UK

4.1   Introduction

The particular hazard posed by burning textiles, especially those based on the natural cellulosic fibres cotton and flax (as linen), was recognised during early civilisations and salts like alum have been used since those times to reduce their ignitability and so confer flame retardancy. These risks remain with us to this day as a consequence of the intimate character of most textiles, primarily as clothing, and in the immediate domestic environment, coupled with the high specific surface area of the fibre-forming polymers present, which enable maximum access to atmospheric oxygen. These factors were highlighted in the recent tragic fire in Saudi Arabia on 16 April 1997 during the haj which killed over 340 pilgrims as fire swept through a tented camp at Mena within the vicinity of Mecca. In addition, many other pilgrims sustained burn injuries as the fire spread quickly through the estimated 70 000 tents. It is most likely that all tents were fabricated from woven cotton and that the fire was a consequence of its flammable character, which was aggravated by the hot, dry environment, the high winds present and the density of the pitched tents.

4.2   Hazards and risks

Across the world very few comprehensive fire statistics exist, especially those which attempt to relate deaths and injuries to cause, such as ignition and burning propagation properties of textile materials.

The annual UK Fire Statistics[1] are some of the most comprehensive available and do attempt to provide information perhaps representative of a European country with a population of about 55 million. For instance, up to 1998, these statistics have demonstrated that while about 20% of fires in dwellings are caused by textiles being the first ignited material, over 50% of the fatalities are caused by these fires. Table 4.1 presents typical data during the last 17 years although since 1993 such detailed data have not

Table 4.1 UK dwelling total and textile-related fire deaths, 1982–1998[1]

Year	Deaths in UK dwelling fires	Textile-related fatalities in dwellings				
		Clothing	Bedding	Upholstery	Floor-coverings	Total
1998[a]	497	62	71	69	11	213
1997[a]	566	59	51	119	8	237
1996[a]	556	60	79	108	11	219
1995[a]	549	85	71	108	8	275
1994[a]	477	65	68	86	5	224
1993	536	51	85	105	19	260
1992	594	71	82	134	22	309
1991	608	59	85	127	10	281
1990	627	61	89	157	20	377
1988	732	92	141	195	20	448
1986	753	69	150	219	17	455
1984	692	59	124	167	22	372
1982	728	86	140	152	23	424

Note: [a] denotes values based on sampling procedure.

been as freely available. This shows that generally deaths from fires in UK dwellings have fluctuated at around 700 per annum between 1982 and 1988; since then they have fallen to the 500–600 level. Fatalities from textile-related fires show a similar pattern and it may be concluded that legislation associated with the mandatory sale of flame retarded upholstered furnishing fabrics into the domestic UK market since 1989 has played a significant factor in these reductions.[2] Plotting the data as in Fig. 4.1 shows the relative risks posed by different types of textiles and the way that they may have changed during this period in which not only upholstered furnishing regulations have been enforced, but also where the incidence of smoke alarms in UK dwellings has increased, especially during the last ten years.

One trend which has not shown any change is the fatality rate associated with clothing where neither of these factors would be expected to have influence. Deaths involving clothing usually fluctuate within the 50–90 annual fatality range and as a group have largely been ignored by both government and the textile industry outside of the areas of nightwear[3] and protective clothing.[4] Clothing fires tend to be of an individual nature and so receive little public attention, and hence legislative pressure unless common groups of hazard are identified, such as there has been with nightwear and more recently, sari and similar clothing fire deaths in places like India and Upper Egypt,[5] where cooking over open fires is prevalent.

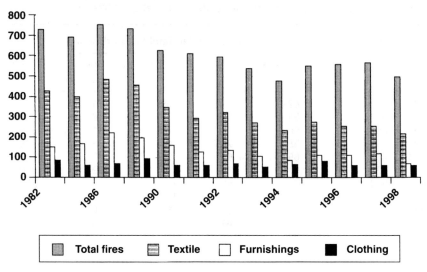

4.1 Fatalities from textile-related UK dwelling fires

Legislation and regulation usually occur only when there is a large loss of life or property and within the UK over the last 20 years a number of significantly well-publicised fires has driven the need to create new or review current fire precautionary and preventative regulations and proce-dures. Table 4.2 lists these major incidents, all of which were associated with textiles being the first ignited material or being responsible for major loss of life or damage. Notable among these is the F.W. Woolworth fire of 1979 from which the first UK cigarette ignition requirement for upholstered fur-nishings derived; subsequently, the more comprehensive legislation of 1990 followed[2] with its demand for match ignition and the mandatory combus-tion modification of foam filling. Similarly, the Manchester Boeing 737 fire brought forward the planned UK Civil and US Federal Aviation Authori-ties' requirement for fire resistant seating materials in all passenger aircraft designed to carry more than 30 passengers.[6]

All the fires in Table 4.2 have the common feature that the textiles present at each scene functioned as the material first ignited by the relevant igniting source. Secondly and subsequently, the speed with which this caused the fire to grow and spread to adjacent materials was a significant feature in the inability of victims to escape or the fire fighters to bring the fires under control. Therefore, these catastrophic fires serve to demonstrate more obvi-ously the ignitability of textiles in the first place followed by the associated speed with which the resulting fire can grow. It is rarely the direct causes of the fire, such as burn severity which are the prime causes of death, however, but the effects of the smoke and emitted fire gases which cause disorienta-tion and impede escape initially followed by subsequent incapacitation,

Table 4.2 Major textile-related fires in the UK and Ireland, 1979–present

Fire	Cause	Consequences
Rail sleeper fire, Taunton, 6 July 1978	Sacks of soiled and clean laundry adjacent to electric heater.	12 fatalities 15 non-fatal casualties
F.W. Woolworth Store Fire, Manchester, 8 May 1979	Ignition by smoker's material of a stack of polypropylene fabric-covered, polyurethane-filled furniture in restaurant area.	10 fatalities 53 non-fatal casualties
Stardust Disco Fire, Dublin, 14 February, 1981	Ignition of PVC-covered, foam-filled furnishings leading to flashover of multi-eating array.	48 fatalities 128 non-fatalities
Boeing 737 Fire, Manchester Airport, 22 August 1985	Punctured fuel tank causing external pool fire which broke through the fuselage into the cabin. Cabin engulfed in toxic fumes and smoke from burning seating materials.	55 fatalities 15 serious non-fatalities
Windsor Castle, 1992	Flood lighting ignition of upper region of large curtain funnelling growing fire into the wooden ceiling and structure.	No casualties £40 million damage and loss of heritage

asphyxiation and death.[1] Only in clothing-related fires are injury and death caused primarily by burns, especially when garments are loose-fitting and worn directly over the body as are nightwear and summer dresses.

4.3   Burning behaviour of textiles

4.3.1   Fibres

The burning behaviour of fibres is influenced by and often determined by a number of thermal transition temperatures and thermodynamic parameters. Table 4.3[7] lists the commonly available fibres with their physical glass (T_g) and melting (T_m) transitions, if appropriate, which may be compared with their chemically related transitions of pyrolysis (T_p) and ignition and the onset of flaming combustion (T_c). In addition, typical values of flame temperature and heats of combustion are given. Generally, the lower the respective T_c (and usually T_p) temperature and the hotter the flame, the more flammable is the fibre. This generalisation is typified by the natural

Table 4.3 Thermal transitions of the more commonly used fibres[7]

Fibre	T_g, °C (softens)	T_m, °C (melts)	T_p, °C (pyrolysis)	T_c, °C (ignition)	LOI, %	ΔH_c, kJ g⁻¹
Wool			245	600	25	27
Cotton			350	350	18.4	19
Viscose			350	420	18.9	19
Nylon 6	50	215	431	450	20–21.5	39
Nylon 6.6	50	265	403	530	20–21.5	32
Polyester	80–90	255	420–447	480	20–21	24
Acrylic	100	>220	290 (with decomposition)	>250	18.2	32
Polypropylene	–20	165	470	550	18.6	44
Modacrylic	<80	>240	273	690	29–30	–
PVC	<80	>180	>180	450	37–39	21
Oxidised acrylic	–	–	≥640	–	–	45
Meta-aramid (eg Nomex)	275	375	410	>500	29–30	30
Para-aramid (eg Kevlar)	340	560	>590	>550	29	–

cellulosic fibres cotton, viscose and flax as well as some synthetic fibres like the acrylics.

In Table 4.3 respective Limiting Oxygen Index (LOI) values are listed, which are measures of the inherent burning character of a material and may be expressed as a percentage or decimal.[8] Fibres having LOI values of 21% or 0.21 or below ignite easily and burn rapidly in air (containing 20.8% oxygen). Those with LOI values above 21 ignite and burn more slowly and generally when LOI values rise above approximately 26–28, fibres and textiles may be considered to be flame retardant and will pass most small flame fabric ignition tests in the horizontal and vertical orientations. Nearly all flammability tests for textiles, whether based on simple fabric strip tests, composite tests (e.g. BS5852: 1979, ISO 8191/2, EN1021-1/2 and EN597-1/2) and more product/hazard related tests (e.g. BS6307 for carpets, BS6341 for tents and BS6357 for molten metal splash) are essentially ignition-resistance tests.

Within the wider community of materials fire science, it is widely recognised that under real fire conditions, it is the rate of heat release that determines burning hazard. While the heats of combustion, ΔH_c, in Table 4.3 indicate that little difference exists between all fibres and indeed some fibres like cotton appear to have a low heat of combustion compared to less flammable fibres like the aramid and oxidised acrylic fibres, it is the speed at which this heat is given out that determines rate of fire spread and severity of burns.

Table 4.4 Peak heat release rates for a number of fibres and blends

Fibre/blend	OSU; 35 kW m^{-2} [11]	Cone calorimeter, 25 kW m^{-2} [12]	Cone calorimeter, 25 kW m^{-2} [13]
Cotton	—	310	115 (as 1.5 mm multi-layer)
Cotton/wool			102 (as above plus wool 4.5 mm interlayer)
Cotton/polyester	167	—	
FR cotton:			
Proban	103		
Ammonium salt		125	
63% oxidised acrylic/17% aramid/ 20% pvc	34	—	
80% oxidised acrylic/ 20% aramid	38	—	
33% modacrylic/ 35% FR viscose/ 32% aramid	47	—	
61.5% Mohair/38.5% polyester (as warp)	58[a]		

[a] Results are from commercial data supplied by Dalton Lucerne Fabrics, UK.

Currently only textiles used in building materials and aircraft and transport interiors and seating are required to have minimal levels of rate of heat release which are measured using instruments such as the cone calorimeter[9,10] and the Ohio State University calorimeter[6] (used to assess aircraft interior textile performance at an incident heat flux of 35 kW m^{-2}). There is currently very little published heat release data for textiles and Table 4.4 presents a selection of that available.[11–13] These results show that the flammable characters of cotton and blends with polyester are determined by respective heat release rates at given incident heat fluxes and that the presence of flame retardant, natural protein fibres like wool or mohair or inherently flame retardant fibres significantly reduce both peak and average values. It is likely that rate of heat release will become a more important textile fire parameter during the next 10 years.

4.3.2 Fabrics and yarn structures

The burning behaviour of fabrics comprising a given fibre type or blend is influenced by a number of factors including the nature of the igniting source and time of its impingement, the fabric orientation and point of ignition (e.g. at the edge or face of the fabric or top or bottom), the ambient temperature and relative humidity, the velocity of the air and last but not least fabric structural variables. Fabric orientation, point of ignition source and time and the atmospheric variables are controlled in any standard test (see below). However, notwithstanding these, and as shown by Backer et al,[14] low fabric area density values and open structures aggravate burning rate and so increase the hazards of burn severity more than heavier and multi-layered constructions.

Hendrix et al[15] have related limiting oxygen index, LOI, linearly with respect to area density and logarithmically with air permeability for a series of cotton fabrics, although correlations were poor. Thus fabric flammability is determined not only by the fibre behaviour but the physical geometry of fibrous arrays in fabrics. Miller et al[16] considered that an alternative measure of flammability was to determine the oxygen index at which the burning rate was zero. The resulting intrinsic oxygen index $(OI)_0$ value for cotton is 0.13 and considerably less than the quoted LOI value of 0.18–0.19 (see Table 4.3). Subsequently Stuetz et al,[17] using a modified oxygen index technique, determined the so-called critical oxygen concentrations (COC) for sustained burning of a number of polymer samples ignited from the top (COC-T) or bottom (COC-B). The former were similar to respective LOI values and were considered to be influenced by extrinsic factors such as polymer geometry, char and melt-dripping. The latter, however, are independent of these and so represent an intrinsic polymer property. The COC-B value for cellulose of 0.135 compares with the above $(OI)_0$ value and it was considered that non-flammable (in air) polymers would have COC-B values above 0.21.

Building on these works, Horrocks et al[8,18–20] defined the extinction oxygen index, EOI, as the oxygen concentration at which the fabric just will not sustain any flame for a finite observable time when subjected to an LOI ignition source at the sample top for a defined ignition time. For simple flammable fabrics like cotton, nylon and polyester, respective EOI values decreased with decreasing igniter application time. Extrapolation enabled EOI values at zero time, $[EOI]_0$, to be defined. For a single layer of a typical cotton fabric, again a value of 0.14 was derived which was considered to be independent of igniter variables. Similar $[EOI]_0$ values for the thermoplastic fibres were determined in the absence of ignition problems caused by shrinkage and melt-dripping. $[EOI]_0$, like LOI values, increased with area density of single and layered fabrics as shown in Fig. 4.2. Repeating the experiments for various flame-retarded cottons produced similar results except that EOI values increased as ignition times reduced because longer times promoted larger areas of char which reduced burning time after the igniter flame was removed.

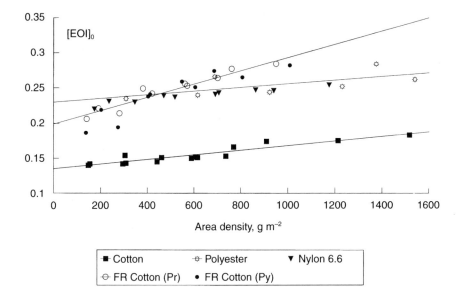

4.2 $[EOI]_0$ versus area density for mono- and multi-layered fabric samples[19]

Table 4.5 Analyses of linear dependences of $[EOI]_0$ with respect to area density, $M(g\,m^{-2})$ and respective LOI values[20]

Fabric	E_0	$E_1/10^{-5}$	LOI
Cotton	0.135	3.28	0.19
Proban cotton	0.199	9.60	0.31–0.33
Pyrovatex cotton	0.187	10.70	0.29–0.30
Polyester	0.226	3.18	0.20–0.215
Nylon 6.6	0.221	2.89	0.20–0.215

Extrapolation of these linear trends defined by the equation $[EOI]_0 = E_0 + E_1M$ enabled the intrinsic fibre extinction oxygen index values, E_0 to be determined as in Table 4.5 along with respective fabric area density sensitivity, E_1 values.

These results show that the E_0 value for cotton of 0.135 is the same as the COC-B value and so is an intrinsic fibre property. Both flame-retardant cottons with LOI values of about 0.30 have E_0 values still less than but close to 0.21 and so may be considered to be intrinsically flame retardant. Both polyester and nylon 6.6 have E_0 values close to respective LOI values, which is probably a consequence of the effect of melting and dripping. The fabric sensitivity values, E_1 are similar for the non-flame retarded fabrics indicating that the effect of area density is independent of fibre type. However, for the flame retardant cottons, the area density dependences are significantly higher, which may be a consequence of char formation. This means that in the presence of flame retardant, the behaviour of the final cotton fabric may be determined by a balance of flame retardant concentration present and area density; thus lightweight fabrics require higher levels of retardant than heavier fabrics. Similar relationships were observed between $[EOI]_0$ values and fabric thickness and the logarithm of air permeability.

The effect of yarn geometry and structure on burning behaviour has not been studied in depth, although the above referenced works on fabric structure infer that coarser yarns will have a greater resistance to ignition. This assumes that fibre type and area density remain constant (for coarser yarns, the cover factor will reduce and the air permeability will increase which will have the converse effect). Recent work by Garvey et al[21] has examined the burning behaviour of blended yarns comprising modacrylic/flame retardant viscose and wool/flame retardant viscose, where the flame retardant viscose is Visil (Sateri Fibres, Finland), produced by both ring-spinning and rotor-spinning methods, having the same nominal linear densities and knitted into panels. Figure 4.3 shows the LOI results for all blends and Fig. 4.4 shows char lengths for modacrylic/Visil blends only since the wool/Visil blended fabrics all failed the test used, BS 5438:1989, Test 2 (face ignition).

Given that each blend contains flame retardant fibre components, then Fig. 4.3 shows that the difference in yarn structure significantly influences the fabric burning behaviour. The more flammable rotor spun yarns are believed to be a consequence of the improved fibre component randomisation that occurs using this spinning method; in ring-spun yarns, component fibre aggregation is known to be a feature.

Figure 4.4 shows results following ignition application to the face and back of the single jersey fabrics and indicates that while they are little influenced by the mode of ignition, there is an effect of yarn structure although not as clear as that from the LOI results. At lowest Visil contents the rotor-spun blends show lower char lengths and hence superior flame retardance; the converse is seen at the highest Visil blends. Research is continuing in this area in order to clarify the position.

This same research has shown that combining two 100% yarns each of different fibre content and half the previous linear densities during the knitting process to give a plated yarn having a 50:50 composition can give improved flame retardancy relative to blended yarns of the same linear density. It is evident, therefore, that yarn structure can influence fabric burning behaviour although in a complex manner.

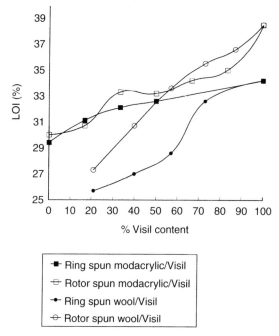

4.3 LOI values of knitted fabrics from blended yarns

4.3.3 Particularly hazardous textiles

The statistical data in Table 4.1 shows that upholstered furnishings are the most hazardous in terms of fatality frequency with smokers' materials being associated with a major ignition source in these fires[1] and, of course, legislation in the UK has addressed this issue.[2] During the 1980s bedding appeared to be the next most hazardous textile group, although the need to use ignition resistant tickings and mattress covers (and combustion-modified foam and fillings if appropriate) was included in the UK furnishings regulations[2] and their earlier version of 1979, which required cigarette ignition resistance only. While cigarettes are a prime cause of smouldering and flaming ignition, a detailed consideration of these is beyond the scope of this review. However, recent interest in both these areas has been revived.[22]

The effect of using retardant covers and tickings is most probably a factor which has seen bedding-related fatalities fall by over a half during the 15 year period since 1982. At the other end of the spectrum is the relatively low involvement of floor-coverings in fire-related deaths. This is not surprising since carpets are used in the relatively low flame-propagating horizontal geometry and they have heavy area densities.

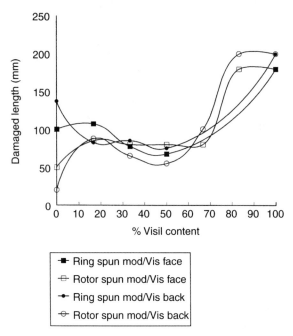

4.4 Maximum damaged or char length of knitted samples tested on both back and face (BS 5438:1989, Test 2, face ignition)

Within the UK the predominance of wool, polyamide and blends in the middle to upper price ranges ensure a low ignition and flammability hazard. This is not always the case with the popular lower-priced polypropylene carpets, which in certain constructions are known to burn quite easily in the horizontal mode. It is not surprising, therefore, that flame-retardant additives for polypropylene are targeted at this market and especially in the contract market where fire regulations demand minimum ignition and fire spread criteria.

In Table 4.1 it is evident that the one set of significant statistics which has shown least change is that concerned with clothing-related fatalities which typically constitute about 10% of total UK fire deaths; this compares with the similar figures of 8–10% determined by Weaver[23] for US clothing fire deaths. The earlier study of Tovey and Vickers,[24] which in 1976 analysed 3087 case histories of textile ignition-caused fire deaths, showed that in the USA loose-fitting clothes such as shirts, blouses, trousers and underwear ranked higher as potential hazards than bedding and upholstered furniture; pyjamas, nightgowns, dresses and housecoats presented very similar hazards to these two latter. While the hazard and severity of burn injury by clothing has been related to sex, age, activity and accident location,[25] major intrinsic textile properties such as time to ignition, high heat release rates and total heat release[26-28] are still of prime importance. An added factor which can increase heat transfer is whether a burning fabric adheres to the skin and burning continues once in contact; this may, of course, be a problem with thermoplastic fibre-containing fabrics especially in blends with non-thermoplastic components like cotton.[27]

Closer analysis of the UK fire statistics[1] as shown in Table 4.6 presents a fatal and non-fatal casualty comparison for the years 1994–98. The ratio of non-fatalities to fatalities for all textile-related dwelling fires is about 15:1 whereas for clothing on the person, excluding nightwear, it is on average 2:1. For nightwear, and assuming the error associated with the relative smallness of figures, this ratio is less than 2:1. These figures suggest that the chances of death in clothing- and nightwear-related fires are significantly higher per incident than the 6–7% chance of fatalities per textile fire.

A recent UK government-sponsored study[29] analysed clothing burn statistics over the period from 1982–92 collected from the UK Home Office,[1] the UK Consumer Unit's Home Accident Surveillance System (HASS) and UK Burns Units in order to identify trends within the 60–80 clothing-related burns fatalities each year (see Tables 4.1 and 4.6). The main conclusions agreed with those of other studies in that the very young and very old are at greatest risk, with more overall fatalities occurring to women (55%) and with loose fitting garments, especially dresses and nightwear posing the highest hazard. With regard to fibre type, natural fibres accounted for 42% of the accidents, synthetic fibres 42% and natural/synthetic blends 16%; the most commonly mentioned and/or identified materials were cotton (29%), polyamide (26%), cotton/polyester blends (13%) and wool (6%), although 'jeans material', presumably cotton in the main, was separately quantified at 4%.

Table 4.6 Casualties from textile-ignited UK dwelling fires

Year	Clothing on person			Nightwear			All textiles		
	Fatalities	Non-fatalities	Fatalities/incident %	Fatalities	Non-fatalities	Fatalities/incident %	Fatalities	Non-fatalities	Fatalities/incident %
1994	25	79	24.0	10	5	66.7	231	3656	5.9
1995	40	78	33.9	4	8	33.3	262	3523	6.9
1996	30	16	65.2	2	9	18.2	256	3738	6.4
1997	29	81	26.4	6	12	33.3	243	3704	6.2
1998	31	76	28.9	1	6	14.3	217	3647	5.6

4.4   Flammability testing of textiles

4.4.1 Attempts to standardise tests

It is probably true to state that nearly every country has its own set of textile fire testing standard methods which are claimed to relate to the special social and technical factors peculiar to each. In addition, test methods are defined by a number of national and international bodies such as air, land, and sea transport authorities, insurance organisations and governmental departments relating to industry, defence and health, in particular. A brief overview of the various and many test methods available up to 1989 is given by Horrocks et al[8] and a more recent list of tests specifically relating to interior textiles has been published by Trevira GmbH in 1997[30] with a focus on Europe and North America. Since 1990 within the EU in particular, some degree of rationalisation has been underway as 'normalisation' of EU member standards continues to occur; for detailed information, the reader is referred to respective national and CEN standard indexes. Because of the process of normalisation, standards are increasingly serving a number of standards authorities; thus for example in the UK most new British Standards are prefixed by BS EN or BS EN ISO. Table 4.7, however, attempts to give an oversight of the complexity of the range of tests available for textile products at the present time.

The complexity of the burning process for any material such as a textile which, because not only is it a 'thermally thin' material, but also has a high specific volume and oxygen accessibility relative to other polymeric materials, proves difficult to quantify and hence rank in terms of its ignition and post-ignition behaviour. Most common textile flammability tests are currently based on ease of ignition and/or burning rate behaviour which can be easily quantified for fabrics and composites in varying geometries. Few, however, yield quantitative and fire science-related data unlike the often maligned oxygen index methods.[8] LOI, while it proves to be a very effective indicator of ease of ignition, has not achieved the status of an official test within the textile arena. For instance, it is well known that in order to achieve a degree of fabric flame retardancy sufficient to pass a typical vertical strip test as defined in Section 4.4.2, an LOI value of at least 26–27% is required which must be measurable in a reproducible fashion. However, because the sample ignition occurs at the top to give a vertically downward burning geometry, this is considered to be unrepresentative of the ignition geometry in the real world. Furthermore, the exact LOI value is influenced by fabric structural variables for the same fibre type and is not single-valued for a given fibre type or blend. However, it finds significant use in developing new flame retardants and optimising levels of application to fibres and textiles.

Table 4.7 Selected test methods for textiles

Nature of test	Textile type	Standard	Ignition source
British Standard based vertical strip method BS 5438	Curtains and drapes	BS 5867:Part 2:1980 (1990)	Small flame
	Nightwear	BS 5722:1991	Small flame
	Protective clothing (now withdrawn)	BS 6249:Part 1:1982	Small flame
ISO vertical strip similar to Tests 1 and 2 in BS 5438	Vertical fabrics	BS EN ISO 6940/1:1995	Small flame
Small-scale composite test for furnishing fabric/fillings	Furnishing fabrics	BS 5852: Pts 1 and 2:1979 (retained pending changes in legislation[2])	Cigarette and simulated match flame (20s ignition)
	Furnishing fabrics	BS 5852:1990 (1998) replaces BS 5852: Pt 2	Small flames and wooden cribs applied to small and full scale tests
		ISO 8191:Pts 1 and 2 (same as BS 5852:1990)	
		BS EN 1021-1:1994	Cigarette
		BS EN 1021-2:1994	Simulated match flame (15s ignition)

Table 4.7 (cont.)

Nature of test	Textile type	Standard	Ignition source
Cleansing and wetting procedures for use in flammability tests	All fabrics	BS 5651:1989	Not applicable but used on fabrics prior to submitting for standard ignition tests
	Commercial laundering	BS EN ISO 10528:1995	
	Domestic laundering	BS EN ISO 12138:1997	
Use of radiant flux	Aircraft seat assemblies, so-called 'Boeing' test	ASTM E9060 1983, uses Ohio State University heat release calorimeter	Irradiate under 35 kW m⁻² with small flame igniter
	All fabrics/composites	NF P 92501-7, French 'M test'	Irradiate with small burner
Protective clothing	Resistance to radiant heat	BS EN 366:1993 (replaces BS 3791:1970)	Exposure to radiant source
		BS EN 367:1992	Determine heat transfer index
	Resistance to molten metal splash	BS EN 373:1993	Molten metal
	Gloves	BS EN 407:1994	Radiant, convective and molten metal
	Firefighters clothing	BS EN 469:1995	Small flame
	General flame spread	BS EN 532:1994 (replaces BS 5438)	Small flame
	General protection	BS EN 533:1997 (replaces BS 6249)	Small flame
	Contact heat transmission	BS EN 702:1994	Contact temps. 100–500 °C

4.4.2 Test categorisation

As Table 4.7 attempts to show, textile flammability tests may be categorised by various means depending whether they are ignition/burning parameter or textile structure/composite-related. At the simplest level, most test procedures are a defined standard procedure (e.g. BS 5438 for vertical fabric strips), the use of a standard test to define a specific performance level for a given product (e.g. BS 5722 uses BS 5438 to test and de- fine performance levels for nightwear fabrics[3]) or a combined test and performance-related set of defining criteria (e.g. BS 5852 Parts 1 and 2:1979 and EN 1021 Parts 1 and 2 for testing upholstered furnishing fabric/filling composites to simulated cigarette and match ignition sources). Ideally, all practical tests should be based on quite straightforward principles which transform into a practically simple and convenient-to-use test method. Observed parameters such as time-to-ignition, post-ignition after- flame times, burning rates and nature of the damage and debris produced should be reproducibly and repeatably measured with an acceptable and defined degree of accuracy. Figure 4.5 shows a schematic representa- tion of a typical vertical strip test such as BS 5438 and BS EN ISO 6941/2 tests in which a simple vertically orientated fabric may be subjected to a standard igniting flame source either at the edge or on the face of the fabric for a specified time such as 10 s. For flame retarded fabrics the properties measured after extinction of the ignition source are the damaged (or char) length (D), size of hole if present, times of after-flame and after- glow and nature of any debris (e.g. molten drips, etc). For slow-burning fabrics, such as are required in nightwear, a longer fabric strip is used across which cotton trip wires connected to timers are placed. In BS 5722, for example, these are at 300 mm and 600 mm above the point of ignition and

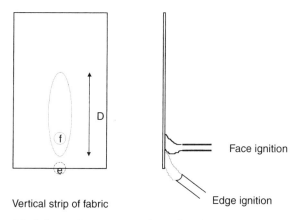

Vertical strip of fabric   Edge ignition

4.5 Schematic representation of a simple vertical strip test the time taken to cut through each thread enables an average burning rate to be determined. Tests of the type shown are simple and give reproducible and repeatable results. Furthermore, for similarly flame retarded fabrics, the length of the damaged or char length can show semi-quantitative relationships with the level of flame retardancy as determined by methods such as LOI.

With the recognition of the hazards posed by upholstered fabrics, the development of the small-scale composite test BS 5852 (see Table 4.7) represented a milestone in the development of realistic model tests which cheaply and accurately indicate the ignition behaviour of full-scale products of complex structure. Figure 4.6 shows a schematic diagram of this and the related EN and ISO tests (see Table 4.7). Again, the test has proved to be a simple to use, cost-effective and reproducible test which may be located in the manufacturing environment as well as formal test laboratory environments.

However, in all flammability test procedures conditions should attempt to replicate real use and so while atmospheric conditions are specified in terms of relative humidities and temperature ranges allowable, fabrics should be tested after having exposed to defined cleansing and after-care processes. Table 4.7 lists BS 6561 and its CEN derivatives as being typical here and these standards define treatments from simple

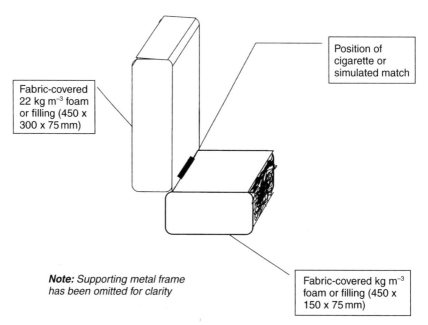

Fabric-covered 22 kg m^{-3} foam or filling (450 x 300 x 75 mm)

Position of cigarette or simulated match

Note: Supporting metal frame has been omitted for clarity

Fabric-covered kg m^{-3} foam or filling (450 x 150 x 75 mm)

4.6 Schematic diagram of BS 5852/EN 1021–1/2 composite test

water soaking, through dry cleaning and domestic laundering to the more harsh commercial laundering processes used in commercial laundries and hospitals, for instance.

As textile materials are used in more complex and demanding environments, so the associated test procedures become more complex. This is especially the case for protective clothing where the garment and its components have to function not only as a typical textile material but be resistant to a number of agencies including heat and flame. Table 4.7 shows a set of tests which has recently been developed across the EU to accommodate the different demands of varying types of protective clothing and the hazards whether open flame, hot surface, molten metal splash or indeed a combination of any of these. One test not yet standardised is that based on the simulation of a human torso and its reaction to a given fire environment when clothed; the original Du Pont 'Thermoman'[31] or instrumented manikin provided the means of recording the temperature profile and simulated burn damage sustained by the torso when clothed in defined garments (usually prototype protective garments) during exposure to an intense fire source. This latter is typically a series of gas burners yielding a heat flux of $80\,kW\,m^{-2}$. Sorensen[32] recently reviewed attempts to establish this and related manikin methods as a standard method. While the test is sometimes specified by fire service purchasing authorities, for example, its as yet poor reproducibility has not given it the robustness necessary for standardisation. One problem with such a test, apart from its sensitivity to garment fit, is that it ignores the consequences of the effect of heat and flame on the exposed head of the wearer. Furthermore, it does not enable the heat fatigue experienced by wearers to be measured during use of garments which can 'pass' the test.

This last test perhaps demonstrates the change in general philosophy associated with material fire testing in general in that post-ignition behaviour is of crucial importance in defining the hazard posed. The measurement of ease of ignition under a high heat flux and the associated heat release have been used to define both the ignition and fire propagation of textiles used in commercial aircraft seatings since the late 1980s (see Table 4.2 and the Manchester Airport disaster, 1985); Table 4.7 lists the relevant 'Boeing' specification using the OSU heat release calorimeter and the current specification demands that seating composites and all interior textiles shall have peak heat release rates $\leq 65\,kW\,m^{-2}$ and average rates $\leq 65\,m^{-2}\,min^{-1}$ when exposed to a heat flux of $35\,kW\,m^{-2}$. The more recently available cone calorimeter[9] has yet to make a significant impact in the assessment of textile fire behaviour apart from in the defence and extreme protective clothing-related sectors (see Table 4.4).

4.5   Burning and flame-retardant mechanisms

In Section 4.3 and Tables 4.3 and 4.4, the fundamental thermal parameters which determine the intrinsic burning behaviour of textile fibres were discussed. In order to understand how currently available flame retardants for textiles function and, more importantly, how future retardants may be developed, it is essential that the burning mechanisms of fibre-forming polymers are more fully explored.

4.5.1   Flame-retardant strategies

Figure 4.7[7] presents the combustion of any textile as a feedback mechanism in which fuel (from thermally degraded or pyrolysed fibres), heat (from ignition and combustion) and oxygen (from the air) feature as the main components. In order to interrupt the mechanism, five modes (a)–(e) are proposed and flame retardants may function in one or more of these. Each stage with a relevant flame retardant action is listed below:

a) Removal of heat. : High heat of fusion and/or degradation and/or dehydration (e.g. inorganic and organic phosphorus-containing agents, aluminium hydroxide or 'alumina hydrate' in back-coatings).

b) Enhancement of decomposition temperature. : Not usually exploited by flame retardants; more usual in inherently flame- and heat-resistant fibres (e.g. aramids).

c) Decreased formation of flammable volatiles, increase in char. : Most phosphorus- and nitrogen-containing flame retardants in cellulose and wool; heavy metal complexes in wool.

d) Reduced access to oxygen or flame dilution. : Hydrated and some char-promoting retardants release water; halogen-containing retardants release hydrogen halide.

e) Interference with flame chemistry and/or increase fuel ignition temperature (T_c). : Halogen-containing flame retardants, often in combination with antimony oxides.

From the above, it is seen that some generic flame retardants function in more than one mode and this is true of the most effective examples. Some flame retardant formulations, in addition, produce liquid phase intermediates which wet the fibre surfaces thereby acting as both thermal and oxygen barriers – the well-established borate–boric acid mixtures act in this manner

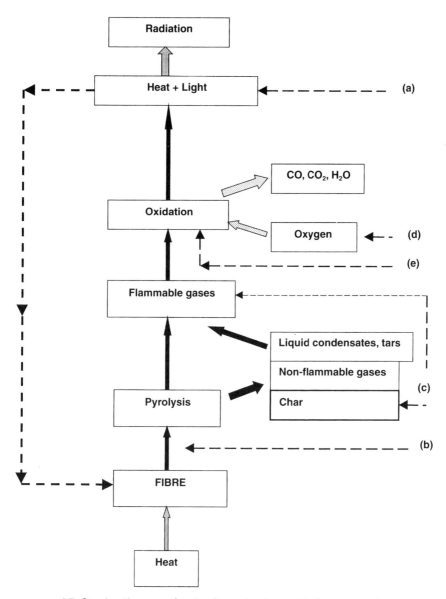

4.7 Combustion as a feedback mechanism with flame retardant actions

as well as promoting char. In order to simplify the classification of different modes of chemical flame retardant behaviour, the terms 'condensed' and 'gas or vapour' phase activities may be used to distinguish them. Both are composite terms and the former will include modes (a–c) above and the latter (d) and (e). Physical mechanisms often operate simultaneously, and these include exclusion of oxygen and/or heat by formation of a coating (mode d), increased heat capacity (mode a) and dilution or blanketing of the flame by non-flammable gases (mode d).

4.5.2 Thermoplasticity

Whether or not a fibre softens and/or melts (as defined by physical transitions in Table 4.3) determines whether it is thermoplastic or not. Thermoplasticity can influence considerably how a flame retardant behaves because of the associated physical change. Conventional thermoplastic fibres like polyamide, polyester and polypropylene will shrink away from an ignition flame and avoid ignition: this can give the appearance of flame retardancy when in fact, if the shrinkage was prevented, they would burn intensely. This so-called scaffolding effect is seen in polyester–cotton and similar blends where the molten polymer melts on to the non-thermoplastic cotton and ignites. Similar effects are seen in composite textiles comprising thermoplastic and non-thermoplastic components.

Added to the above is the problem of molten and often flaming drips which, while removing heat from a flame front and encouraging flame extinction (and hence achieve a 'pass' in vertical flame tests), can lead to burns or secondary ignition of underlying surfaces, such as carpets or human skin.

Most flame retardants applied to conventional synthetic fibres during manufacture or as finishes usually function by increasing melt dripping and/or promoting extinction of flaming droplets. None to date reduce their thermoplasticity and promote significant char formation as is the case in flame-retarded cellulosics, including viscose fibres.[33]

4.5.3 Flame retardant mechanisms and char formation

Flame retardants which function in the vapour phase by modes (d) and/or (e) share the advantage that they will reduce ignition propensity and aid in flame extinction of any textile fibre-forming polymer. This is because once the volatile products or fuels formed from thermal degradation enter into the oxidative reaction with oxygen in the flame, their chemistries are similar. Thus starvation of oxygen (mode (e)) or generation of interfering free radicals (mode (f)), for example, will assure the flame retardant's effectiveness.

Antimony–halogen flame retardants are the most successful within both the bulk polymer sectors and back-coated textile areas based on both cost and effectiveness. Unlike the fibre-reactive, durable phosphorus- and nitrogen-containing retardants used for cellulosic fibres (see below), they can only be applied topically in a resin binder, usually as a back-coating.[34]

For textiles, most antimony-halogen systems comprise antimony III oxide and bromine-containing organic molecules such as decabromodiphenyl oxide (DBDPO) or hexabromocyclododecane (HBCD). On heating, these release HBr and Br$^\bullet$ radicals which interfere with the flame chemistry by the following general scheme where R$^\bullet$, CH$_2$$^\bullet$, H$^\bullet$ and OH$^\bullet$ radicals are part of the flame oxidative chain reaction[34,35] which consumes fuel (R.CH$_3$) and oxygen:

$$R.Br \longrightarrow R^\bullet + Br^\bullet \qquad [4.1]$$

$$Br^\bullet + R.CH_3 \longrightarrow R.CH_2^\bullet + HBr \qquad [4.2]$$

$$HBr + OH^\bullet \longrightarrow H_2O + Br^\bullet \qquad [4.3]$$

$$H^\bullet + HBr \longrightarrow H_2 + Br^\bullet \qquad [4.4]$$

The recent concern regarding bromine-containing molecules is causing end-users of Sb–Br formulations to demand reduced flame retardant concentrations or alternatives (see Table 4.10 and Section 4.11 below).

Depending on the nature of the resin binder, often an acrylic copolymer or ethylene–vinyl acetate copolymer,[34] these coated systems may have some char-forming character. This enables them to be used successfully on synthetic fibre-containing furnishing fabrics, for example, which must have a means of counteracting the effects of fibre thermoplasticity if they are to pass such composite tests as BS5852, ISO 1891/2, EN 1021, and others.

However, without doubt, the most effective flame retardants are those which promote char formation by converting the organic fibre structure to a carbonaceous residue or char and hence reduce volatile (i.e. fuel) formation (mode (c)). Indirectly, these flame retardants, which require absorption of heat for them to operate, will offer the additional mode (a) and, by releasing non-flammable molecules like CO$_2$, NH$_3$ and H$_2$O during char formation, mode (d). In addition, the char behaves as a carbonised replica of the original fabric, which continues to function as a thermal barrier, unlike flame retardant thermoplastic fibres, for example.

Char-forming flame retardants, therefore, offer both flame and heat resistance to a textile fibre and so can compete with many of the so-called high performance flame and heat resistant fibres like the aramids and similar fibres (see Table 4.8).

For char-formation to be most effective, the polymer backbone must comprise side-groups, which on removal lead to unsaturated carbon bond formations and eventually a carbonaceous char following elimination of most of the non-carbon atoms present. Most phosphorus- and nitrogen-containing retardants, when present in cellulose, reduce volatile formation and catalyse char formation. While this is a considerably oversimplified view of the actual chemistry involved,[36] a brief overview of some of the essen-

Table 4.8 Durably-finished and inherently flame-retardant fibres in common use

Fibre	Flame retardant structural components	Mode of introduction
Natural: COTTON	Organophosphorus and nitrogen-containing monomeric or reactive species e.g. Proban CC (Rhodia formerly Albright & Wilson), Pyrovatex CP (Ciba), Aflammit P and KWB (Thor), Flacavon WP (Schill & Seilacher)	F
	Antimony-organo-halogen systems e.g. Flacavon F12/97 (Schill & Seilacher), Myflam (B F Goodrich, formerly Mydrin)	F
WOOL	Zirconium hexafluoride complexes, e.g. Zirpro (IWS); Pyrovatex CP (Ciba), Aflammit ZR (Thor)	F
Regenerated: VISCOSE	Organophosphorus and nitrogen/sulphur-containing species e.g. Sandoflam 5060 (Clariant, formerly Sandoz) in FR Viscose (Lenzing); polysilicic acid and complexes e.g. Visil AP (Sateri)	A A
Inherent Synthetic: POLYESTER	Organophosphorus species: Phosphinic acidic comonomer e.g. Trevira CS (Trevira GmbH, formerly Hoechst); phosphorus-containing additive, Fidion FR (Montefibre).	C/A
ACRYLIC (modacrylic)	Halogenated comonomer (35–50% w/w) plus antimony compounds e.g. Velicren (Montefibre); Kanecaron (Kaneka Corp.)	C
POLYPROPYLENE	Halo-organic compounds usually as brominated derivatives, e.g. Sandoflam 5072 (Clariant, formerly Sandoz)	A
POLYHALOALKENES	Polyvinyl chloride, e.g. Clevyl (Rhone-Poulenc). Polyvinylidene chloride, e.g. Saran (Saran Corp.)	H
High Heat and Flame Resistant (Aromatic): POLYARAMIDS	Poly(m-phenylene isophthalamide) e.g. Nomex (Du Pont), Conex (Teijin). poly (p-phenylene terephthalamide) e.g. Kevlar (Du Pont), Twaron (Acordis, formerly Enka)	Ar Ar
POLY (ARAMID-ARIMID)	e.g. Kermel (Rhone-Poulenc)	Ar
POLYBENZIMIDAZOLE	e.g. PBI (Hoechst-Celanese)	

Key
F : chemical finish
A : additive introduced during fibre production
C : copolymeric modifications
H : homopolymer
Ar : aromatic homo- or copolymer

tial features of the mechanism will provide a model for char formation in general. Most phosphorus-containing retardants act in this double capacity because, on heating, they first release polyphosphoric acid, which phosphorylates the C(6) hydroxyl group in the anhydroglucopyranose moiety, and simultaneously acts as an acidic catalyst for dehydration of these same repeat units. The first reaction prevents formation of laevoglucosan, the precursor of flammable volatile formation[7,36] and this ensures that the competing char-forming reaction is now the favoured pyrolysis route and the rate of this route is increased further by the acidic catalytic effect of the released polyacid. While considerable research has been undertaken into char formation of flame retarded cellulose, the actual mechanisms of both unretarded and retarded cellulose charring are not well understood.[36] Recent work in our own laboratories confirms the competition between volatile and char formation and considers a three-stage process which depends on both temperature and the exact nature of the flame retardant present.[37] Figure 4.8 shows the overall scheme, which builds on previously published mechanisms[36] and our own research based on evolved gas analytical, DTA, GC, pyrolysis-FTIR and temperature oxygen index studies of a range of flame retarded cotton fabrics. Stage I shows the well-established competing mechanisms of char formation and volatilisation within the temperature range 300–400 °C and Stage II, within the range 400–600 °C, shows a competition between char oxidation and conversion of aliphatic char to an aromatic form. Volatiles from Stage I are also oxidised within this range to yield similar products to those formed from char oxidation and aromatisation. During the higher temperature regime of 600–800 °C, some char decomposition to acetylene occurs, while above 800 °C, Stage III follows during which complete combustion of all remaining carbonaceous species to CO and CO_2 takes place. Vapour-phase active bromine-containing species do influence the pyrolysis to the extent that they favour volatile reactions by enhancing the decomposition of laevoglucosan to flammable furans, aldehydes and similar species. Phosphorus-containing flame retardants increase char formation as expected, but evidence suggests that those with a greater dehydrating power, such as ammonium polyphosphate, have a greater tendency to form aromatic chars than those based on organophosphorus (see Section 4.6 below). Furthermore, most of the original phosphorus remains in the char,[38] some of which is believed to combine with the carbon present via P–O–C bonds, for example.[36] This has the effect not only of increasing the oxidation resistance but also of mechanically toughening the structure. Surprisingly, the bromine-containing retardants studied also appeared to have slight char-promoting effects.

Clearly, char formation is not a simple process and the above discussion serves to illustrate that rarely do flame retardants function by a single mode. Furthermore, the general route to char requires the presence of functional

Stage I

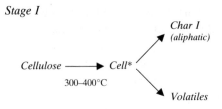

Stage II

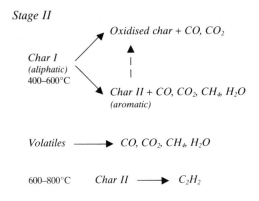

Stage III

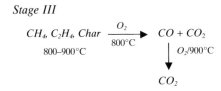

4.8 Pyrolysis of cellulose and char formation

groups which enable both dehydration and cross-linking reactions to occur as precursors to the formation of an aliphatic carbonaceous and finally an aromatic char structure. The presence of elements like nitrogen and sulphur are known to enhance synergistically the performance of phosphorus-containing retardants by further increasing char-forming tendencies. While the chemistry of these actions is not well understood, it is considered that not only are the char-forming chemistries influenced but that the char structures and thermal stabilities are modified by the presence of the elements by formation of P–N and C–N bonds.[36]

Such reactions will also occur in wool fibres as a consequence of their complex protein (keratin) structure and in the non-thermoplastic aromatic fibres (see Section 4.9) which have wholly aromatic chains and which behave as char-precursor structures. The major problem lies, however, with the commonly available synthetic polymers, polyester, polyamide, polypropylene and polyacrylic, which because of their tendencies to pyrolysis by chain scission or unzipping reactions and their general lack of reactive side groups, do not tend to be char-forming. The polyacrylic fibres are the only real exception here (see Section 4.8). This lack is aggravated by their thermoplasticity. An ideal char-promoting flame retardant would have to promote cross-linking reactions before thermoplastic effects physically destroyed the coherent character of the textile. Few, if any, commercially available flame retardants, whether as additives, as treatments or as copolymeric modifications react with the conventional synthetic fibre structures in char-enhancing modes (see Section 4.8), unless a degree of prior cross-linking has been introduced for example by radiation.[39]

4.6   Cellulosic textiles

Table 4.8 compares the main flame retardant cellulosic fibres with the major flame-retardant textile fibres of current importance, with special reference to flame retardant cotton and viscose.

Flame-retardant cellulosic textiles generally fall into three groups based on fibre genus:

1 Flame retardant cotton
2 Flame retardant viscose (or regenerated cellulose).
3 Blends of flame retardant cellulosic fibres with other fibres, usually synthetic or chemical.

4.6.1 Flame-retardant cottons

It is most important that effective flame retardants are also effective afterglow retardants.[36] All flame-retardant cottons are usually produced by after-treating fabrics chemically as a textile finishing process which, depending on chemical character and cost, yields flame-retardant properties having varying degrees of durability to various laundering processes. These may be simple soluble salts to give non-durable finishes (e.g. ammonium phosphates, polyphosphate and bromide; borate-boric acid mixtures); they may be chemically reactive, usually functional finishes to give durable flame retardancy (e.g. alkylphosphonamide derivatives (Pyrovatex, Ciba; Antiblaze TFR1, Rhodia, formerly Albright & Wilson (now discontinued); Aflammit KWB, Thor; Flacavon WP, Schill & Seilacher); tetrakis (hydroxy methyl) phosphonium salt condensates (Proban, Rhodia, formerly Albright & Wilson; Aflammit P, Thor) and back-coatings, which often usually comprise a resin-bonded antimony-bromine flame retardant system. Table 4.9 summarises the currently popularly-used treatments with selected commercial examples.[40]

Levels of flame retardant to be applied depend upon the degree of flame retardancy required and the area density and structure of fabric as discussed above in Section 4.4. Generally, however, phosphorus levels of 1.5

Table 4.9 Commonly-available flame-retardant finishes for cotton

Type	Durability	Structure/formula
Salts:		
(i) Ammonium polyphosphate	Non- or semi-durable (dependent on n)	$HO\left[\begin{array}{c} O \\ \parallel \\ P-O \\ \mid \\ NH_4 \end{array}\right]_n H$
(ii) Diammonium phosphate	Non-durable	$(NH_4)_2HPO_4$
Organophosphorus:		
(i) Cellulose reactive methylolated phosphonamides	Durable to more than 50 launderings	$(CH_3O)_2P.CH_2.CH_2.CO.N\begin{array}{c}{}^{\displaystyle H}\\[-2pt]{}_{\displaystyle CH_2OH}\end{array}$ (with $O \parallel$ above P)
		e.g. Pyrovatex CP (Ciba) Antiblaze TFR 1 (Albright & Wilson) Aflammit KWB (Thor)
(ii) Polymeric tetrakis (hydroxy methylol) phosphonium salt condensates	Durable to more than 50 launderings	THPC – urea – NH_3 condensate e.g. Proban CC (Albright & Wilson), Aflammit P (Thor)
(Back) Coatings:		
(i) Chlorinated paraffin waxes	Semi-durable	$C_nH_{(2n-m+2)}.Cl_m$ e.g. Flacavon FK (Schill & Seilacher)
(ii) Antimony/halogen (aliphatic or aromatic bromine-containing species)	Semi- to fully durable	Sb_2O_3 (or Sb_2O_5) + Decabromodiphenyl oxide or Hexabromocyclododecane + Acrylic resin e.g. Myflam (Mydrin) Flacavon F12 (Schill & Seilacher)

to 4% (w/w) on fabric are used which can give finish add-ons in the range of 5 to 20% (w/w) depending on the finishing agent phosphorus content. In order to maximise phosphorus–nitrogen synergy, P:N molar ratios of between 1:1 and 1:2 are recommended.[40,41] In the case of back-coatings, a total application is typically between 20–30% (w/w) of which 50% (w/w) is the flame retardant Sb–Br system (present in a 1:3 Sb:Br molar ratio). This is equivalent to an effective bromine concentration range of 5–8% (w/w) on fabric.

Most of these treatments have become well-established during the last 30 years and few changes have been made to the basic chemistries since that time. The earlier review by Horrocks[41] and the more recent update[40] provide comprehensive statements of the current state of chemistry and flame retardant finishing of cotton. However, during this same period, many other flame retardants mainly based on phosphorus chemistry, have ceased to have any commercial acceptability for reasons which include toxicological properties during application or during end-use, antagonistic interactions with other acceptable textile properties and cost.[41] The examples cited above may be considered to be those which continue to satisfy technical performance and enable flammability regulatory requirements to be met, while having acceptable costs and meeting health and safety and environmental demands: this last issue is becoming increasingly important. It must also be recognised that the most effective flame retardants contain either phosphorus or antimony–bromine-based systems and this generates a perception of unacceptable environmental hazard in spite of scientific information which indicates the contrary (see Section 4.11.1).[40]

4.6.2 Flame-retardant viscose

These FR fibres usually have flame retardant additives incorporated into the spinning dopes during their manufacture, which therefore yield durability and reduced levels of environmental hazard with respect to the removal of the need for a chemical flame retardant finishing process (see Table 4.8). Additives like Sandoflam 5060[33] are phosphorus-based and so are similar to the majority of FR cotton finishes in terms of their mechanisms of activity (condensed phase), performance and cost-effectiveness. Again, environmental desirability may be questioned and this issue has been minimised by Sateri (formerly Kemira) Fibres, Finland with their poly-silicic acid-containing Visil flame-retardant viscose fibre.[33,42] This fibre not only has removed the need for phosphorus, but also chars to form a carbonaceous and silica-containing mixed residue which offers continued fire barrier properties above the usual 500 °C where carbon chars will quickly oxidise in air.

4.6.3 Flame retarded cellulosic blends

In principle, flame retardant cellulosic fibres may be blended with any other fibre, whether synthetic or natural. In practice, limitations are dictated by a number of technical limitations including:

1   Compatibility of fibres during spinning or fabric formation; fibres must be available with similar dimensions and be processible simultaneously with other types on the same equipment.
2   Compatibility of fibre and textile properties during chemical finishing; for instance, flame retardant cotton treatments must not adversely influence the characteristics of the other fibres present in the blend during their chemical application.
3   Additivity and, preferably synergy, should exist in the flame retardant blend; it is well known that with some flame retardant blends, antagonism can occur and the properties of the blend may be significantly worse than either of the components alone.[41]

Consequently, the current rules for the simple flame retarding of blends are either to apply flame retardant only to the majority fibre present or apply halogen-based back-coatings, which are effective on all fibres because of their common flame inhibiting mechanism.

The prevalence of polyester–cotton blends coupled with the apparent flammability-enhancing interaction in which both components participate (the so-called scaffolding effect, reviewed elsewhere[8,41]) has promoted greater attention than any other blend. However, because of the observed interaction, only halogen-containing coatings and back-coatings find commercial application to blends which span the whole blend composition range; the (1975) Caliban F/R P-44 formulation comprising decabromodiphenyl oxide and antimony III oxide in a 2:1 mass ratio (equivalent to a molar ratio of Br:Sb = 3:1) in a latex binder[43] has been the model for current back-coating formulations for polyester–cotton blends as well as back-coatings in general[34] (see examples from B F Goodrich (formerly Mydrin) and Schill & Seilacher in Tables 4.8 and 4.9).

Most non-durable finishes for cellulosics (see Table 4.9) function on cellulosic-rich blends with polyester although the converse does not hold true unless some bromine is present. Antiblaze FSD (Rhodia (formerly Albright & Wilson)), Flovan BU (Ciba) and Flammentin BL (Thor) are examples of non-durable salt mixtures able to flame retard polyester (and other synthetic fibre)-rich cellulosic blends because they contain ammonium bromide.

In the case of durable, phosphorus-containing cellulose flame retardants, these are generally only effective on cellulose-rich blends with polyester. THP-based systems like Proban CC (Rhodia (formerly Albright & Wilson)) are effective on blends containing no less than 55% cotton if a combination of flame retardation and acceptable handle are required. This is because the THP condensate is substantive only on the cellulose content, which would require over 5% (w/w) phosphorus to be present on this component in order to confer acceptable flame retardancy to the whole blend. High phosphorus and hence finish levels lead to excessive surface deposits on fibres, reduced durability to laundering and create unacceptable harshness of handle. Furthermore, such an application only works well on medium- to heavyweight fabrics ($>200\,g/m^2$) and so is particularly effective for protective clothing applications. The use of a cotton-rich blend here is particularly advantageous because the lower polyester content confers a generally lower thermoplastic character to the fabric with less tendency to produce an adhesive molten surface layer when exposed to a flame.

In order to achieve the high finish levels necessary, often a double pass pad (or foam)-dry stage is required before the THPC-urea-impregnated fabric is ammonia-cured in the normal way. If a lower degree of durability is required then the cheaper semi-durable Antiblaze LR2 (Rhodia (formerly Albright & Wilson)) and similar finishes based on ammonium polyphosphate (at phosphorus levels of about 6%) and 5–6% Antiblaze CU and similar products, a cyclic phosphonate (see Section 4.5) applied to cotton and polyester components respectively in the blend will give a 40 °C, 30 minute water soak-resistant finish.

Application of methylolated phosphonamide finishes (e.g. Pyrovatex CP, Ciba) is effective on blends containing 70% or less cellulose content. This is because the phosphorus present is less effective on the polyester component than in THP-based finishes.[41] The reasons for this are not clear but are thought to be associated with some vapour-phase activity of phosphorus in the latter finish on the polyester component.

4.7   Flame-retarded wool and blends

The dyeing and finishing of wool continues to pose a challenge for textile and protein chemists because the complexity of its chemical and physical structure and the need to find effective processes are in competition in recent years with its almost constant world tonnage production and diminishing share of world fibre markets and textile economy. Within the area of flammability of all so-called conventional fibres, wool has the highest inherent non-flammability and for some end-uses, where high densities of structure and horizontal orientation (e.g. carpets) are required in the product, wool fabrics will often pass the required flame-retardancy tests untreated. Table 4.3 shows it to have a relatively high LOI value of about 25 and a low flame temperature of about 680 °C. Its high ignition temperature of 570–600 °C is a consequence of its higher moisture regain (8–16% depending upon relative humidity), high nitrogen (15–16%) and sulphur (3–4%) contents and low hydrogen (6–7%) content by weight. While organo–sulphur compounds are generally flame retardant to some degree, the disulphide–containing cystine links are easily oxidisable and so this can offset some of the anticipated natural flame retardancy. Pre-oxidation of wool and hence cystine to cysteic acid residues restores this expected retardant activity and oxidised wools can have greater inherent flame retardancy as a consequence.

From the above, it will be obvious that char-promoting flame retardants will be particularly beneficial, although bromine-containing, vapour phase-active surface treatments are effective. The review by Horrocks[41] discusses comprehensively developments in flame retardants for wool up to 1986 and very little has changed since that time. Given that a number of traditional non-durable finishes based on boric acid–borax (1 : 2 w/w) mixtures and sulphamic acid (as the ammonium salt) are still used, those currently listed by a selection of major flame retardant manufacturers are given in Table 4.10

It is significant that ammonium phosphates and derivatives will function as Lewis acids on any functional polymer which has pendant –OH groups and so will promote char formation in wool. Released phosphorus acids will probably promote the deamination of wool protein and so further encourage char promotion. These salts, when dried and cured at temperatures up to 130 °C, will give dry-clean durability up to as many as 10 cycles. Even the highly water-soluble ammonium bromide gives some degree of durability on wool as seen in Table 4.10 (see Antiblaze FSD and possibly Flovan BU).

In spite of considerable research into the use of functional phosphorus-based finishes, including the more recent study of the effectiveness of methylolated phosphonamides (e.g. Pyrovatex CP) by Hall and Shah,[44] and substantive halogenated species like chlorendic, tetrabromophthalic anhydride and dibromo-maleic anhydrides and brominated salicylic acid derivatives, the most commonly used durable flame retardants are based on Benisek's Zirpro (IWS) system[41] (see Table 4.8). Major advantages of this treatment are the absence of any discoloration or other effect on wool aesthetics, coupled with its application via a simple exhaust process.

The Zirpro process is based upon the exhaustion of negatively charged complexes of zirconium or titanium onto positively charged wool fibres under acidic conditions at a relatively low temperature of 60 °C. Zirpro treatments can be applied to wool at any processing stage from loose fibre to fabric using exhaustion techniques either during or after dyeing. The relatively low treatment temperature is an advantage because this limits the felting of wool.

Table 4.10 Non- and semi-durable flame-retardant finishes for wool and wool blends

Trade Name	Chemical Constitution	Durability
Rhodia, formerly Albright & Wilson		
Antiblaze FSD	Ammonium polyphosphate (APP) + ammonium bromide	Dry cleaning
Antiblaze RD1	Ammonium salt of phosphonic acid	—
Antiblaze LR3	APP powder (30% w/w P) for use in coatings	Dry cleaning
Antiblaze LR4	APP powder (27% w/w P) for use in coatings	Dry cleaning
Ciba		
Flovan BU	Inorganic (ammonium?) halide (bromide?)	Dry cleaning
Flovan CGN	Ammonium acid phosphonate	—
Thor		
Flammentin ASN	Ammonium phosphate (APP or DAP?)	Dry cleaning
Flammentin HM	Ammonium salts (phosphates?); useful up to 30% wool in blends	Dry cleaning
Flammentin KRE	Organic phosphorus-nitrogen compound	—
Flammentin MCFC	Cross-linking silicone + P and N– containing compounds	40 °C water soak, dry cleaning
Schill & Seilacher		
Flacavon RNEU	Organic P- and N-containing compound	Dry cleaning

The processor has the choice of potassium hexafluorozirconate ($K_2 Zr F_6$) or a mixture of this and potassium hexafluorotitanate ($K_2 Ti F_6$). The simple chemistry of application is:

$$\text{Wool–NH}_2 + H^+ \longrightarrow \text{Wool–NH}_3^+ \qquad [4.5]$$

$$[Zr F_6]^{2-} + 2[\text{Wool. NH}_3^+] \longrightarrow [\text{Wool.NH}_3^+]_2 [Zr F_6]^{2-} \quad [4.6]$$

It is important to maintain a low pH (≤ 3) in order to maximise penetration and wash-fastness to as many as 50 washes at 40 °C or 50 dry cleaning cycles in perchloroethylene. Acids like hydrochloric and formic acid are preferred buffers because, unlike sulphuric acid, for example, they do not have anions which compete with the metal fluoride ions for protonated amino groups in wool. However, the general simplicity of the whole process enables it to be used either concurrently with 1:1 premetallized and acid levelling dyes or after dyeing when applying acid milling reactive 1:2 premetallized and chrome dyes. Furthermore, the treatments are compatible with shrink-resist, insect-resistant and easy-care finishes.

The effectiveness of the Zirpro treatment is not fully understood from the mechanistic point of view and while Benisek[45] attributes it to enhanced intumescent char formation, Beck et al contest this view.[46] Clearly, however, its ability to create extremely effective flame and heat barrier properties at high heat fluxes is associated with the char structure generated.

Recently the process has come under the critical eye of environmentalists (see Section 4.11) as a consequence of the release of heavy metal ions into effluent discharges. In attempts to reduce effluent problems, replacement of exhaust by padding methods has not been successful because both potassium metal fluoride complexes are not very soluble (~10 g/l) at room temperature.

More recent research during the last five years or so has been limited to that of Lewin and Mark[48] who have demonstrated that sulphation with ammonium sulphamate followed by curing at 180–200 °C in the presence of urea can give a 50 hard water wash-durable finish for wool fabrics with little change in handle. Clearly there is an opportunity to develop a commercial process not dependent on heavy metal complexes. More recent research by Horrocks et al[49] has show that intumescents may also offer effective flame retardancy to wool and wool-containing blends and this work is reviewed in Section 4.10.2.

Wool blends pose different challenges, but given the complexity of wool and the position of the Zirpro process as the currently major durable FR treatment, its specificity ensures that little if any transferability occurs to other fibres present. Furthermore, antagonisms between Zirpro and other flame retardant fibres were reported by Benisek in 1981.[47] In the absence of any back-coating treatment, acceptable flame retardancy of Zirpro-treated blends is obtainable in 85:15 wool:polyester or polyamide combinations. For lower wool contents in blends and without the possibility of using alternative FR treatments, flame retardance can be maintained only if some of the Zirpro-treated wool is replaced by certain inherently flame retardant fibres, except for Trevira CS polyester.[49] Chlorine-containing fibres such as PVC and modacrylics are particularly effective in this respect.

4.8   Flame-retardant synthetic fibres

4.8.1   Inherently flame-retardant synthetic fibres

The conventional synthetic fibres may be rendered flame retardant during production by either incorporation of a flame retardant additive in the polymer melt or solution prior to extrusion or by copolymeric modification. Synthetic fibres produced in these ways are often said to be inherently flame retardant. However, problems of compatibility, especially at the high temperatures used to extrude melt-extruded fibres like polyamide, polyester and polypropylene, have ensured that only a few such fibres are commercially available. Table 4.8 lists examples of inherently FR synthetic fibres and the absence of polyamides reflects their high melt reactivities and hence poor flame retardant compatabilities. Flame retardant acrylics are usually so highly modified in terms of comonomer content that they are termed modacrylics. This latter group has been commercially available for 40 years or so but at present few manufacturers continue to produce them. This is largely because of the success of back-coatings applied to normal acrylic fabrics which create high levels of flame retardancy more cost-effectively. On the other hand, one group which continues to be successful is FR polyester (see Table 4.8) typified by the well-established Trevira CS,[33] which contains the phosphinic acid comonomer shown in Table 4.11. Other flame retardant systems, both based on phosphorus-containing additives, are also shown although only the Toyobo GH (and variants) are commercially available. The Rhodia (formerly Albright & Wilson) Antiblaze P45 additive is the former Mobil Chemical Antiblaze 19 compound which is available in dimeric form as a melt additive and in monomeric form as a polyester textile finish (Antiblaze CU, see Table 4.13). All three of these FR polyester vari-

Table 4.11 Flame-retardant modifications for polyester fibres

Generic type	Nature	Structure
Phosphinic acid derivative (Trevira CS)	Comonomer	
Bisphenol-S oligomer (Toyobo GH)	Additive	
Cyclic phosphonate (Amgard/ Antiblaze P45)	Dimeric additive	

Table 4.12 LOI values and char production of flame retarded acrylic polymers (pure polyacrylonitrile and as a copolymer with 10% methyl acrylate, MA)[49]

Polymer	Flame Retardant (present at 15 pph)	LOI, %	Char at LOI, %
100% PAN	None	19.0	17.3
100% PAN	Ammonium dihydrogen phosphate	27.0	34.4
10% MA	None	20.4	20.8
	Ammonium polyphosphate	29.0	39.1
	Ammonium dihydrogen phosphate	28.0	36.7
	Diammonium hydrogen phosphate	27.0	41.7
	Antiblaze/Antiblaze CU	27.0	34.3
	Red phosphorus	26.5	22.0
	Phosphonitrilic chloride trimer	26.0	27.3
	Sandoflam 5060	26.0	35.8
	Flacavon TOC (Sb–Br system)	26.0	29.6
	Proban CC polymer	24.0	28.9
	Ammonium sulphate	24.0	24.7
	Ammonium chloride	24.0	23.1
	Melamine	24.0	25.0
	Urea	23.0	17.0
	Sodium dihydrogen phosphate	23.0	21.9
	Thiourea	23.0	26.0
	Zinc phosphate	22.7	23.0
	Zinc borate	22.6	24.5
	Ammonium thiocyanate	21.0	24.0
	Antimony III oxide	20.0	23.0
	Hexabromododecane (10 pph) + Sb_2O_3 (5 pph)	27.5	27.8
	Hexabromododecane	26.5	26.4
	Decabromodiphenyl oxide (10 pph) + Sb_2O_3 (5 pph)	25.2	27.6
	Decabromodiphenyl oxide	25.0	26.2

ants do not promote char but function mainly by reducing the flaming propensity of molten drips normally associated with un-modified polyester. As yet, no char-promoting flame retardants exist for any of the conventional synthetic fibres and this must constitute the real challenge for the next generation of acceptable inherently FR synthetic fibres.

Recently, we have studied ways of increasing char-forming tendency of the polyacrylics.[50-52] Of all simple synthetic fibre-forming polymers, this group can cross-link via its pendant nitrile substituents to give a carbonised structure as evidenced by its carbon fibre precursor suitability. We have shown[50] that the high flammability of acrylic fibres is associated with the rapid heating rate associated with the burning process which favours the volatilisation (probably by unzipping) of the polyacrylonitrile chains. Slow heating rates favour the oligomerisation and cross-linking reactions more usually associated with carbon fibre production conditions. Any effective char-promoting flame retardant should therefore reduce the former volatilisation tendency at high heating rates and enhance oligomerisation in the first instance.

Incorporation of a range of selected flame retardants as mixtures in a range of fibre-forming acrylic polymers provided the respective LOI and residual char results in Table 4.12.[51] Plotting LOI versus residual char (% w/w) after burning at respective LOI conditions for these and all other results gave the relationship:

$$LOI = 14.6 + 0.36 \, [char \, w/w\%]_{LOI} \tag{4.7}$$

which showed that the most effective flame retardants were those which promoted highest residual char levels. Results in Table 4.12 show that the most effective flame retardants are not those containing halogen but phosphorus-containing species and in particular, the ammonium phosphates and polyphosphate (APP). Simple Lewis acids like ammonium chloride are ineffective and the presence of antimony III oxide does not seem to raise significantly the LOI when in the presence of bromine at a constant total concentration of flame retardant. Subsequent studies[50] proposed a mechanism for APP in which it functions:

1 As a physical barrier to oxygen following release of polyphosphoric acid.
2 As a nucleophilic agent which promotes oligomerisation of the adjacent, pendant nitrile groups to form a ladder polymer.
3 As a catalyst for the dehydrogenation of the ladder structure to a fully carbonaceous char to which polyphosphate groups are firmly bonded and with an empirical formula $C_{30}H_{13}N_7P_2$.

The proposed char structure is shown in Fig. 4.9. While there are problems which are associated with the commercial exploitation of APP as an effective flame retardant for acrylics, such as its relatively high solubility in water, the above research has demonstrated that high levels of char formation are achievable in some aliphatic synthetic fibre-forming polymers.

4.8.2 Flame-retardant finishes for synthetic fibres

Polyamide, polyester, polyacrylic and polypropylene are also candidates for semi-durably and durably flame retarding if suitable finishes are available. Table 4.13 lists examples of those currently available for polyester, polyamide and blends. In the case of acrylics, because of the difficulty in finding an effective flame retardant finish, modacrylic fibres are preferred, unless a back-coating is considered as an acceptable solution, as it would be for finishing fabrics to be tested to BS 5852:1979 or EN 1021–1/2. While back-

4.9 Proposed char structure for APP in acrylic polymer

coatings may be similarly effective on other synthetic fibre-containing fabrics and may offer sufficient char-forming character and char coherence to offset fibre thermoplastic and fusion consequences (see Flacavon H12 and H14 examples in Table 4.13), this is less easily achieved for polypropylene fabrics.

The low melting point, non-functionality and high hydrocarbon fuel content (see Table 4.3) of polypropylene are three factors that have created problems in finding an effective durable flame retardant finish and also pose difficulties in the design of effective back-coatings.

This leaves only polyamides and polyesters as major candidates for durable flame retardant treatments. While the scientific literature contains possible solutions,[41] few have entered the commercial arena as examples in Table 4.13 show.[53]

Table 4.13 Durable finishes for synthetic fibre-containing textiles

Trade Name	Chemical Constitution/Comments
Rhodia (formerly Albright & Wilson) Antiblaze CU/CT	(See Table 4.11) $n = 1$; cyclic oligomeric phosphonate; pad-dry(110–135 °C)-cure(185–200 °C) Primary use: polyester Secondary uses: polyamide, polypropylene
Thor Aflammit PE Aflammit NY	As above for Antiblaze CU/NT, polyester Organic nitrogen and sulphur compound (probably a thiourea derivative) and a reactive cross-linking compound; polyamide. Cure at 150–170 °C for 45–60 s.
Schill & Seilacher Flacavon AM	Nitrogen and sulphur-containing compound (thiourea derivative?); polyamide; 100–110 °C dry only; durable to dry cleaning
Flacavon AZ	Organic phosphorus compound (as for Antiblaze CU?); polyester
Flacavon H12/10	Organic phosphorus and nitrogen-containing compound (+ binder)
Flacavon H14/587	Antimony oxide + bromine compound (+ binder); all fibres especially polyester-cotton blends
Apex Chemicals (US) Apex Flameproof 334 Apex Flameproof 1510	Organohalogen compounds: polyester Organohalogen compounds: polyester, polyamide
Emco Services (US) Flame Out PE-60 Flame Out PE-19 Flame Out N-15	Organohalogen compounds: polyester Cyclic phosphorus compound: polyester Organic nitrogen compound: polyamide
Glo-Tex International (US) Guardex PFR-DPH	Organohalogen–phosphorus compound: polyester, polyamide, polypropylene
Guardex FR-MEHN	Organophosphorus compound: polyamide
Sybron Chemicals (US) Flame Gard PE conc Flame Gard 908	Organophosphorus compound: polyester Organic nitrogen compound: polyamide

The Antiblaze CU product (formerly Antiblaze 19[41] and now known as Antiblaze N or NT in the US) mentioned above is claimed to be effective on polyamides and polypropylene as well as polyester, for which it was initially developed. Antiblaze CU has a high phosphorus content (21.5% w/w) and is a clear viscous liquid which is applied to polyester at 3 to 6% (w/w) add-on buffered at pH 6.5 with disodium phosphate and a small amount of wetting agent. After padding at about 40–60% expression, fabric is dried at 110–135 °C followed by thermofixation at 185–205 °C for 1–2 minutes. Thermofixation usually only results in about 80% retention of original finish because of the finish volatility at high temperature. After rinsing and drying, the finish should resist 50 washes at 60 °C or 10 dry cleaning cycles with 90% retention.

The finish may be incorporated in a resin for coating polyester and its blends. Durability is not as great but loss does not occur during processing, as is the case of Antiblaze CU above. Inclusion of melamine increases the finish effectiveness on 100% polyester. Thor's Aflammit PE and Schill & Seilacher's Flacavon AZ are believed to be similar if not the same as Antiblaze CU/NT.

A number of US-manufactured flame retardants, which are able to confer BS 585:1979:Part 1 or EN 1021-1/2 passes on polyester fabrics, presumably after exposure to the required 40 °C water soak requirement, are included in Table 4.13 and tend to be halogen-based.[53]

For the flame-retardant treatment of nylon fabrics there are few that are satisfactory. Application of 10% (w/w) ammonium bromide or 18% (w/w) ammonium dihydrogen phosphate by a pad-dry route is effective but nondurable. The use of urea–formaldehyde resins or aminotrazine–aldehyde condensates can be used with ammonium bromide using a pad-dry-cure process to improve the durability of the finish. Durable but stiff flame retardant finishes based upon methylated urea–formaldehyde with thiourea–formaldehyde have been successfully applied to nylon nets for evening wear and underskirts. Some 15–20% (w/w) thiourea–formaldehyde precondensate is padded with ammonium chloride (1% on the weight of the resin) as a latent catalyst followed by low temperature drying and then curing at 170 °C for 1 minute.

Examples of these finishes are probably included in Table 4.13, although exact chemical constitutions of polyamide–specified retardants are not available.

4.9   High heat and flame resistant fibres and textiles

In the UK, Europe and the USA, the majority of flame and heat resistant fibres and textiles are still chemically after-treated; this probably constitutes a figure of about 80% by weight. Inherently flame and heat resistant fibres and textiles, including the inherently FR-viscose and synthetic fibres above comprise the remaining percentage. Table 4.8 includes the main members of the group of high heat and flame resistant fibres, which have fundamentally combustion-resistant all-aromatic polymeric structures. Table 4.3 shows that most of these decompose above 375 °C or so. Their all-aromatic structures are responsible for their low or non-thermoplasticity and high pyrolysis temperatures. In addition, their high char-forming potentials are responsible for their low flammabilities and, as established by van Krevelen,[54] their high LOI values. Table 4.14 lists the major aromatic fibre-forming polymeric structures currently available with respective char and LOI values.

This group is typified by the polyaramids, poly(aramid-arimids) and polybenzimidazole, which may be employed in end-uses where high levels of heat resistance are required in addition to flame retardancy. However, also included in Table 4.14 are the novoloid and carbonised acrylic fibres, which while having poorer fibre and textile physical properties, do have significant char-forming potentials and flame and heat resistance. These fibres tend to be used in nonwoven structures or in blends with other fibres in order to offset their less desirable textile characteristics. All these fibres compete with flame retarded cotton and wool fabrics, which also are heat resistant, although their chars are more brittle than those formed from this inherently heat-resistant group of fibres. The high cost and high temperature performance of these heat-resistant synthetic fibres restricts their use in applications where performance requirements justify the price to be paid. In practice, these fibres find use in high performance protective clothing and barrier fabrics, particularly in fire-fighting, transport and defence areas. However, the recently reported intumescent systems developed and described in Section 4.10.2 demonstrate heat barrier characteristics similar to and in some cases superior to those of the heat-resistant synthetic fibres.

Finally, and with respect to this group of fibres, it is interesting to note that the recent development of high modulus and high strength fibres based on linear and symmetrical aromatic structures also gives rise to heat and flame resistance. This was exemplified by the development of the poly (para-aramids) in the 1970s and typified by Kevlar (Du Pont) and Twaron (AKZO). Of more recent interest is the development of fibres such as the polybenzoxazoles (PBO) of which Zylon (Toyobo) is a recent example and comprises poly (p-phenylene-2, 6-benzobisoxazole). While having been developed as a high modulus, high tenacity fibre, an LOI value of 68% is claimed.

Table 4.14 Heat and flame resistant fibres based on aromatic polymeric structures

Generic type	Structure	LOI %	Char %, 850°C
Polyaramid e.g., poly(m-phenylene isophthala-mide) (Nomex)	$-NH\!\!-\!\!C_6H_4\!\!-\!\!NH\!\!-\!\!CO\!\!-\!\!C_6H_4\!\!-\!\!CO-$	30	35[a]
Polyamide-imide (Kermel)	(structure)	30	55[b] 30[c]
Polybenzi-midazole (PBI)	(structure)	41	66[b] 59[c]
Novoloid (Kynol)	(structure)	36	45[a]
Carbonised acrylic (Panox)	(structure)	55	93[c]

Notes: After Krevelen[54]: [a]experimental value; [b]calculated; [c]graphically estimated.

4.10   Methods of flame retardancy

4.10.1 Direct application methods

Successful flame retardant finishes are those which combine acceptable levels of flame retardancy at an affordable cost and are applicable to textile fabrics using conventional textile finishing and coating equipment. Such established techniques are well-documented in the general textile literature, so excessive detail here is superfluous.

Figure 4.10 attempts to present an overall summary of four basic processes shown schematically and as they would be used on open-width textile fabrics. Each relates to one or more of the flame-retardant finishes for cellulosic textiles identified above in Table 4.9, wool finishes in Table 4.10 and synthetic textile finishes in Table 4.13. It is of interest to note that alternative application methods to padding may be used in processes (i)–(iii) such as foam application; padding perhaps represents the most commonly used technique. Each process, (i)–(iv), relates to finish type as follows:

Process (i):   This simple pad/dry technique is applicable with most non-durable and water-soluble finishes such as the ammonium phosphates.

Process (ii):   This sequence is typical of those used to apply crease-resistant and other heat curable textile finishes. In the case

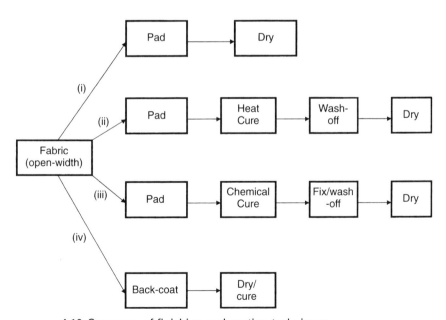

4.10 Summary of finishing and coating techniques

of flame retardant finishes it finds best use for application of the phosphonamide systems such as Pyrovatex (Ciba), Afflamit (Thor) and the now obsolete Antiblaze TFR1 (Rhodia (formerly Albright & Wilson)) which are applied with resin components like the methylolated melamines (see Section 4.6.1). Because the process requires the presence of acidic catalysts (e.g. phosphoric acid), the wash-off stage will include an initial alkaline neutralisation stage.

This same sequence without the washing-off stage may be used to apply semi-durable finishes where a curing stage allows a degree of interaction to occur between the finish and the cellulose fibre; a typical example is given by the ammonium phosphates which during curing at about 160 °C give rise to phosphorylation of the cellulose. Thus the finish develops a degree of resistance to water soak and gentle laundering treatments.

Process (iii):   This is best exemplified by the Proban process described in Section 4.6 above and which requires an ammonia gas curing process in order to polymerise the applied finish into the internal fibre voids. In this way the Proban CC condensate of tetrakis (hydroxy methyl) phosphonium chloride and urea after padding and drying onto the fabric is passed through a patented ammonia reactor which cross-links the condensate to give an insoluble polymeric finish. In order to increase the stability and hence durability of the finish, a subsequent oxidative 'fixation' stage is required before finally washing off and drying.

Process (iv):   Back-coating describes a family of application methods where the flame-retardant formulation is applied in a bonding resin to the reverse surface of an otherwise flammable fabric. In this way the aesthetic quality of the face of the fabric is maintained while the flame retardant property is present on the back or reverse face. Flame retardants must have an element of transferability from the back into the whole fabric and so they almost always are based on the so-called vapour-phase active antimony–bromine (or other halogen) formulations as typified by Myflam (B F Goodrich, formerly Mydrin) and Flacavon (Schill & Seilacher) products which comprise brominated species such as decabromodiphenyl oxide or hexabromocyclododecane and antimony III oxide (see Tables 4.8 and 4.9). Application methods include doctor blade or knife coating methods and the formulation is as a paste or foam. These processes and finishes are used on fabrics where aesthetics of the front face are of paramount importance such as furnishing fabrics and drapes.

The relatively high application levels required (see Sections 4.6–4.9) can adversely influence fabric handle, drape and appearance; these effects are minimised by ensuring that finishing application is carried out so that in processes (i) to (iii) minimal finish remains on fibre and fabric surfaces and in process (iv) the coating is applied solely to the surface fibres of the fabric reverse face. In addition, softening agents may be included within the formulations during application; careful selection of these is essential if compatibilty with the formulation is to be assured and they are to have minimal effect on the resulting flame retardant property.

4.10.2 Intumescent application to textiles

Clearly any enhancement of the char barrier in terms of thickness, strength and resistance to oxidation will enhance the flame and heat barrier performance of textiles. Generation or addition of intumescent chars as part of the overall flame retardant property will also reduce the smoke and other toxic fire gas emissions. The application of intumescent materials to textile materials has been reviewed[33] and is exemplified in the patent literature by the following fibre-intumescent structures which offer opportunities in textile finishing:

1   More conventional, flexible textile fabrics to which an intumescent composition is applied as a coating have been reported.[55] In one example of this patent, the glass-fibre-cored yarns used in the woven or knitted structure complement the flame and heat resistance of the intumescent coating. Presence of sheath fibres of a more conventional generic type ensures that the textile aesthetic properties may be optimised. More recently the Flammentin IST flame retardant from Thor Chemicals, UK is based on the use of intumescents as replacements for antimony–bromine systems in coating and back-coating formulations. This has been demonstrated to be effective on polyester-based fabrics.[56]

2   The recent development of a back-coating for technical nonwovens by Schill & Seilacher has been reported and is based on exfoliated graphite.[57] This seems to be particularly effective on polyamides and polyester.

3   Recently, Horrocks et al[58] have patented a novel range of intumescent-treated textiles that derive their unusually high heat barrier properties from the formation of a complex char that has a higher than expected resistance to oxidation. These require the intumescent to be in intimate contact with the surfaces of flame retarded, char-forming fibres and for respective char-forming mechanisms to be physically and chemically similar. Exposure to heat promotes simultaneous char formation of both intumescent and fibre to give a so-called 'char-bonded' structure. This integrated fibrous-intumescent char structure has a physical integrity superior to that of either charred fabric or intumescent alone and, because of reduced oxygen accessibility, demonstrates an unusually high resistance to oxidation when exposed to temperatures above 500 °C and even as high as 1200 °C. Furthermore, these composite structures show significantly reduced rates of heat release when subjected to heat fluxes of $35 \, kW \, m^{-2}$, thus demonstrating additional significant fire barrier characteristics.[59]

More recent work has been reviewed elsewhere[60] and has shown that the intumescents, which are based on ammonium and melamine phosphate-containing intumescents applied in a resin binder, can raise the fire barrier properties of flame retarded viscose and cotton fabrics to levels associated with high performance fibres such as aramids. Table 4.15 lists the cellulosic fibre-intumescent combinations which have been studied to date using thermal analysis, scanning electron microscopy and mass calorimetry. We have shown that these systems are compatible in their char-forming physical and chemical mechanisms to yield greater-than-expected char residues which have unusually high resistances to oxidation above 500 °C. The complex char structures have been examined by scanning electron microscopy[60] and more recently by EDAX[61] which demonstrate the uniqueness of the physical and chemical characteristics. Furthermore, the heat barrier performance of these fabrics is similar to that for similarly structured, commercially available nonwoven fabrics comprising aramid and carbonised acrylic fibres. Figure 4.11 shows the coherent fabric char residues after exposure for 10 minutes in air in a Fire Testing Technology mass loss calorimeter at $50 \, kW \, m^{-2}$ heat flux.

Current research is now focused upon enabling these intumescent formulations to be applied to other fibres such as wool and to more conventional fabric structures using normal application technologies (see Fig. 4.10). Initial studies with wool and wool-containing blended fabrics[48] show that the presence of intumescents applied as fibre surface coatings can enhance the flame retardancy of the underlying fibres and that evidence of char-forming interactions similar to those

Table 4.15 Compatible flame retardant cellulosic fibres and intumescents[60]

Fibre	Flame retardant	Intumescent (applied to all fibres)
Visil, hybrid viscose (Sateri (formerly Kemira) Fibres, Finland)	Polysilicic acid, 30% by weight as silica.	(i) Ammonium polyphosphate, pentaerythritol, melamine (3:1:1 mass ratio) (MPC 1000, Rhodia (formerly Albright & Wilson)).
Viscose FR (Lenzing, Austria)	Sandoflam 5060 (Sandoz) (2, 2-oxybis) 5,5-dimethyl – 1, 2, 3 – dioxaphosphorinane – 2, 2 – disulphide), 10–15% by weight.	(ii) Melamine phosphate, dipentaerythritol as Antiblaze NW (formerly MPC 2000), Rhodia (formerly Albright & Wilson).
Cotton	Ammonium polyphosphate (APP) and urea (Antiblaze LR2, Rhodia (formerly Albright & Wilson)), applied 1.7% P (w/w), heat cured.	
Cotton	Tetrakis (hydroxy methyl) phosphonium chloride (THPC)-urea condensate applied via an ammonia cure (Proban CC, Rhodia (formerly Albright & Wilson)), 2.5–4% P (w/w).	
Cotton	N-Methyeol dimethyl phosphono-propionamide (Antiblaze TFRl, Rhodia (formerly Albright & Wilson)) applied with trimethylolated melamine resin, 2.3–2.7% P (w/w).	

observed for cellulosic fibre substrates exists. Furthermore, these same studies are shedding light on the little known char-forming processes occurring in wool fibres.

Clearly there is an increasing interest in the development and use of intumescent flame retardants across the whole spectrum of flame retardant polymeric materials. This is driven by the need to reduce the concentrations and usage of the common Sb–Br formulations coupled with the superior fire barrier and reduced toxic combustion gas properties which they generally confer. In the next few years, there will be increased use of these materials in the textile and related sectors.

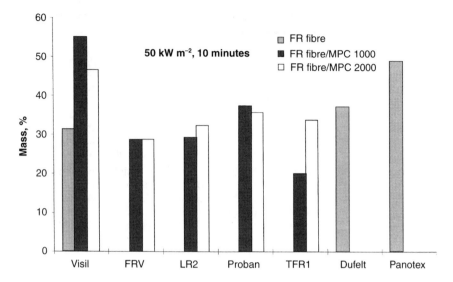

4.11 Residual coherent fabric chars following exposure to 50 kW m-2 heat flux for 10 minutes in air (Key: FRV = Lenzing Viscose FR, LR2 = APP-treated cotton, Proban = Proban-treated cotton, TFR1 = phosphonamide-treated cotton, Dufelt = aramid fabric, Panotex = carbonised acrylic fabric. MPC 1000 = APP/Melamine/Pentaerythritol intumescent, MPC 2000 = Antiblaze NW intumescent)

4.11   Environmental issues

Apart from the inevitable pressure to reduce expenditure throughout the textile industry and use the most cost-effective finishes and application processes available, a major issue is the influence of environmental factors and the related current concerns levelled at the use of flame retardants in general.

While Holme[62] has addressed and reviewed environmental concerns of textile processing in general, Horrocks et al[63] have attempted to quantify the environmental impact of the currently available flame retardant fibres including finished and inherently flame retardant types. A simple model has been devised which identifies each stage in the processing and end-use history of each fibre from 'cradle to grave'. Each stage is ranked from 0 to 5 for zero to maximum environmental impact respectively and then summed and expressed as a percentage environmental impact index. Figure 4.12 shows the results for the most common and high performance, flame retardant fibres and indicates that in spite of very different process histories they span a very small range of 39–51%. Assuming that the model is valid, then the similarity of values might suggest that since all fibres and

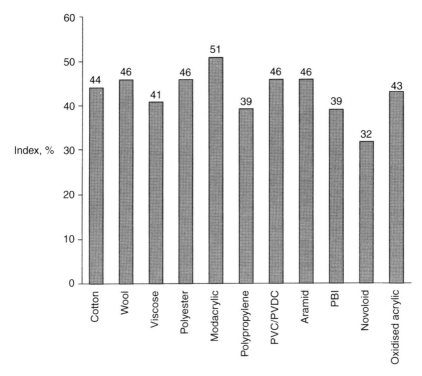

4.12 Environmental indices for flame retardant fibres

textiles are produced by the most economically efficient processes, including raw material production, then environmental impact minimisation is already a factor that determines economic success. Obviously as process efficiency, energy and waste minimisation and recycling of by-products and even end-products improves, such environmental impact will reduce further.

4.11.1 The bromine – antimony question

However, notwithstanding the above, considerable concern has been shown regarding the possible formation of polybrominated dioxins associated with incineration of organobromine compounds, especially those based on poly-brominated diphenyl oxides (PBDPO).[40] Without wishing to enter into very great detail, following the initial concern in Germany in 1986, the EU published in 1991 a draft amendment to EC Directive 76/769/EEC, which would essentially ban use of all polybrominated diphenyl oxides or ethers within five years. In 1994 this Directive was withdrawn as subsequent studies cast doubt on the earlier concerns. Simultaneously, other organisations (e.g. US Environmental Protection Agency, OECD) initiated risk analyses of these compounds. At the same time the World Health Organisation initiated an evaluation of the risk to health of PBDPO which in 1994 indicated that it did not pose a significant hazard. While the full details of the OECD programme are complex,[64] one outcome has been an industrial commitment to address environmental exposure and purity of PBDPOs and minimisation of the presence of non-commercial congeners.

As a consequence of these concerns, European companies like Schill and Seilacher now offer DBDPO alternatives such as hexabromocyclododecane and Wragg[57] has reviewed the current position. Even more recently, the role of antimony III oxide in 'cot deaths' or sudden infant death syndrome (SIDS) was raised on UK Television in 1994 and although refuted[57] and subsequently shown to be without foundation,[65] the image of Sb–Br finishes in general is poor within the media and environmental circles.

Efforts to reduce Sb–Br concentrations and eventually replace them in back-coating formulations have continued since the early 1990s. A number of ammonium polyphosphate formulations (e.g. Antiblaze LR3/LR4 (Rhodia (formerly Albright & Wilson)), Pyrovatim SB (Ciba) and Flammentin UCR (Thor)) are claimed to be able to replace partially or even wholly Sb–Br systems in back-coating formulations and are recommended for cellulosic furnishing fabrics. Their effectiveness is enhanced by use of chlorine-containing binding resins. One drawback lies in the solubility of APP, however, and so they need to be used with care if they are to pass the 40 °C water soak test (BS 5651) as required by UK furnishing fabric regulations.[2] In addition, because these phosphorus-based formulations are not vapour-phase active, the need to increase and control coating penetration becomes of importance. An alternative to APP is the use of other phosphorus compounds such as Antiblaze CU/NT (Rhodia (formerly Albright & Wilson)) and the Flammentin NAH range of products from Thor.

The recent innovative approach, reported by Wragg,[57] which uses a modified carbon pigment (possibly an intumescent exfoliated graphite (see Section 4.3)) has been mentioned in Section 4.10.2: Flacavon DPL from Schill & Seilacher is an example although other grades are available.

4.11.2 Effluent and water minimisation

As a consequence of current environmental legislation at UK (Environmental Protection Act, 1990) and EU levels strict controls over effluent discharge are demanded. Particular problems associated with flame retardant applications are:

- Discharge to effluent of unused flame retardant liquors.
- Emissions of formaldehyde to the atmosphere, especially during curing (currently required to be ≤20 ppm).
- Emissions of volatile organic compounds (VOCs) (currently ≤50 ppm).
- Use and emission of ammonia in THP-based treatments.
- Discharge of unfixed flame retardants from washing-off effluent.

In order for most commercial textile finishers to achieve acceptable formaldehyde and VOC emissions when applying formaldehyde-based finishes such as Pyrovatex CP and its analogues, gaseous exhausts from the drying and curing stages pass through scrubbers before release into the environment. Liquid effluents require neutralisation and dilution before release. Not surprisingly, use of techniques such as controlled impregnation technologies, low formaldehyde finishes and recycling of wash waters not only reduce effluents but save money and so are economically attractive to finishers. Furthermore, recent research at Bolton Institute under a UK Environmental Technology Best Practice Programme[66] has shown that application of chemometrics to flame retardant finishing chemistry reduces formaldehyde emissions by up to 75% and reduces levels of phosphorus in effluent by improved finish fixation.

Clearly a better understanding and optimisation of the process chemistry, if used with minimum add-on (e.g. foam application) and wastewater recycling systems can minimise waste even further, thus achieving environmental and economic savings.

4.12   Conclusions

It is evident, therefore, that it is not sufficient any longer to have an effective flame retardant textile *per se*, but that the flame retardant, its application technology and the product must be environmentally as well as technically acceptable. In addition, of course, the toxicological properties should be fully understood if the safety factor that they confer upon a textile is to be truly quantifiable. In a recent study, Stevens and Mann[67] have considered the risk benefits of using flame retardants in consumer goods in general and have arrived at the conclusion that the benefits in lives saved outweigh the risks. However, there is currently considerable debate in both the EU[68] and the US[69] regarding the overall acceptability of flame retardants with respect to the possible introduction of EU nightwear and US furnishing fabric flame retardant legislation.

In addition to the environmental issues is the need to enhance FR textile performance and this will arise by maximising char formation in both natural and synthetic fibre-containing textiles. This will not only increase the protective efficiency and hence safety of flame retardant textiles, but will also reduce problems associated with thermoplasticity and toxic gas emissions formed in fires. The role of intumescents is only now being explored within the textile area and it is probable that they will play an increasing part in improving textile fire performance while also addressing environmental and toxicological concerns.

References

1 *Fire Statistics, United Kingdom, 1997*, London, The Home Office, The Government Statistical Office, UK, ISSN 0143 6384, 1998.
2 *Consumer Protection Act (1987), the Furniture and Furnishings (Fire) (Safety) Regulations, 1988*, SI 1324 (1988), London, HMSO, 1988.
3 *The Nightwear (Safety) Regulations 1985*, SI 1985/2043, London, HMSO, 1985.
4 *Health and Safety at Work Act (1974)*, London, HMSO, 1974.
5 Hammad F H and Rashad S M, 'Review and analysis of fire events in Egypt 1980–1991', in *Materials and Fire*, proceedings of an international conference on materials science, Alexandria, Egypt, The Arab Society of Materials Science, 1998, 319.
6 *US Federal Aviation Regulation FAR 23:853*, Appendix F, Part IV.
7 Horrocks A R, 'An introduction to the burning behaviour of cellulosic fibres', *J. Soc. Dyers. Col.*, 1983, **99**, 191.
8 Horrocks A R, Price D and Tunc D, 'The burning behaviour of textiles and its assessment by oxygen index methods', *Text. Prog.*, 1989, **18** (1–3), 1–205.
9 Babrauskas V and Grayson S J, *Heat Release in Fires*, London and New York, Elsevier Applied Science, 1992.
10 ASTM E9060 1983, *Standard Test Method for Heat and Visible Smoke Release Rates for Materials and Products*.
11 Saville N, private communication, Universal Carbon Fibres, Cleckheaton, UK, 1993.
12 Hshieh F-Y and Beeson H D, 'Flammability testing of pure and flame-retardant-treated cotton fabrics', *Fire Mat.*, 1995, **19**, 233–9.
13 Patel M, *Rate of Heat Release from Multi-layer Clothing using the Cone Calorimeter*, TCR7/91, Garston, UK, Building Research Establishment, 1991.
14 Backer S, Tesoro G C, Toong T Y and Moussa N A, *Textile Fabric Flammability* Cambridge, Mass., MIT Press, 1976.
15 Hendrix J E, Drake G L and Reeves W A, 'Effects of fabric weight and construction on OI values for cotton cellulose', *J. Fire Flamm.*, 1972, **3**, 38.
16 Miller B, Goswami B C and Turner R, 'The concept and measurement of extinguishability as a flammability criterion', *Text. Res. J.*, 1973, **43**, 61.
17 Stuetz D E, DiEdwardo A H, Zitomer F and Barnes B P, 'Polymer combustion', *J. Polym. Sci., Polym., Polym. Chem. Edn.*, 1980, **18**, 967, 987.
18 Horrocks A R and Ugras M, 'The persistence of burning of textiles in different oxygen environments and the determination of the extinction oxygen index', *Fire Mat.*, 1983, **7**, 119.
19 Horrocks A R, Price D and Tunc M, 'Extinction oxygen index as a means of quantifying minimum burning behaviour', in *Fundamental Aspects of Polymer Flammability*, IOP Short Meeting Series No. 4, Bristol, UK, Institute of Physics, 1987, 165.

20 Horrocks A R, Price D and Tunc M, 'Studies on the temperature dependence of extinction oxygen index values for cellulose', *J. Appl. Polym. Sci.*, 1987, **34**, 1901–16; 1989, **37**, 1051–61.

21 Garvey S J, Anand S C, Rowe T and Horrocks A R, 'The effect of fabric structure on the flammability of hybrid viscose blends', in *Fire Retardancy of Polymers – The Use of Intumescence* (eds. M J Le Bras, G Camino, S Bourbigot and R Delobel), London, Royal Society of Chemistry, 1998, 376.

22 Dyakonov A J and Grider D A, 'Smolder of cellulosic fabrics. I Development of a framework', *J. Fire Sci.*, 1998, **16**, 297–322.

23 Weaver W J, 'Mortality and flammable fabrics', *Text. Chem. Col.*, 1978, **10** (1), 42.

24 Tovey H and Vickers A, 'Hazard analysis of fires involving blankets', *Text. Chem. Col.*, 1976, **8** (20), 19.

25 Laughlin J M, Parkhurst A M, Reagan B M and Janecek C M, 'Midwest regional burn injury study: characteristics of clothing-related burn accidents', *Text. Res. J.*, 1985, **55**, 285.

26 Miller B and Meiser C H, 'Heat emission from burning fabrics; potential harm ranking', *Text. Res. J.*, 1978, **48**, 238–43.

27 Krasny J F and Fisher A L, ' Laboratory modelling of garment fires', *Text. Res. J.*, 1973, **43**, 272–83.

28 Krasny J F, 'Apparel flammability: accident simulations and bench-scale tests', *Text. Res. J.*, 1986, **56**, 287.

29 *Clothing Flammability Accidents Survey*, London, Consumer Safety Unit, Department of Trade and Industry, 1994.

30 *Standards for Testing of Interior Textiles*, Frankfurt, Trevira (formerly Hoechst) GmbH, 1997.

31 Chouinard MP, Knodel D C and Arnold H W, 'Heat transfer from flammable fabrics', *Text. Res. J.*, 1973, **43**, 166–75.

32 Sorensen N, 'A manikin for realistic testing of heat and flame protective clothing', *Technical Text. Int.*, 1992, June, 8–12.

33 Horrocks A R, 'Developments in flame retardants for heat and fire resistant textiles – the role of char formation and intumescence', *Polym. Degrad. Stab.*, 1996, **54**, 143–54.

34 Dombrowski R, 'Flame retardants for textile coatings', *J. Coated Fabrics*, 1996, **25**, 224.

35 Hastie J R, 'Molecular basis of flame inhibition', *J. Res. Nat. Bureau Stds.*, 1973, **77A** (6), 733.

36 Kandola B, Horrocks A R, Price D and Coleman G, 'Flame retardant treatments of cellulose and their influence on the mechanism of cellulose pyrolysis', *Revs. Macromol. Chem. Phys.*, 1996, **C36**, 721–94.

37 Price D, Horrocks A R, Akalin M and Faroq A A, 'Influence of flame retardants on the mechanism of pyrolysis of cotton (cellulose) fabrics in air', *J. Anal. Appl. Pyrol.*, 1997, **40/41**, 511–24.

38 Lewin M, 'Flame retardance in fabrics', in *Chemical Processing of Fibers and Finishes. Part B; Handbook of Fiber Science and Technology: Vol. II*, (eds. M Lewin and S B Sello), New York, Dekker, 1984, 1–143.

39 Elton S, 'Reduction of the thermoplastic melt hazard of polyester fabrics through the application of a radiation cross-linking technique', *Fire Mater.*, 1998, **32**, 19–23.

40 Horrocks A R, 'Flame retardant finishes and finishing', in *Textile Finishing, Volume 2*, (ed. D Heywood), Bradford, UK, Society of Dyers and Colourists, in press.

41 Horrocks A R, 'Flame retardant finishing of textiles', *Revs. Prog. Colouration*, 1986, **16**, 62–101.

42 Heidari S and Kallonen, 'Hybrid fibres in fire protection', *Fire Materials*, 1993, **17**, 21–4.

43 Mischutin V, 'New FR system for synthetic/cellulosic blends', *Text. Chem. Colorist.*, 1975, **7** (3), 40–2.

44 Hall M E and Shah S, 'The reaction of wool with N-hydroxymethyl dimethyl phosphonopropionamide', *Polym. Degrad. Stab.*, 1991, **33**, 207.

45 Benisek L, 'Use of titanium complexes to improve the natural flame retardant of wool', *J. Soc. Dyers. Col.*, 1971, **87**, 277–8.

46 Beck P P, Gordon P G and Ingham P E, 'Thermogravimetric analysis of flame-retardant-treated wools', *Text. Res. J.*, 1976, **46**, 478.

47 Lewin M and Mark H F, 'Flame retarding of polymers with sulfamates; Part 1 Sulfation of cotton and wool,' 8th annual conf *Recent Advances in Flame Retardancy of Polymer Materials*, Co, Norwalk, USA, Business Communications Company, 1997.

48 Horrocks A R and Davies P J, 'Char formation in flame-retarded wool fibres; Part 1 Effect of intumescent on thermogravimetric behaviour,' *Fire Mater.*, 2000, **24**, 151–7.

49 Benisek L, 'Antagonisms and flame retardancy', *Text. Res. J.*, 1981, **51**, 369.

50 Horrocks A R, Zhang J and Hall M E, 'Flammability of polyacrylonitrile and its copolymers II. Thermal behaviour and mechanism of degradation', *Polym. Int.*, 1994, **33**, 303–14.

51 Hall M E, Zhang J and Horrocks A R, 'Flammability of polyacrylonitrile and its copolymers III. Effect of flame retardants', *Fire Mater.*, 1994, **18**, 231–41.

52 Zhang Z, Horrocks A R and Hall M E, 'Flammability of polyacrylonitrile and its copolymers IV. The flame retardant mechanism of ammonium polyphosphate', *Fire Mater.*, 1994, **18**, 307–12.

53 Anon, 'Flame retardant buyers' guide', *Amer. Dyest. Rep.*, 1998, January, 13–29.

54 Van Krevelen R W, 'Some basic aspects of flame resistance of polymeric materials', *Polymer*, 1975, **16**, 615.

55 Tolbert T W, Dugan J S, Jaco P and Hendrix, Springs Industries, Fire Barrier Fabrics, US Patent Office, 333 174, 4 April 1989.

56 Caze C, Devaux E, Testard G and Reix T, 'New intumescent systems: an answer to the flame retardant challenges in the textile industry', 6th European Conference on Flame Retardant Polymers *Fire Retardancy of Polymers: The Use of Intumescence* (eds M Le Bras, G Camino, S Bourbigot and R Delobel), London, Royal Society of Chemistry, 1998, 363–75.

57 Wragg P J, 'Where now with FR?', 2nd int conf *Ecotextile'98 – Sustainable Development* (ed A R Horrocks), Cambridge, UK, Woodhead Publishing, 1999, 247–58.

58 Horrocks A R, Anand S C and Hill B J, *Fire and Heat Resistant Materials*, UK Patent Office, GB 2 279 084 B, 20 June 1995.

59 Horrocks A R, Anand S C and Sanderson D, 'Novel textile barrier fabrics which retain coherence and flexibility up to 900 deg C', int conf *Interflam'93*, London, Interscience Communications, 1993, 689–98.

60 Horrocks A R and Kandola B K, 'Flame retardant cellulosic textiles', 6th European Conference on Flame Retardant Polymers *Fire Retardancy of Polymers: The Use of Intumescence* (eds M Le Bras, G Camino, S Bourbigot and R Delobel), London, Royal Society of Chemistry, 1998, 343–62.

61 Horrocks A R and Kandola B K, 'Complex char formation in flame-retarded fibre-intumescent combinations – III Physical and chemical nature of the char', *Text. Res. J.*, 1999, **69**, 347–81.

62 Holme I, 'Flammability – the environment and the green movement', *J. Soc. Dyers Col.*, 1994, **110**, 362–6.

63 Horrocks A R, Hall M E and Roberts D, 'Environmental consequences of using flame-retardant textiles – a simple life cycle analytical model', *Fire Mater.*, 1997, **21**, 229–34.

64 Hardy M C, 'Status of regulatory activity in Europe and the United States impacting brominated flame retardants', 8th annual conf *Recent Advances in Flame Retardancy of Polymer Materials*, Stamford, USA, Business Communications Co, Norwalk, USA, 1997.

65 Anon, 'No evidence for cot death link', *Chem. Brit.*, 1998, **34** (7), 8; *Expert Group to Investigate Cot Death Theories: Toxic Gas Hypothesis*, London, UK Department of Health; http://www.open.gov.uk/doht.limer.htm.

66 Edmonds A and Horrocks A R, 'Optimised process reduces formaldehyde', *Environmental Best Practice Programme, FP 70*, Harwell, UK, ETSU, 1997.

67 Stevens G C and Mann A H, *Risks and Benefits in the Use of Flame Retardants in Consumer Products*, Consumer Safety Unit, Department of Trade and Industry, 1999.

68 *Mandate/263 to CEN in the Field of Standardisation Relative to the Safety of Consumers for a Feasibility Study on Possible Standardisation on Fire Resistance of Nightwear.* Directorate General XXIV, Brussels, 8 December 1997. Report to EU 16 June 1999.

69 *Toxicological Risks of Selected Flame Retardant Chemicals*, Report by the Subcommittee on Flame Retardant Chemicals, National Research Council, National Academy for Sciences, National Academy Press, Washington, DC, 2000.

5
Composites

BALJINDER K KANDOLA AND
A RICHARD HORROCKS

Faculty of Technology, Bolton Institute,
Bolton, UK

5.1 Introduction

A composite material is a heterogeneous mixture of two or more homogeneous phases which have been bonded together. In the finished form sometimes these two phases are not visibly distinguishable. It is, however, different from an engineered structure containing more than one material, such as alloys, in the sense that although the composite material may seem to be homogeneous, there are still different constituent phases. The term 'composite' originally started in engineering science when two or more materials were combined together in order to rectify some shortcomings of a particular useful component. A composite comprises a large number of strong stiff fibres called the reinforcement, embedded in a continuous phase of a second material known as the matrix. The resulting product has the advantage of lower weight, greater strength and higher stiffness than individual constituents and even conventional load-bearing structures such as steel and aluminium.

The idea of reinforcement is not new. Over the centuries natural fibres, such as grass or animal hair, have been used to improve the strength and to lessen shrinking of pottery prior to firing and increase the strength in mud bricks. This idea in the present form has been exploited with the development of glass, carbon and later of aramid fibres.[1] Initially the glass-reinforced plastic materials have been used only for defence structures such as aircraft radomes, boathulls and seaplane floats, but now they have wider daily uses in the lives of most people in industrialised societies. In aircraft and automobiles, they are exploited to make body components, structural members, tyres and interior furnishings of vehicles. Many types of sporting and leisure goods such as boats, gliders, sailboards, skis and racquets utilise composite material extensively. Even in homes, many plastic-bodied appliances incorporate reinforcement in the form of short chopped fibres.

The major advantages of composite materials are low density, high specific strength and stiffness, good corrosion resistance and improved fatigue properties. Because of these characteristics, they have successfully replaced many conventional metals and other polymeric materials in load-bearing structures in aircraft, automobiles, ships, pipelines, storage tanks, and so forth as bulkhead, framework and panel components. Flexibility of manufacturing is also a unique characteristic. Large complex structures can be fabricated in one piece thus minimising tooling costs and the need for joints and fastenings. The fact that during manufacture the material itself is being made at the same time as the component, required properties can be attained by suitable choice of the constituents. For example, it is possible to design components with low or even zero thermal expansivity. Using carbon fibres fatigue characteristics can be improved, whereas glass fibres are excellent electric insulators.

All matrix materials are highly flammable in comparison with metals such as aluminium or steel, so they can burn vigorously with evolution of smoke and this partly offsets their many advantages. Even if inorganic fibres like E-glass are the reinforcing structures, the composite fire resistance will be determined by that of the organic matrix and the relatively low melting point of these fibres in comparison with typical flame temperatures. Flame retardancy of these materials is a major issue these days because, depending on applications, they must pass some type of regulatory fire test in order to assure public safety. For these reasons, it is important to understand how individual components of the end-products burn and how best to modify materials to make them flame resistant without compromising their uniquely valuable low weight to high mechanical property ratios.

5.2   The properties of the constituents of composites

As discussed earlier a composite material is a heterogeneous material having two or more components: these comprise the reinforcing elements (the fibres) that provide necessary mechanical characteristics to the material and a matrix that allows uniform deformation of reinforcing elements. This definition is very wide. It permits the fibres to be natural or man-made, metallic, inorganic or organic. Likewise, the matrix can be metal or metal alloy, an inorganic cement or glass or a natural or synthetic high polymer. The most commonly used fibres and resins for polymeric composites are shown schematically in Table 5.1. Adhesion between two dissimilar phases is necessary to allow uniform load distribution between them. The mechanical and other properties of a composite material depend upon the properties of the reinforcement and the matrix as well as upon the adhesive strengths at their interfaces. The properties also depend upon the methods used to combine these components into one material. The structure of the reinforcement also affects the mechanical properties, such as whether the fibres are unidirectional or multidirectional, continuous or

Table 5.1 Constituents of fibre-reinforced polymeric composite materials

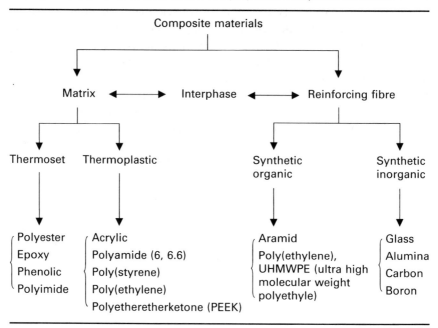

chopped, or whether they are in fabric form, which may be of the woven, nonwoven or knitted type.

5.2.1 Fibres

Fibres are the dominant constituents of most composite systems as they act as ideal reinforcing elements providing composites with mechanical performance (excellent stiffness and strength) as well good thermal, electrical and chemical properties, while offering significant weight savings over metals. The main objective of any design is to place fibres in positions and orientations in which they are able to contribute efficiently to load-carrying capabilities. The most widely used fibre aggregate for advanced structural applications is in the form of continuous tow. Such tows produce highly anisotropic materials of very high stiffness and strength in the direction of the reinforcement. Many layers of plies called laminae can be used, each with different orientation with respect to each other to form a laminate. Fibrous arrays can be in the form of woven cloths. They may have equal numbers of fibres in warp and weft directions, hence equal properties in both directions. Different layers with different orientation can be used to create equal properties in all directions. Discontinuous fibres can

Table 5.2 Physical and mechanical properties of fibres[2,3]

Fibre	Diameter (μm)	Tensile strength (GPa)	Young's modulus (GPa)	Density (kg m^{-3})	Thermal properties, max temp in service, °C
Glass	3–20	2–6	50–100	2400–2600	250
Carbon	5–6	1.5–7.0	150–800	1500–2000	400–450 (LOI 55–60)
Aramid	10	2–4	70–150	1410–1450	200 (LOI 30)
Boron	100–200	2–4	370–430	2500–2700	350
UHMPE	10–30	1.5	70	990–1020	100
Alumina	10–20	0.5	310	3800–4000	1000

also be used e.g. chopped fibres about 30–50 mm long, distributed in a random manner in a plane and held together with a resin binder. Both tows and cloths can be preimpregnated with resin, processed and then used as 'prepregs' during composite manufacture. Some of the most commonly used fibres are detailed below.

5.2.1.1   Glass fibres

Glass is the common name given to a number of mutually soluble inorganic oxides which can be cooled below their melting points without crystallisation. The main component is silica (SiO_2), while other oxides such as those of calcium, sodium, and aluminium, among others reduce the melting temperature and hinder crystallisation.[2] Based on different chemical compositions, various grades of glass are available commercially e.g. E-, S-, R- and C-glass.

By pulling swiftly and continuously from the melt, glass can be drawn into very fine filaments and such continuous glass fibres are usually 3–20 × $10^{-6}\mu$m in diameter: their physical properties are given in Table 5.2.[2,3] The advantages of glass fibres are in their high tensile and compressive strengths, low cost, good compatibility and good processibility. The disadvantages are associated with their low modulus and physical thermal stability. When heated they soften at relatively low temperatures so have limited temperature performance ranges.

Glass/epoxy materials have a variety of applications. For example, they are used to manufacture pressure vessels, reservoirs, ships, boats and yachts, parts for car bodies, and aircraft propellers.

5.2.1.2   Carbon fibres

Carbon fibres are manufactured by controlled pyrolysis and cyclisation of certain organic precursors such as polyacrylonitrile (PAN). Carbon fibres have such physical characteristics as low density, high strength and stiffness and these are presented in Table 5.2.[3] Their stiffness is high compared to that of glass fibres. Mechanical characteristics of carbon fibres do not deteriorate with temperature increases up to 450 °C, so they can be used for both polymeric and metal matrices. They are used for manufacturing load-carrying panels of aircraft wings and fuselages, drive shafts of cars and parts operating under intense heating.

5.2.1.3 Aramid fibres

Fibres which are based on aromatic polyamides, where at least 85% of the amide groups are connected directly to an aromatic group, are generically called aramid fibres. The general chemical formula for a para-aramid is typified by that for poly(p-phenylene terephthalamide) (PPT):

PPT fibres are generally available in two forms: low and high modulus. Their main advantage is their low density (lower than glass or carbon) giving high values of specific strength and stiffness and excellent toughness. Low modulus fibres are useful for textile applications like ropes, cables and flexible protective clothing: high modulus fibres are used for rigid laminates of maximum stiffness such as in boats, sports-car bodies and aircraft components. Some examples of commercially available PPT aramid fibres are Kevlar (Du Pont), Twaron (AKZO) and Technora (Teijin). The thermal stability of PPT fibres is discussed in detail in Chapter 4.

5.2.1.4 Boron fibres

Boron fibres are obtained by high temperature reduction of boron trichloride vapour on a tungsten or carbon substrate. With rise in temperature, fibres start to degrade in air at 400 °C. In order to prevent their oxidative degradation, they are covered with a refractory silicon or boron carbide coating. They are typically $100–200 \times 10^{-6} \mu m$ in diameter. Because of their large diameter and high stiffness, it is not possible to carry out normal textile processes such as weaving, and so these are used in the form of single-thickness, parallel-laid, pre-impregnated sheets or narrow continuous tapes.[2] Their main advantages are high stiffness and compression strength, but they are rather expensive.

5.2.1.5   Polyethylene fibres

Fibres from ultra high molecular weight polyethylene (UHMWPE) may be produced to have similar tensile properties as aramids. Their very low density (~30% lower than aramids) means that their specific tensile properties are considerably higher. Polyethylene has good chemical, abrasion resistance and low moisture absorption.[1] Their main disadvantage is their low melting point, 130–150 °C and hence low maximum service temperatures. Examples of UHMWPE are Spectra (Allied fibres) and Dyneema (DSM).

5.2.1.6   Alumina fibres

Fibres of polycrystalline alumina can be made by extruding a thickened mixture of fine alumina powder suspended in an alginate binder and then sintering the fibrous mass at high temperature.[2] Alumina fibres are very strong and are resistant to temperatures as high as 900–1000 °C. These fibres are used with epoxy, polyimide and maleimide resins.

All the fibres used for composites except UHMWPE are relatively non-flammable and LOI values for aramid and carbon are 30 and 55 respectively (see Table 5.2), hence they do not need further treatment. However, as discussed earlier, at high temperatures they soften or melt and their mechanical strength is reduced.

5.2.2   Matrix polymers

The most common matrix materials for composites, and the only kinds discussed here, are polymeric and they can be thermoset or thermoplastic. Thermoset matrices are fabricated from the respective resin, a curing agent, a catalyst or curing initiator and a solvent sometimes introduced for lowering the viscosity and improving impregnation of reinforcements.

In thermosets, solidification from the liquid phase takes place by the action of an irreversible chemical cross-linking reaction which produces a tightly bound three-dimensional network of polymer chains. The molecular units forming the network and the length and density of the cross-links of the structure will influence the mechanical properties of the material. The network and length of the units depend upon the chain segment lengths of the relevant monomers or oligomers. However, the level of cross-linking between functional groups depends on the degree of cure and this usually involves application of heat and pressure although some resins cure at room temperature.

The second type of polymers are thermoplastic in nature and have the advantage that they can be formed by the physical processes of heating and cooling. Thermoplastics readily flow under stress at elevated temperatures, can be fabricated into required components and become solid and retain their shape when cooled to room temperature. However, the reversibility of this process generates composites having a thermoplastic property and hence, poor physical resistance to heat.

The most widely used matrix materials are now described.

5.2.2.1   Polyester resins

Polyesters are probably the most commonly used of polymeric resin materials. The advantages of polyester matrices include their ability to cure over a wide range of temperatures under moderate pressures, their low viscosities providing their good compatibility with fibres and their ability to be readily modified by other resins. Essentially, they consist of a relatively low molecular weight unsaturated polyester chain dissolved in styrene. Curing occurs by the polymerisation of the styrene, which forms cross-links across unsaturated sites in the polyester. Curing reactions are highly exothermic which can affect processing rates due to the excessive heat generated and can damage the final laminate.[1]

The general formula for a typical resin[4] is:

Among the drawbacks of polyester resins are poor mechanical characteristics, low adhesion, relatively large shrinkage and the presence of toxic components of the styrene type.

Because of their structure, polyesters have LOI values of 20–22 and hence flame readily, and sometimes vigorously, after ignition. Unsaturated polyesters, cross-linked with styrene, burn with heavy sooting. These can be flame retarded by addition of inorganic fillers, addition of organic flame retardants, chemical modification of the acid, alcohol or unsaturated monomer component and the chemical combination of organo-metallic compounds with resins.[5]

It is common practice to add inert fillers to polyester resins to reinforce the cured composite, to lower cost and to improve flame retardance. Glass fibre and calcium carbonate often increase the burning rate of the composition,[5] but other fillers such as antimony trioxide for halogenated compositions and hydrated alumina are quite effective flame retardants. Ammonium polyphosphate is also used as a filler.[6]

Modification of the saturated acid component has been by far the most successful commercial method of preparing flame-retardant unsaturated polyesters. Examples are halogenated carboxylic acids, such as chlorendic acid or their anhydrides, tetrachloro- or tetrabromophthalic anhydride.[6] Halogenated alcohols or phenol can also be incorporated into the polymeric chain. Examples are tribromo-neopentyl glycol, tetrabromobisphenol-A and dibromophenol.

The cross-linking partner may also be flame retardant, as in the case of monochloro- or dichlorostyrene and hexachloropentadiene. Examples of halogenated additive compounds are tetrabromo-*p*-xylene, pentabromobenzyl bromide, pentabromoethyl benzene, pentabromotoluene, tribromocumene, decabromodiphenyl oxide and brominated epoxy resins.[6] The effectiveness of halogenated components is enhanced by simultaneous addition of antimony trioxide. Phosphorus-containing flame retardants like phosphonates and dialkyl phosphites can be incorporated into the polyester chain. In addition allyl or diallyl phosphites may act as cross-linking agents.[6]

5.2.2.2 Epoxy resins

These resins are extensively used in advanced structural composites particularly in the aerospace industry. They consist of an epoxy resin and a curing agent or hardener. They range from low-viscosity liquids to high melting point solids, can be easily formulated to give suitable products for the manufacture of prepregs by both the solution and hot-melt techniques, and are modified easily with a variety of different materials. They are manufactured by the reaction of epichlorohydrin with materials such as phenols or aromatic amines.

Epoxy resins contain the epoxy or glycidyl group shown below:

$$R \left[CH_2 - CH \overset{O}{\underset{\diagdown}{\overbrace{}}} CH_2 \right]_n$$

The resin can exist in the uncured state for quite a long time and this property allows the manufacture of so-called prepregs, where the fibres are impregnated with resin and are partially cured.[3]

Glass transition temperature of epoxies ranges from 120–220 °C,[2] hence they can be safely used up to these temperatures. Some of the epoxy resins used in advanced composites are N-glycidyl derivatives of 4,4′-diaminodiphenylmethane and 4-aminophenol, and aromatic di- and polyglycidyl derivatives of bisphenol A, bisphenol F, phenol novolacs and tris (4-hydroxyphenyl) methane.[2] Cross-linked epoxy resins are combustible and their burning is self-supporting with LOI values in the range 22–23. They mainly require reactive flame retardants, such as tetrachloro- or tetrabromobisphenol-A and various halogenated epoxides. Even the cross-linking agent may be flame retardant, as in the case of chlorendic anhydride, tetrabromo- or tetrachlorophthalic anhydride[7] or possibly phosphorus compounds.[8] Halogenated agents can be supplemented with antimony trioxide.[6] Additive flame retardants like ammonium polyphosphate, tris(2-chloroethyl) phosphate or other phosphorus-containing plasticizers are also used. Alumina trihydrate used as a filler is an effective flame retardant for epoxy resins.[6]

5.2.2.3 Phenolic resins

Phenolic resins are manufactured from phenol and formaldehyde. Reaction of phenol with less than equimolar proportions of formaldehyde under acidic conditions gives novolac resins containing aromatic phenol units linked predominantly by methylene bridges. Novolac resins are thermally stable and can be cured by cross-linking with formaldehyde donors such as hexamethylenetetramine. However, resoles are the most widely used phenolic resins for composites: they are manufactured by reacting phenol with a greater than equimolar amount of formaldehyde under alkaline conditions and are essentially hydroxymethyl functional phenols or polynuclear phenols. Unlike novolacs, they are low-viscosity materials and easier to process. Phenolic resins can also be prepared from other phenols such as cresols or bisphenol. The general formula is:

Phenolics are of particular interest in structural applications owing to their inherent fire-resistant properties yielding LOI values of 25 or so, although they tend to increase smoke generation. Their main disadvantages are low toughness and a curing reaction that involves the generation of water. This water can remain trapped within the composite and during a fire steam can be generated, which can damage the structure of the material.

Cured phenolic resins do not ignite easily because of their high thermal stability and high charring tendency on decomposition. The principal volatile decomposition products are methane, acetone, carbon monoxide, propanol and propane. In a few cases, where phenolic resins require flame-retardant treatment, additive and reactive flame retardants can be used. Tetrabromobisphenol A, various organic phosphorus compounds, halogenated phenols and aldehydes (e.g., *p*-bromobenzaldehyde) are some of the reactive flame retardants used for phenolics. Phosphorus can be introduced by direct reaction of the phenolic resin with phosphorus oxychloride and similarly inorganic compounds such as boric acid may be incorporated into phenolic resin by chemical reaction.[9]

Chlorine compounds (e.g. chloroparaffins) and various thermally stable aromatic bromine compounds are utilised as additive flame retardants. Antimony trioxide is usually added as a synergist. Suitable phosphorus compounds include halogenated phosphoric acid esters such as tris(2-chloroethyl)phosphate, halogenated organic polyphosphates, calcium and ammonium phosphates. Zinc and barium salts of boric acid and aluminium hydroxide also find frequent application.[9] In order to suppress the afterglow of phenolic resins, use is made of such compounds as aluminium chloride, antimony trioxide and organic amides.

5.2.2.4 Maleimide and polyimide resins

Thermosetting bismaleimide resins are used widely in advanced composites and are prepared by reaction of maleic anhydride with the corresponding primary amine. There are other polyimide matrices in addition to the bismaleimide resins. Some are thermoplasic resins prepared by condensation reactions of an aromatic tetracarboxylic acid dianhydride, such as benzophenone tetracarboxylic acid dianhydride (BTDA) and an aromatic amine.[2] Others are thermosetting resins that cross-link on cure. Examples are PMR 15 matrix and LARC 160 systems. The general formula for these is:

The processing conditions required to manufacture composite components from bismaleimide and other polyimide resins are more severe than used for epoxy systems and the resulting composites are more brittle than those of epoxy matrices. They cure at about 250–350 °C for several hours.[2] However, the glass transition temperature of cured resin is about 100 °C higher than cured epoxy matrices and so they retain better their mechanical properties at higher temperatures.

Polyimides are characterised by high char formation on pyrolysis, low flammability (LOI > 30) and low smoke production when subjected to a flame in a non-vitiated atmosphere.

5.2.2.5 Vinyl ester resins

Vinyl ester resins, like unsaturated polyesters, cure by a radical initiated polymerisation. They are mainly derived from reaction of an epoxy resin, for instance, bisphenol A diglycidyl ether with acrylic or methacrylic acid. Their general formula is:

$$R \left[CH_2CHCH_2O - \underset{O}{\overset{OH}{\underset{\|}{C}}} - \underset{}{\overset{R'}{C}} = CH_2 \right]_n$$

Like unsaturated polyesters they are copolymerised with diluents such as styrene using similar free-radical initiators. They differ from polyesters in that the unsaturation is at the end of the molecule and not along the polymer chain. When methacrylates are used they offer better chemical resistance than unsaturated polyesters. Their burning behaviour falls between that of polyester and epoxy resins (LOI 20–23).

5.2.2.6 Thermoplastic resins

Thermoplastic resins are high molecular weight linear chain molecules often with no functional side groups (e.g. linear polyesters, polyamides) or with side groups which do not easily cross-link at processing temperatures. They are fundamentally different from the thermosets in that they do not undergo irreversible cross-linking reactions, but instead melt and flow on application of heat and pressure and resolidify on cooling. The main advantages of thermoplastics are improved damage tolerance, infinite storage life, easy, rapid and low cost processability as no curing is required. However, as the temperatures required for processing are high, expensive tooling is required where particularly complex shapes are needed. The temperature performance of any individual resin depends upon its glass transition temperature. Some of the commonly used thermoplastic resins are poly(phenylene sulphide), poly(ether ether ketone), poly(ether ketone), poly(sulphone), poly(ether imide), poly(phenyl sulphone), poly(ether sulphone), poly(amide imide) and poly(imide). Their glass transition temperatures are 85, 143, 165, 190, 216, 220, 230, 249–288 and 256 °C, respectively.[1]

Thermoplastic composites are almost always processed in the form of prepreg materials. Impregnation of the fibres to form the prepreg can be difficult owing to high vicosities of the thermoplastic melt or the requirement to use high boiling point, polar solvents.[1]

As their name denotes, thermoplastics soften when heated. In a fire, such materials can soften enough to flow under their own weight and drip or run.

The extent of dripping depends upon such factors as thermal environment, polymer structure, molecular weight, presence of additives and fillers. Dripping can increase or decrease the fire hazard depending upon the fire situation. With small ignition sources, removal of heat and flame by the dripping away of burning polymer can protect the rest of the material from the flame spreading. In other situations, the flaming molten polymer might flow and ignite other materials. Since thermoplastics are not used for rigid composites, the methods to impart flame retardancy are not discussed here.

5.2.3 Resin–matrix interface

The properties of composite materials depend upon those of the matrix system and reinforcement and upon the interaction between the two. The interface is an important region and is required to provide adequate and stable bonding, both chemical and physical, between the fibres and the matrix. For example, aminosilane is used to bond glass fibre with epoxy matrix systems. Carbon fibre is both surface-treated in order to improve the mechanical properties of the composite, and coated with a sizing agent in order to aid processing of the fibre. Surface treatment creates potentially reactive groups such as hydroxyl and carboxyl groups upon the surface of the fibres, which are capable of reaction with the matrix. While epoxy-based sizing agents are quite common, they may not be suitable for the resin matrix.[2]

The mechanical properties of the composite structure are extremely dependent upon the interface properties. Those composite materials which have weak interfaces have low strengths and stiffness but have high resistances to fracture: those with strong interfaces have high strength and stiffness but are very brittle. This effect is a function of the ease of debonding and 'pull-out' of the fibres from the matrix material during crack propagation.[10] In addition, the nature of the interface will affect the burning of the material as well. If the binding material is highly flammable, it will increase the fire hazard of the whole structure. However, if the interface is weak and the two phases (fibre and matrix) are pushed apart in case of fire, the matrix will burn more vigorously and the inorganic fibres, which can no longer act as insulators, sometimes act rather as heat conductors thus increasing flammability.

5.3   Flammability of composite structures

As discussed previously, composite structures contain two polymeric structures, fibre and resin. Both these components behave differently in a fire depending upon their respective thermal stabilities. Composite structures which are layered tend to burn in layers. When heated, the resin of the first layer degrades and any combustible products formed are ignited. The heat penetrates the adjacent fibre layer. If inorganic fibre is used it will melt or soften: if organic fibre is used, it will degrade into smaller products depending upon its thermal stability. The heat then penetrates further, reaches the underlying resin, causing its degradation and any products formed will then move to the burning zone through the fibrous char. Burning will slow down at this stage. If the structure is multilayered, it will burn in distinct stages as the heat penetrates subsequent layers and degradation products move to the burning zone through the fibrous layers. In general, the thickness of a structure can affect the surface flammability characteristics up to a certain limiting value, after which the full depth of the material is not involved in the early stages of burning and the material is said to be 'thermally thick'.[11] Scudamore[12] has shown by cone calorimetric results that this effect decreases as the external heat flux increases. At 35 and $50 kW m^{-2}$, thin samples (3 mm) ignited easily compared to thick samples (9.5 mm), but at 75 and $100 kW m^{-2}$ there was not much difference and both sets of samples behaved as if they were 'thermally thin'.

When compared with the flammability of different resins, phenolic resins behave very well. LOI values of various thermoplastic, thermoset resins and their composites at 23 °C conducted by Kourtides et al[13,14] are given in Table 5.3. Brown et al[15] have also conducted and reviewed cone calorimetric studies for the flammability of various composite materials. All the studies indicate that ranking of fire resistance of thermoset resins is:

Phenolic > Polyimide > Bismaleimide > Epoxy

Table 5.3 Limiting oxygen index values for polymers and composites at 23 °C[13,14]

Resin	LOI (%)	
	Resin	40% resin/ 181 glass cloth
Thermoplastic resins		
Acrylonitrile–butadiene–styrene	34	
Polyaryl sulfone (PAS)	36	
Polyether sulfone (PES)	40	
9,9 bis-(4-hydroxyphenyl) fluorene/polycarbonate – poly(dimethyl siloxane) (BPFC-DMS)	47	
Polyphenylene sulphide (PFS)	50	
Thermoset resins		
Epoxy	23	27
Phenolic	25	57
Polyaromatic melamine	30	42
Bismaleimide	35	60

As discussed above, the inherent fire retardant behaviour of the phenolics is due to the char-forming tendency of the cross-linked chemical structure. A recent review by Brown[16] discusses the fire performance and mechanical strengths of phenolic resins and their composites. Owing to the large proportion of aromatic structures in the cross-linked cured state, phenolic resins carbonise in a fire and hence extinguish once the source of fire is removed: they may thus be said to encapsulate themselves in char and therefore do not produce much smoke.[17] Epoxy and unsaturated polyesters, on the other hand, carbonise less than phenolics and continue to burn in a fire, and structures based on these aromatic compounds produce more smoke. Although phenolics have inherent flame retardant properties, their mechanical properties are inferior to other thermoset polymers, such as polyester, vinyl ester and epoxies.[16] For this reason, they are less favourable for use in load-bearing structures. Epoxies, on the other hand, because of their very high mechanical strength are the more popular choice.

It is well known that for char-forming polymers, flame retardants acting in the condensed phase are very successful. These flame retardants enhance the formation of polymer char at the expense of combustible volatiles. In other words, there is a direct relationship between the flammability of a polymer and its char yield as discussed comprehensively by van Krevelen[18] and observed in our own laboratories for flame retarded polyacrylic fibre-forming polymers.[19] Gilwee et al[20] have found that, in a nitrogen atmosphere, a linear relationship exists between limiting oxygen index and char yields from TGA results, as shown in Table 5.4. Also shown in this table is

Table 5.4 Limiting oxygen index and char yields from TGA curves in nitrogen[20,21]

	Char yield at 800 °C	LOI (%)
Resins[20]		
Epoxy	10	23
Polyimide	53	27
Phenolic	54	25
Melamine	58	27
Benzyl	63	43
Graphite composites[21]		
Epoxy	79	41
Phenolic – Xyloc	83	46
Bismaleimide A	82	47
Phenolic – Novolac	86	50
Polyether sulphone	77	54
Polyphenyl sulphone	81	52

a similar relationship for graphite-reinforced composite laminates from data provided by Kourtides,[21] demonstrating that composite structures behave similarly to bulk resin polymers, while char formation determines the flammability of the composite and the presence of inorganic fibre does not improve the flame retardancy of the structure.

Brown et al[22] have studied the fire performance of extended-chain polyethylene (ECPE) and aramid fibre-reinforced composites containing epoxy, vinyl ester and phenolic matrix resins by cone calorimetry. Various parameters were determined for ECPE and aramid fabrics only, matrix resins only and their composites. Maximum rates of heat release (max RHR) are plotted in Fig. 5.1. Typically, ECPE reduced the flammability of epoxy but increased that for vinyl ester matrix resins. Aramid, on the other hand, had little effect on time to ignition (compared to resin alone) except for the phenolic, but reduced the RHR. In general, resin and reinforcement contributions to the composite rate of heat release behaviour as a function of time are generally discernible and depend on the respective flame-retardant mechanisms operating, the levels of their transferability and possibly synergisms and antagonisms. This indicates that a flame- or heat-resistant fibre can be effective for one type of resin but not necessarily for another.

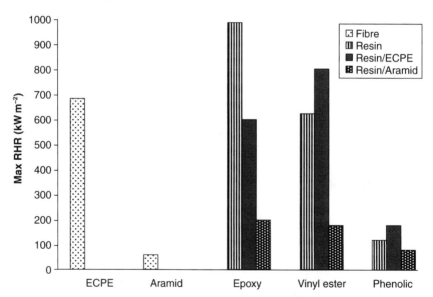

5.1 Max RHR values for fibre reinforcements, matrix resins and composite materials at 50 kW m^{-2} cone irradiance[22]

5.4   Methods of imparting flame retardancy to composites

Considerable research in this area is being done at various laboratories around the world. Usually most means of imparting flame retardancy to composites result in a reduction of their mechanical strengths. Therefore, achieving a certain level of flame retardancy while maintaining other properties of the composite such as light weight and high mechanical strength is a major challenge. Although some flame-retardant systems are used commercially, most work in this area is still at an experimental stage. We have recently published a review[4] of the work done in this field and selected examples are discussed here. In general, the following methods for imparting flame retardancy to composites are used:

5.4.1   Use of mineral and ceramic wool

This method is quite popular for naval applications to flameproof conventional composite hull, deck and bulkhead structures.[23] The main disadvantages of using mineral and ceramic wool are that they occupy space, add significant weight and can act as an absorbent for spilled fuel or flammable liquid during a fire. When this occurs, extinguishing the fire will be more difficult and the insulating property of the ceramic wool is lost.

5.4.2   Chemical or physical modification of resin matrix

Additives like zinc borate and antimony oxide have been used with halogenated polyester, vinyl ester or epoxy resins.[16,24] Alumina hydrate and bromine compounds are other examples. However, many of these resins and additives are ecologically undesirable and in a fire increase the amount of smoke and toxic fumes given off by the burning material. Furthermore, this method usually results in a reduction in the mechanical properties of the composite structure.

As stated above, Scudamore[12] has studied the fire performance of glass-reinforced polyester, epoxy and phenolic laminates by cone calorimetry and the effect of flame retardants on these composites was also observed. The FR polyester examined consisted of brominated resin whereas FR epoxy and phenolic resins contained ATH (alumina trihydrate). ATH was used in the FR phenolic laminate. It was concluded that the fire properties depend on the type of resin and flame retardant, the type of glass reinforcement, and, for thin laminates, the thickness. Flame retardants for all resins seem to be effective in delaying ignition and decreasing heat release rates. Phenolic laminates have lower flammability than FR polyester or epoxy resins,

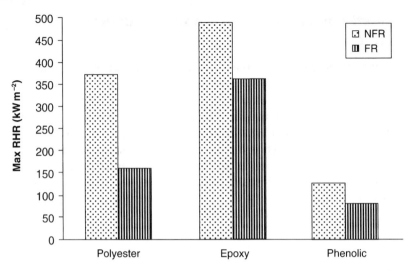

5.2 Max RHR values for glass-reinforced materials at 50 kW m^{-2} cone irradiance and effect of flame retardant additives[12]

but addition of ATH further enhances flame retardancy. Max RHR values for these structures at 50 kW m^{-2} are plotted in Fig. 5.2 from which it is seen that the presence of ATH in a phenolic composite can reduce max RHR values to less than 100 kW m^{-2} at 50 kW m^{-2} heat flux.

Morchat[26] and Hiltz[25] have studied the effect of the FR additives antimony trioxide, alumina trihydrate and zinc borate on the flammability of FR resins – polyester, vinyl ester and epoxy – by TGA, smoke production, toxic gas evolution, flame spread and OI methods. Except for epoxy resin, they contained halogenated materials from which they derived their fire-retardancy characteristics through the vapour-phase activity of chlorine and/or bromine. In most cases, with a few exceptions, the additives lowered the flame spread index (2–70%), increased LOI (3–57%) and lowered specific optical density (20–85%) of smoke, depending on the fire retardant and the resin system evaluated. However, for the majority of resins, the addition of antimony trioxide resulted in an increase in smoke production. The best performance was observed upon addition of zinc borate to the epoxy resin.

Nir et al[27] have studied the mechanical properties of brominated flame retarded and non-brominated epoxy (tris-(hydroxyphenyl)-methane triglycidyl ester)–graphite composites. The incorporation of bromine did not change the mechanical properties within ± 10% of those of the non-brominated resin. The addition of bromine helped in decreasing water absorption and increasing environmental stability, thereby indicating that

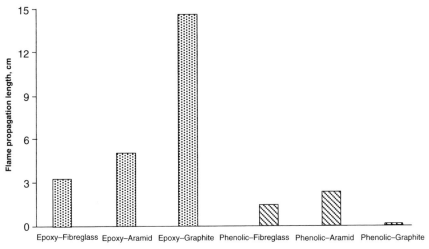

5.3 Flame propagation lengths of composites[28]

this is an easy method to flame retard and increase impact strength of graphite-reinforced composites.

Kovlar and Bullock[23] have used an intumescent component as an additive in the phenolic matrix. A novel composite structure was developed consisting of phenolic resin and intumescent in 1:1 ratio, reinforced with glass fabric. Upon exposure to fire the intumescent composite panel immediately began to inflate, foam, swell and char on the side facing the fire, forming a tough, insulating, fabric-reinforced carbonaceous char that blocks the spread of fire and insulates adjacent areas from the intense heat. When tested for their fire performance, the intumescent-containing samples showed marked improvement in the insulating properties than control phenolic or aluminium panels.

5.4.3 Use of high performance fibres

The reinforcing fibre phase can be rendered flame retardant by appropriate treatment or by the use of high heat- and flame-resistant fibres, such as aramids, although the flame-retardancy levels desired should really match those of the matrix if high levels of fire performance are to be realised. Hshieh and Beeson[28] have tested flame-retarded epoxy (brominated epoxy resin) and phenolic composites containing fibre glass, aramid (Kevlar) and graphite fibre reinforcements using the NASA upward flame test and the controlled atmosphere, cone-calorimeter test. The upward flame propagation test showed that phenolic–graphite had the highest and epoxy–graphite composites had the lowest flame resistance as shown in Fig. 5.3. The controlled-atmosphere, cone calorimeter test showed that phenolic

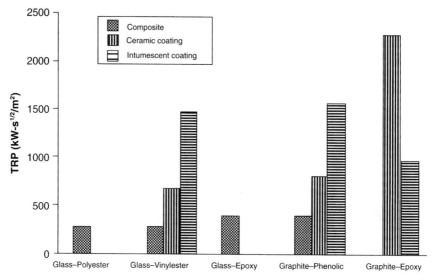

5.4 Thermal response parameter (TRP) values for composites[30]

composites had lower values of time of ignition, peak heat release rate, propensity to flashover and smoke production rate.

5.4.4 Use of flame-retardant coatings

Another way of flame retarding or fire-hardening composite structures is to use flame-retardant (usually intumescent based) paints or coatings. Intumescent systems are chemical systems, which by the action of heat evolve gases and form a foamed char. This char then acts as an insulative barrier to the underlying structural material against flame and heat. One very effective intumescent coating is fluorocarbon latex paint.[29]

Tewarson and Macaione[30] have evaluated the flammability of glass–resin composite samples and the effect of intumescent and ceramic coatings by Factory Mutual Research Corporation (FMRC) 50 kW-scale apparatus. From the FMRC test, thermal response parameter (TRP) values were calculated and are plotted in Fig. 5.4, which show that ceramic and intumescent coatings are quite effective in improving fire resistance.

Sorathia et al[31] have expored the use of integral, hybrid thermal barriers to protect the core of the composite structure. These barriers function as insulators and reflect the radiant heat back towards the heat source. This delays the heat-up rate and reduces the overall temperature on the reverse side of the substrate. Thermal barrier treatments evaluated include ceramic fabrics, ceramic coatings, intumescent coatings, hybrids of ceramic and intumescent coatings, silicone foams and a phenolic skin. The composite

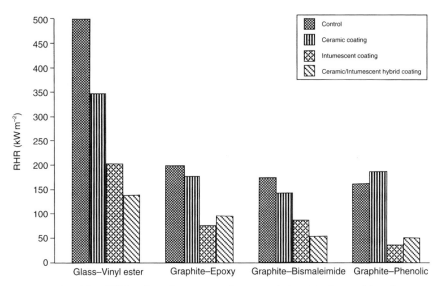

5.5 Max RHR values for composite materials with thermal barrier treatment at 75 kW m^{-2} cone irradiance[31]

systems evaluated in combination with thermal barrier treatments include glass–vinyl ester, graphite–epoxy, graphite–bismaleimide and graphite–phenolic. All systems were tested for flammability characteristics and max RHR values at 75 kW m^{-2} cone irradiance are plotted in Fig. 5.5. Without any barrier treatment, all composites failed to meet the ignitability and peak heat release requirements, whereas all treated ones passed. Ceramic–intumescent hybrid coating seems to be very effective.

5.5   Conclusions: some important considerations

From the above discussion it can be concluded that the choice of resin and fibre is important in determining the flammability properties of the whole structure. Inorganic fibres such as glass and carbon do not help in reducing flammability. If high performance fibres are used their compatibility with the resin matrix needs to be considered. For a flame-retarding resin matrix, reactive additive agents are generally used. If flame-retardant chemicals, which are compatible with both fibres and resin matrix, are selected, the effect can be synergistic. Char-forming agents seem to be the best choice and the way forward in developing effective fire-resistant composites.

References

1 Eckold G, *Design and Manufacture of Composite Structures*, Cambridge, Woodhead, 1994.

2 Phillips L N (ed), *Design with Advanced Composite Materials*, London, Springer-Verlag, 1989.

3 Vasiliev V V, Jones R M and Man L I (eds), *Mechanics of Composite Structures*, Washington, Taylor & Francis, 1988.

4 Kandola B K and Horrocks A R, 'Flame retardant composites, a review: the potential for use of intumescents', in Bras M L, Camino G, Bourbigot S and Delobel R (eds), *Fire Retardancy of Polymers – The Use of Intumescence*, Cambridge, The Royal Society of Chemistry, 1998.

5 *Fire Safety Aspects of Polymeric Materials' Vol 1– Materials State of Art*, Chapter 6, A Report by National Materials Advisory Board, National Academy of Sciences, Washington, Technomic Publ., 1977.

6 Pal G and Macskasy H, *Plastics – Their Behaviour in Fires*, Chapter 5, Amsterdam, Elsevier, 1991.

7 Lo J and Pearce E M, 'Flame-retardant epoxy resins based on phthalide derivatives', *J. Polym. Sci. Polym. Chem. Ed.*, 1984, **22**, 1707.

8 Mikroyannidis J A and Kourtides D A, 'Fire resistant compositions of epoxy resins with phosphorus compounds', *Polym. Mat. Sci. Eng. Proc.*, 1983, **49**, 606.

9 Troitzsch J, *International Plastics Flammability Handbook*, Chapter 5, New York, Hanser, 1990.

10 Holloway L, *Polymer Composites for Civil and Structural Engineering*, London, Blackie Academic & Professional, 1993.

11 Mikkola E and Wichman I S, 'On the thermal ignition of combustible materials', *Fire Mater.*, 1989, **14**, 87–96.

12 Scudamore M J, 'Fire performance studies on glass-reinforced plastic laminates', *Fire Mater.*, 1994, **18**, 313–25.

13 Brown J E, Loftus J J and Dipert R A, *Fire Characteristics of Composite Materials – A Review of the Literature*, Report 1986, NBSIR 85–3226.

14 Kourtides D A et al, 'Thermochemical characterisation of some thermally stable thermoplastic and thermoset polymers', *Polym. Eng. Sci.*, 1979, **19** (1), 24–29.

15 Brown J E, Braun E and Twilley W H, *Cone Calorimetric Evaluation of the Flammability of Composite Materials*, Report 1988, NBSIR-88-3733.

16 Brown J R and St John N A, 'Fire-retardant low-temperature-cured phenolic resins and composites', *TRIP*, 1996, **4** (12), 416–20.

17 Gabrisch H-J and Lindenberger G, 'The use of thermoset composites in transportation: their behaviour', *SAMPE J.*, 1993, **29** (6), 23–7.

18 Van Krevelen D W, 'Some basic aspects of flame resistance of polymeric materials', *Polymer*, 1975, **16**, 615–20.

19 Hall M, Zhang J and Horrocks A R, 'The flammability of polyacrylonitrile and its copolymers part III: Effect of flame retardants', *Fire Mater.*, 1994, **18**, 231–41.

20 Gilwee W J, Parker J A and Kourtides D A, 'Oxygen index tests of thermosetting resins', *J. Fire Flamm.*, 1980, **11** (1), 22–31.

21 Kourtides D A, 'Processing and flammability parameters of bismaleimide and some other thermally stable resin matrices for composites', *Polym. Compos.*, 1984, **5** (2), 143–50.

22 Brown J R, Fawell P D and Mathys Z, 'Fire-hazard assessment of extended-chain polyethylene and aramid composites by cone calorimetry', *Fire Mater.*, 1994, **18**, 167–72.

23 Kovlar P F and Bullock D E, 'Multifunctional intumescent composite fire-barriers' in *Recent Advances in Flame Retardancy of Polymeric Materials*, M Lewin (editor), Vol. 4, Business Communications Co, Norwalk, 1993, 87–98.

24 Stevart J L, Griffin O H, Gurdal Z and Warner G A, 'Flammability and toxicity of composite materials for marine vehicles', *Naval Engn. J.*, 1990, **102** (5), 45–54.

25 Morchat R M and Hiltz J A, 'Fire-safe composites for marine applications', Proc 24th int conf *SAMPE Tech.*, 1992, **24**, T153–T164.

26 Morchat R M, *The Effects of Alumina Trihydrate on the Flammability Characteristics of Polyester, Vinylester and Epoxy Glass Reinforced Plastics*, Techn. Rep. Cit. Govt. Rep. Announce Index (US), 1992, **92** (13), AB NO 235, 299.

27 Nir Z, Gilwee W J, Kourtides D A and Parker J A, 'Rubber-toughened polyfunctional epoxies: brominated vs nonbrominated formulated for graphite composites', *SAMPE Q.*, 1983, **14** (3), 34–8.

28 Hshieh F Y and Beeson H D, 'Flammability testing of flame-retarded epoxy composites and pheniolic composites', Proc int conf *Fire Safety*, 1996, **21**, 189–205.

29 Ventriglio D R, 'Fire safe materials for navy ships', *Naval Engn. J.*, October 1982, 65–74.

30 Tewarson A and Macaione D P, 'Polymers and composites – an examination of fire spread and generation of heat and fire products', *J. Fire Sci.*, 1993, **11**, 421–41.

31 Sorathia U, Rollhauser C M and Hughes W A, 'Improved fire safety of composites for naval applications', *Fire Mater.*, 1992, **16**, 119–25.

32 Kourtides D A et al, 'Thermal response of composite panels', *Polym. Eng. Sci.*, 1979, **19** (3), 226–317.

6

Nanocomposites

Faculty of Technology, Bolton Institute, Bolton

6.1   Introduction

Nanocomposites constitute a new development in the area of flame retardancy and offer significant advantages over conventional formulations where high loadings are often required. In general, when composites are formed two or more physically and chemically distinct phases (usually polymer matrix and reinforcing element) are joined and the properties of the resulting product differ from and are superior to those of the individual components. The structures and properties of the composite materials are greatly influenced by the component phase morphologies and interfacial properties. Nanocomposites are based on the same principle and are formed when phase mixing occurs at a nanometer dimensional scale (compare the microscopic scale, μm – mm, in conventional composites). As a result, nanocomposites show superior properties over their micro counterparts or conventionally filled polymers.

Polymer-layered silicates are the commonest group of nanocomposites. Although first reported by Blumstein[1] in 1961, the real exploitation of this technology started in the 1990s. Because of their nanometer size dispersions, nanocomposites exhibit superior properties in comparison with pure polymer constituents or conventionally filled polymers. The main advantages are light weight, high modulus and strength, decreased gas permeability, increased solvent resistance and increased thermal stability. Their mechanical properties are superior to unidirectional fibre-reinforced polymers because reinforcement from the inorganic layers will occur in two rather than in one dimension.[2] Because of the length scale involved that minimises scattering, nanocomposites are usually transparent.[3] They also exhibit significant increases in thermal stability as well as a self-extinguishing character.

In polymer-layered silicates, composite properties are achieved at a much lower volume fraction of reinforcement in comparison with conventional fibre or mineral-reinforced polymers. They can be processed by such techniques as extrusion and casting common to polymers which are superior to the costly and cumbersome techniques used for conventional fibre and mineral-reinforced composites and furthermore are adaptable to films, fibres and monoliths.

Most of the work in this area is at present at the experimental stage, although some commercial exploitation has been reported.[3] For example, the Toyota Motor Company is using an automotive timing-belt cover made from a nylon-layered silicate nanocomposite.[3] Potential applications are barrier films for food packaging, aeroplane interiors, fuel tanks and components in electrical or electronic parts, brakes and tyres.

6.2   The structure and properties of layered silicates

Layered silicate clays, because of their chemically stable siloxane surfaces, high surface areas, high aspect ratios and high strengths are most widely used for the formation of organic-inorganic nanocomposites.[4] Their high aspect ratios and high strengths make them very good reinforcing elements as well. Their two particular characteristics exploited for the formation of nanocomposites are:

- The rich intercalation chemistry used to facilitate exfoliation of silicate nanolayers into individual layers. As a result, an aspect ratio between 100–1000 can be obtained (compared to 10 for poorly dispersed particles). Layer exfoliation maximises interfacial contact between organic and inorganic phases.[4]
- The ability to modify finely their surface chemistries through ion exchange reactions with organic and inorganic cations.

The silicates most commonly used in nanocomposites are layered silicates (clay minerals) or phyllosilicates (rock minerals). Clay minerals are built of two structural units. One is a sheet of silica tetrahedra arranged as a hexagonal network in which the tips of the tetrahedra all point in the same direction;[5] this is the same unit as for phyllosilicates. The other structural unit consists of two layers of closely packed oxygen or hydroxyl groups in which aluminium, iron or magnesium atoms are embedded so that each is equidistant from six oxygens or hydroxyls.

Most clay minerals are sandwiches of two structural units, the tetrahedral and octahedral. The simplest type of sandwich is made of a single layer of silica tetrahedra with an aluminium octahedral layer on top: these are called 1:1 minerals and are of the kaolinite family. The other main type of sandwich is that of the 2:1 structure (smectite minerals), consisting of an octahedral filling between two tetrahedral layers. In smectite minerals the octahedral sites may be occupied by magnesium, iron or small metal ions as well as by aluminium. The structure of a 2:1 clay mineral (smectite) is

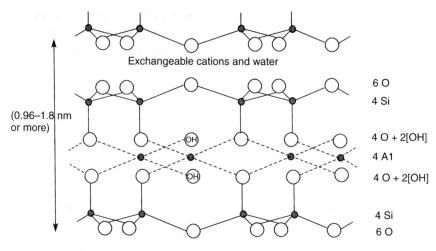

Exchangeable cations and water

(0.96–1.8 nm
or more)

6 O
4 Si

4 O + 2[OH]
4 Al
4 O + 2[OH]

4 Si
6 O

6.1 The structure of a 2:1 clay mineral (smectite)[5]

shown in Fig. 6.1.[5] Montmorillonite clay minerals of this group make a very popular choice for nanocomposites because of their small particle size ($<2\,\mu m$) and hence easy polymer diffusion into the particles. They also possess high aspect ratios (10–2000) and high swelling capacity, which are essential for an efficient intercalation of the polymer.[6] Phyllosilicates include muscovite ($KAl_2(AlSi_3O_{10})(OH)_2$), talc ($Mg_3(Si_4O_{10})(OH)_4$) and mica.

Stacking of the layers leads to a regular van der Waals gap between the layers called the interlayer or gallery. Isomorphic substitution within the layers generates negative charges that are normally counterbalanced by cations residing in the interlayer space. The interlayer cations are usually hydrated Na^+ or K^+, which can be exchanged with various organic cations e.g. alkylammonium, rendering the normally hydrophilic silicate surface organophilic. The organic cations lower the surface energy of the silicate surface and improve wetting with the polymer matrix which makes organosilicates more compatible with most engineering plastics. The organic cations may contain various functional groups that react with the polymer to improve adhesion between the inorganic phase and the matrix.[3]

Layered silicic acids can also be used for the preparation of nanocomposites. The family of layered silicic acids includes five members: kanemite ($NaHSi_2O_5 \cdot nH_2O$), makatite ($Na_2Si_4O_9 \cdot nH_2O$), octasilicate ($Na_2Si_8O_{17} \cdot nH_2O$), magadiite ($Na_2Si_{14}O_{29} \cdot nH_2O$) and kenyaite ($Na_2Si_{20}O_{41} \cdot nH_2O$).[4] They can be easily synthesized by hydrothermal methods and their acidic analogues can be obtained by proton exchange reactions. The intercalation chemistry of layered silicic acids is similar to that of smectite clays.[4] Burkett

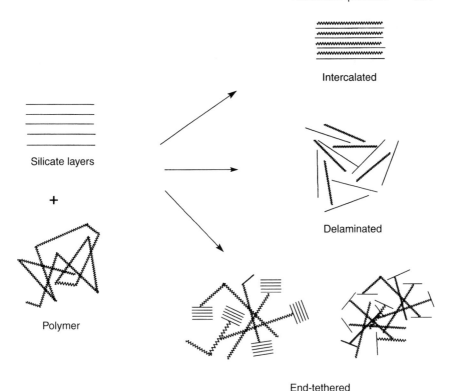

6.2 A schematic view of polymer-layered silicate nanocomposite structures

et al[7] have synthesised layered 2:1 trioctahedral phyllosilicates (magnesium organosilicates) for preparation of nanocomposites.

6.3   The structure of nanocomposites

There are three structurally different types of nanocomposites and they are shown schematically in Fig. 6.2. These three types are:

- Intercalated structures – the individual monomers and polymers are sandwiched between silicate layers.
- Delaminated or exfoliated – the silica is exfoliated to produce 'a sea of polymer with rafts of silicate'. The exfoliated structure can be ordered, where the silicate layers are more or less in one direction, or disordered, where they are dispersed randomly.
- End-tethered structure – the whole silicate or a single layer of the silicate is attached to the end of the polymer chain.[8,9]

In most materials two or more of these structures are combined, and one structure dominates.

Intercalated structures are those in which a single extended polymer chain is intercalated between the silicate host layers resulting in a well-ordered multilayer with fixed interlayer spacings. Delaminated or exfoliated structures are formed when the individual silicate layers are no longer close enough to interact with the adjacent layers' gallery cations.[10] The silicate nanolayers are individually dispersed in the polymer matrix with the average distance between the segregated layers (20–200 nm) being dependent on the clay loading. The separation between the nanolayers may be regular or disordered. Exfoliated nanocomposites show greater homogeneity than intercalated nanocomposites, and so exfoliated structures possess properties that are superior to those of intercalated ones.[4] Two types of end-tethered structures can be produced, one where the end of the polymer is attached to the outside of the silicate sheet and the other where the end of the polymer is attached to an exfoliated layer of the silicate. The second type is similar to a delaminated structure with polymer surrounding exfoliated layers of silicate.[8,9]

6.4   Synthesis methods

Not all physical mixtures of polymer and silicate will form a nanocomposite: the compatibility between the two phases is important. Kawasumi et al[11] have synthesised nanocomposites from various polymers – nylon 6, polyimide, epoxy resin, polystyrene, polycaprolactone and acrylic. The exfoliated and homogeneous dispersion of the silicate layers, however, could be achieved only in few cases, such as polymers containing polar functional groups such as amides and imides. This is due to the fact that silicate layers of clay have polar hydroxy groups and are compatible with polymers containing polar functional groups.[11]

Silicate clay layers are bound together by a layer of Na^+ or K^+ ions and are naturally hydrophilic. Ion exchange reactions with cationic surfactants including primary, tertiary and quaternary ammonium ions render the normally hydrophilic silicate surface organophilic, which makes intercalation of many engineering polymers possible. The role of the alkyl ammonium cations in the organosilicates is to lower the surface energy of the inorganic host and improve the wetting characteristics and, therefore, miscibility with the polymer.[2]

Nanocomposites can be formed in one of three ways:

- Melt blending synthesis.
- Solvent based synthesis.
- *In-situ* polymerisation.

6.4.1 Melt blending synthesis

The melt blending process involves mixing the layered silicate by anneal-ing, statically or under shear, with the polymer while heating the mixture above the softening point of the polymer. During the annealing process, the polymer chains diffuse from the bulk polymer melt into the galleries between the silicate layers. Giannelis[3] has used the 'direct polymer melt' method to intercalate poly(ethylene oxide), PEO, by heating the polymer and silicate at 80 °C for 6 h. Other polymer nanocomposites comprising polystyrene,[12] polyamides, polyesters, polycarbonate, polyphosphazene and polysiloxanes[13] can also be synthesised with this method.[3]

In some cases the polymer–silicate mixture can be extruded as well. Gilman et al[14] have prepared polystyrene-layered silicate nanocomposites using (a) static melt intercalation by mixing and grinding dried powders of polystyrene and organic silicate in a pestle and mortar and then heating the mixture at 170 °C for 2–6 h in vacuum, and (b) extrusion melt intercalation by extruding the mixture under nitrogen at 150–170 °C for 2–4 min. Kawasumi et al[11] and Hasegawa et al[15] used a twin screw extruder to produce a polypropylene nanocomposite from a modified polypropylene oligomer and modified clay.

6.4.2 Solvent based synthesis

The solvent based synthesis involves mixing a preformed polymer solu-tion with clay. A polystyrene–clay hybrid can be prepared by mixing a polystyrene-toluene solution and silicate to yield a suspension and then evaporating the solvent.[14] Polyimide–clay hybrids can be prepared by dissolving clay in dimethylacetamide (DMAC) and mixing with precursor solution of polyimide and then removing the solvent.[16] Jeon et al[17] have pre-pared polymer–polymer nanocomposites by solution blending of HDPE (high density polyethylene) and nitrile copolymer.

6.4.3 Polymerisation

In the case of polymerisation, the clay is dispersed in the monomer and the polymerisation reaction is carried out. Polystyrene clay nanocomposites can be prepared by the polymerisation of styrene in the presence of clay.[18] Moet[19] have achieved chemical grafting of polystyrene onto montmoril-lonite interlayers by addition polymerisation reactions. Kojima et al[20,29] have prepared nylon 6–clay hybrids by intercalating montmorillonite with ε-caprolactam (the cyclic lactam used in nylon 6 production). ε-caprolactam and 6-aminocaproic acid (accelerator) were polymerised with the interca-lated montmorillonite at 260 °C for 6 h, yielding a nylon 6-clay hybrid.

Most nanocomposites are produced by either of the first two methods.[18]

6.5   Characterisation

The most commonly used techniques for characterisation of nanocomposites are X-ray diffraction (XRD) and transmission electron microscopy (TEM). X-ray diffraction allows the determination of the spaces between structural layers of silicate utilising Bragg's law: $\sin \theta = n\lambda/2d$. Intercalation and delamination change the dimensions of the gaps between the silicate layers, so an increase in layer distance indicates that a nanocomposite has formed. A reduction in the diffraction angle corresponds to an increase in the silicate layer distance.[8] Generally, diffraction peaks observed in the low angle region ($2\theta = 3$–$9°$) indicate the d-spacing (basal spacing) of ordered-intercalated and ordered-delaminated structures.[2,14] If the nanocomposites are disordered, no peaks are observed in the XRD due to loss of structural registry of the layers, the large d-spacings (>10 nm), or both.[14] In general, the following relationship between the composite and the X-ray diffraction pattern holds:

composite	X-ray diffraction pattern
conventional	d-space as original
intercalated	d-space expands
ordered exfoliated	d-space further expands
disordered exfoliated	X-ray amorphous

The other technique used to characterise nanocomposites is that of transmission electron microscopy (TEM). When nanocomposites have formed, the intersections of the silicate sheets are seen as dark lines which are the cross-sections of the silicate layers, measuring 1 nm thick.[8,11] Sometimes other analytical techniques like differential scanning calorimetry (DSC), thermal gravimetric analysis (TGA) and (Fourier Transform infrared) (FTIR) spectroscopy are also used to characterise polymer-nanocomposite structures by comparing the results for polymer alone and polymer–nanocomposite structures.[8,18] Porter et al[8] have used X-ray diffraction with pyrolysis-mass spectrometry to analyse interlayer bonding in nanocomposites.

6.6   Properties of nanocomposites

The Toyota research group first observed that exfoliation of layered silicates in nylon 6 greatly improved the thermal, mechanical and barrier properties of the polymer.[20,29] The nylon 6 layered-silicate nanocomposites with a silicate mass fraction of 5% exhibited increases of 40% in tensile strength, 68% in tensile modulus, 60% in flexural strength and 126% in flexural modulus. The heat distortion temperature showed an increase from 65 to 152 °C.[14,20,29] These composites are now used in under-the-bonnet applications in the automobile industry.[4]

After this success the nanocomposite technology has been extended to other polymers – polypropylene,[11] polystyrene,[14] polyimide,[16] epoxy[4,10] and unsaturated polyester resins,[6] which showed similar results. For example, aliphatic amine-cured, epoxy nanocomposites containing 15 wt% (7.5 vol%) organoclay showed more than a ten-fold increase in tensile strength and modulus.[10] The toughness properties of montmorillonite–unsaturated polyester nanocomposites showed significant improvement.[6] The nanocomposite containing only 1.5 vol% clay exhibited the fracture energy 138 J/m^2 compared to 70 J/m^2 for pure unsaturated polyester.[6] These polymer-layered silicates often exhibit increased thermal stability and hence reduced flammability as discussed in the following section.

6.6.1 Flame-retardant properties of nanocomposites

6.6.1.1 Thermal stability

Nanocomposites were first reported in literature by Blumstein in 1961[1] and his subsequent studies in 1965[21] also demonstrated the improved thermal stability of polymethyl methacrylate (PMMA) – layered silicate nanocomposites. These PMMA nanocomposites were prepared by free radical polymerisation of methyl methacrylate (MMA). TGA results showed that both linear PMMA and cross-linked PMMA intercalated into Na$^+$ montmorillonite have 40 to 50 °C higher decomposition temperatures than does the respective PMMA. According to Blumstein[21] the stability of PMMA–nanocomposite is due not only to its particular structure but also to steric factors restricting the thermal motion of the segments sandwiched between the two lamellae. The unzipping of the chain starts when the temperature is high enough to bring about this motion.

Similar results were shown for dimethyl siloxane and polyimide nanocomposites by the research group at Cornell University.[2,3,13] PDMS–nanocomposite was prepared by melt intercalation of silanol-terminated PDMS into dimethyl ditallow ammonium treated montmorillonite.[13] PDMS-nanocomposite containing only 10% mass fraction of clay showed an increase of 140 °C in decomposition temperature compared to pure PDMS elastomer, which normally decomposes into volatile cyclic oligomers. For PDMS–nanocomposites the permeability also decreased dramatically, hence the increased thermal stability was attributed to hindered diffusion of volatile decomposition products within the nanocomposite.

The TGA data for several aliphatic polyimide nanocomposites (both intercalated and delaminated structures) containing different organically modified montmorillonite and fluorohectorite clays[2] also showed higher decomposition temperatures, indicating their increased thermal stabilities.

Table 6.1 The residual char yields from isothermal
TGA experiments at different temperatures for PEI
and PEI-intercalated nanocomposites[2]

	Residual char yield (%)	
	PEI	PEI-intercalated nanocomposites
At 450°C		
after 20 minutes	45	90
after 120 minutes	15	45
At 500°C		
after 40 minutes	0	55

The intercalated nanocomposites showed much higher char yields than
any of the other (polymer, immiscible and delaminated) systems. The
residual weights after different times at different temperatures from
isothermal TGA experiments for PEI and PEI-intercalated nanocomposites are given in Table 6.1. There was, however, no difference between the
montmorillonite- and fluorohectorite-based nanocomposites containing
the same nanostructure, suggesting that the particle size of the silicates
is not an important factor. Thermal stability was independent of the
cation in the organosilicate with the nanostructure again being the predominant variable.

Both polyimide[2] and PDMS nanocomposites,[3] when exposed to an open
flame, stopped burning after the flame was removed. They retained their
integrity and this contrasted with the respective pure polymer which kept
burning. It is considered that the silicate layer probably acts as a barrier
inhibiting gaseous products from diffusing to the flame and shielding the
polymer from the heat flux, leading to self-extinguishing behaviour.

6.6.1.2 Flammability reduction

The other research group that is very active in this field is at the US
National Institute of Standards and Technology (NIST). The workers there
have prepared polypropylene-graft maleic anhydride (PPgMA), polystyrene (PS),[14] nylon 6, nylon 12, vinyl ester and epoxy-layered silicate
nanocomposites using montmorillonite and fluorohectorite.[22] Workers
studied flammability properties by cone calorimetry. They observed a
common mechanism of flammability reduction and found that the type of
layered-silicate studied, the level of dispersion and processing degradation

Table 6.2 Selected cone calorimetry results for polymer-layered silicate nanocomposites at 35 kW/m² heat flux[22]

Sample	Peak HRR[a] kW m⁻²	Av HRR kW m⁻²	Mean H_c[b] MJ/kg	Residue yield (%)
Nylon 6	1010	603	27	1
Nylon 6–nanocomposite 2% (delaminated)	686	390	27	3
Nylon 6–nanocomposite 5% (delaminated)	378	304	27	6
Nylon 12	1710	846	40	0
Nylon 12–nanocomposite 2% (delaminated)	1060	719	40	2
PS	1120	703	29	0
PS silicate – mix 3% (immiscible)	1080	715	29	3
PS silicate – nanocomposite 3% (intercalated)	567	444	27	4
PS – DBDPO/Sb₂O₃ 30%	491	318	11	3
PPgMA	2030	861	38	0
PPgMA–nanocomposite 5% (intercalated)	922	651	37	8
Mod-bis–A vinyl ester (A)	879	598	23	0
A – nanocomposite 6% (intercalated)	656	365	20	8
Bis-A/novolac vinyl ester (B)	977	628	21	2
B – nanocomposite 6% (intercalated)	596	352	20	9

[a] HRR = Heat release rate.
[b] H_c = Heat of combustion.

influenced the magnitude of flammability reduction.[14] Selected cone calorimetry results at 35 kW m⁻² are given in Table 6.2.

Cone calorimetry results indicated that for nanocomposites peak and average heat release rates (HRR) were considerably reduced. While char yields are not significantly increased (after taking into account silica present in the residue), specific heat of combustion (H_c), specific extinction areas (SEA) and CO yields were unchanged. From this the workers concluded that the source of improved flammability properties of these materials is due to differences in condensed-phase decomposition processes and not to a gas-phase effect. For comparison they also studied polystyrene flame retarded by decabromodiphenyloxide (DBDPO) and Sb₂O₃, which, because it functions in the gas phase, showed lower heat of combustion and higher

CO yields typical of incomplete combustion. They also studied charred residues with TEM and XRD analysis, which indicated a common mechanism of fire retaradancy in all the materials.

6.7   The mechanism of flame retardance in nanocomposites

According to Gilman et al[22] the nanocomposite flame-retardant mechanism is a consequence of high performance carbonaceous–silicate char build-up on the surface during burning, which insulates the underlying material and slows down the mass loss rate of decomposition products. This residue layer forms as the polymer burns away and the silicate layers reassemble. Since there was little improvement in residue yields, once the presence of silicate is accounted for, this indicated that reduced flammability of these materials is not via retention of a large fraction of carbonaceous char in the condensed phase.

Gilman and his coworkers also studied the effect of nanocomposite structure on the flammability of polystyrene-layered silicates and found that only delaminated polystyrene-nanocomposites have reduced flammability[14] whereas, for epoxy and vinyl esters, the intercalated structure produced such an effect.[22] Furthermore, the flammability of polystyrene-nanocomposites was also affected by processing conditions during their preparation. When polystyrene-nanocomposites were prepared via melt blending in an extruder (at 170 °C under nitrogen or vacuum) or by solvent (toluene) blending, the nanostructure had reduced flammability. However, if the extrusion conditions included high temperatures and air was not excluded, the nanocomposite formed showed no improvement in the flammability.[14]

Zhu and Wilkie[18] prepared nanocomposites of polystyrene with several organophilic clays at various levels of clay loading using a bulk polymerisation technique. The resulting nanocomposites showed enhanced thermal stability by TGA and cone calorimetry and even 0.1 % clay loading reduced peak heat release by 40%. However, the char yields as determined from both TGA and cone calorimetry were not much enhanced, which is in agreement with the results of Gilman et al.[22]

Lomakin et al[23] and Ruban et al[24] have studied combustion behaviour of PVA, polystyrene and nylon 6 intercalated nanocomposites. They also observed that triphenylphosphine (TPP), which itself is a very effective flame retardant, when intercalated using kaolin (TPP-i), became even more effective. TPP works in the gas phase releasing P$^\bullet$, which acts as a free radical trap in the gas phase. But when TPP is intercalated, the mechanism of degradation of the TPP changes to the condensed phase. The presence of char formation during polymer combustion gave evidence of solid phase cross-linking and aromatisation reactions, which occur normally in TPP-i-

Table 6.3 LOI and char yield values from TGA experiments at 600 °C for polymer-layered nanocomposites[23]

Sample	LOI	Char yield at 600 °C, % (-clay fr.)*
Polyvinyl alcohol (PVA)	20.7	3.0
PVA–nanocomposite (5% kaolin) (intercalated)	23.7	12.5
Nylon 6	23.0	0
Nylon 6–nanocomposite (5% kaolin) (intercalated)	27.5	6.8
Nylon 6 + 10% kaolin intercalated/modified by TPP (1 : 1)	26.3	
PS	18.0	
PS–nanocomposite (5% montmorillonite) (intercalated)	23.0	
PS–7% of kaolin intercalated/modified by TPP (1 : 1)	30.0	

* Char yield is in addition to clay fraction residue.

PS systems. The combustion properties of some polymer nanocomposites using mineral kaolin are given in Table 6.3.

Bourbigot et al[25] have used the char-forming nylon 6 clay nanocomposite as carbonising agent in an intumescent formulation (an intumescent formulation contains three components – an acid source, carbon source and blowing agent, see Chapter 10). This formulation increased both mechanical and flame retardancy properties of ethylene vinyl acetate (EVA) based materials. They suggest that the clay increases the thermal stabilisation of the phosphocarbonaceous structure in the intumescent char and hence its thermal insulation. The silicate layers also act as thermal barriers.

Sometimes layered silicate–nanocomposites are used in combination with other flame retardants as a means of improving the mechanical properties of the polymer as recently reviewed by Gilman et al.[14] The General Electric Company has used this approach for polybutylene terephthalate (PBT, Valox 315) [26](cited in reference 14). The treated MMT (2%, dimethyl di(hydrogenated tallow) ammonium montmorillonite) in combination with poly(tetrafluoroethylene) (PTFE) dispersed on a styrene–acrylonitrile copolymer (50% PTFE) is used to replace 40% of the brominated polycarbonate–Sb_2O_3 flame retardant in PBT. According to Takekoshi et al,[26] there is a synergistic interaction between PTFE and organo-MMT because, without any of these additives, the flame retardancy is not achieved. It is anticipated that because of the low loading level of the silicate and its

Table 6.4 Percentage elongation-at-break and peak heat release (PHR) values from cone calorimetric studies at 50 kW/m² for polyethylene (PE) samples containing polymer-layered silicates and traditional flame retardants[27] (cited in ref 14)

Sample	PHR (kW/m²)	Elongation (%)
Polyethylene (PE)	1327	980
PE + SBAN N-400 (clay, 10 phr)	687	900
PE + (DBDPO + Sb$_2$O$_3$), (10 phr)	1309	830
PE + (DBDPO + Sb$_2$O$_3$), (15 phr)	1189	720
PE + APP, (10 phr)	1272	590
PE + APP, (15 phr)	989	490
PE + SBAN N-400 (10 phr) + APP, (5 phr)	493	900
PE + SBAN N-400 (10 phr) + phenyl phosphate, (5 phr)	543	930

PE = polyethylene; SBAN N-400 is an organic modified layered silicate; DBDPO = decabromo diphenyl oxide; Sb$_2$O$_3$ = antimony trioxide; APP = ammonium polyphosphate; and phr = parts per hundred resin

nanocomposite characteristics, the mechanical properties of the system are also improved.

Okada has shown the effect of organic modified layered silicates in polyethylene in combination with a variety of conventional flame retardants and metal oxides [27](cited in reference 14). As an example, peak heat release (PHR) values from cone calorimetry at 50 kW/m² heat flux for polyethylene and polyethylene containing either one type of clay (SBAN N-400) and/or a traditional flame retardant are given here in Table 6.4. The peak heat release for polyethylene containing 10 parts per hundred resin (phr) clay is about 50% lower than the value for the control sample, whereas, the same level of DBDPO + Sb$_2$O$_3$ and of ammonium polyphosphate did not have much effect on the peak value. However, when 5% ammonium polyphosphate or phenyl phosphate is added to a polyethylene sample containing 10 phr clay, the peak heat release value is reduced further by 30% and 20%, respectively. Furthermore, the mechanical properties observed by percentage elongation-at-break of these samples remain the same compared to samples containing an equivalent loading (15 parts per hundred resin) of conventional flame retardants, where the breaking elongation values are reduced by half.

Inoue and Hosokawa [28](cited in ref 14) have used melamine salts and polymer-layered silicates to render polymeric (nylon 6, PBT, polyoxymethylene (POM) and polyphenylene sulphide (PPS)) composites flame retardant. Melamine salts were pre-intercalated into the synthetic silicate, fluorinated-mica (FSM) and 10 to 15% total mass fraction of additives were used. In the resulting products, along with enhancement of flame retardant properties, the bending modulus and heat distortion temperatures also increased.

6.8   Conclusion

Polymer-layered silicate nanocomposites may be considered to be environmentally friendly alternatives to some traditional flame retardants. Not only does this fact give rise to a promising means of producing flame-retarding polymers, but it does not have the usual drawbacks associated with other additives. For instance, relatively low concentrations of silicates are necessary compared with the amounts used for conventional additive flame retardants in order to achieve similar or indeed, superior levels of flame retardancy. Moreover, polymer nanocomposites can be processed with normal techniques used for polymers like extrusion, injection moulding and casting. Furthermore, the physical properties of the polymer are not degraded but are greatly improved. An additional advantage is that during combustion, the silicates remain intact at very high temperatures and act as insulating layers against the heat. They slow down the release of volatile decomposition products from the polymer, and thus impart a self-extinguishing character to it.

Clearly, nanocomposites offer novel means of developing flame-retardant polymeric materials and increased research in this area continues to signify their potential.

References

1 Blumstein A, 'Etude des polymerisations en couche adsorbee I', *Bull Chim. Soc.*, 1961, 899.
2 Lee J, Takekoshi T and Giannelis E P, 'Fire retardant polyetherimide nanocomposites', *Mat. Res. Soc. Symp. Proc.*, 1997, **457**, 513.
3 Giannelis E P, 'Polymer layered silicate nanocomposites', *Adv. Mater.*, 1996, **8** (1), 29.
4 Wang Z and Pinnavaia T J, 'Hybrid organic–inorganic nanocomposites: exfoliation of magadiite nanolayers in an elastomeric epoxy polymer', *Chem. Mater.*, 1998, **10**, 1820.
5 Raiswell R W, Brimblecombe P, Dent D L and Liss P S, *Environmental Chemistry*, Edward Arnold, London, 1992, Chapter 3.
6 Kornmann X, Berglund L A, Sterte J and Giannelis E P, 'Nanocomposites based on montmorillonite and unsaturated polyester', *Polym. Eng. Sci.*, 1998, **38** (8), 1351.
7 Burkett S L, Press A and Mann S, 'Synthesis, characterization and reactivity of layered inorganic–organic nanocomposites based on 2:1 trioctahedral phyllosilicates', *Chem. Mater.*, 1997, **9**, 1071.

8   Porter D, Metcalfe E and Thomas M J K, 'Nanocomposite fire retardants – a review', *Fire Mater.*, 2000, **24**, 45.

9   Krishnamoorti R and Giannelis E P, 'Rheology of end-tethered polymer layered silicate nanocomposites', *Macromolecules*, 1997, **30**, 4097.

10 Lan T and Pinnavaia T J, 'Clay-reinforced epoxy nanocomposites', *Chem. Mater.*, 1994, **6**, 2216.

11 Kawasumi M, Hasegawa N, Kato M, Usuki A and Okada A, 'Preparation and mechanical properties of polypropylene-clay hybrids', *Macromolecules*, 1997, **30**, 6333.

12 Vaia R A, Ishii H and Giannelis E P, 'Synthesis and properties of two dimensional nanostructures by direct intercalation of polymer melts in layered silicates', *Chem. Mater.*, 1993, **5**, 1694.

13 Burnside S D and Giannelis E P, 'Synthesis and properties of new poly(dimethylsiloxane) nanocomposites', *Chem. Mater.*, 1995, **7** (9), 1597.

14 Gilman J W, Jackson C L, Morgan A B, Harris R, Manias E, Giannelis E P, Wuthenow M, Hilton D and Phillips S, 'Flammability properties of polymer layered-silicate (clay) nanocomposites' *Flame Retardants 2000*, London, Interscience, 2000, 49.

15 Hasegawa N, Kawasumi M, Kato M, Usuki A and Okada A, 'Preparation and mechanical properties of polypropylene–clay hybrids using a maleic anhydride-modified polypropylene oligomer', *J. Appl. Polym. Sci.*, 1998, **67**, 87.

16 Yano K, Usuki A, Okada A, Karauchi T and Kamigaito O, 'Synthesis and properties of polyimide–clay hybrid', *J. Polym. Sci. Part A: Polym. Chem.*, 1993, **31**, 2493.

17 Jeon H G, Jung H-T, Lee S W and Hudson S D, 'Morphology of polymer/silicate nanocomposites. High density polyethylene and a nitrile copolymer', *Polymer Bulletin*, 1998, **41**, 107.

18 Zhu J and Wilkie C, 'Thermal and fire studies on polystyrene–clay nanocomposites', *Polymer International*, 2000, **49** (10) 1158.

19 Moet A S and Akelah A, 'Polymer–clay nanocomposites: polystyrene grafted onto montmorillonite interlayers', *Mat. Lett.*, 1993, **18**, 97.

20 Kojima Y, Usuki A, Kawasumi M, Okada A, Kurauchi T and Kamigaito O, 'Synthesis of nylon 6–clay hybrid by montmorillonite intercalated with ε-caprolactam', *J. Polym. Sci. Part A: Polym. Chem.*, 1993, **31**, 983.

21 Blumstein A, 'Polymerization of adsorbed monolayers. II. Thermal degradation of the inserted polymer', *J. Polym. Sci: Part A*, 1965, **3**, 2665.

22 Gilman J W, Kashiwagi T, Nyden M, Brown J E T, Jackson C L, Lomakin S, Giannelis E P and Manias E, 'Flammability studies of polymer layered silicate nanocomposites: polyolefin, epoxy and vinyl ester resins', Chapter 14 in Al-Malaika S, Golovoy A and Wilkie C A (eds), '*Chemistry and Technology of Polymer Additives*', Blackwell Science, Oxford, UK, 1999.

23 Lomakin S M, Usachev S V, Koverzanova E V, Ruban L V, Kalinina I G and Zaikov G E, 'An investigation of thermal degradation of polymer flame retardant additives: triphenylphosphine and modified/intercalated trimethylphosphine', 10th annual conf. *Recent Advances in the Fire Retardancy of Polymeric Materials*, Business Communication Co, Norwalk, USA, 1999.

24 Ruban L, Lomakin S and Zaikov G, 'Polymer nanocomposites with participation of layer aluminium silicates' in Zaikov G E and Khalturinski N A (eds),

'*Low Flammability Polymeric Materials*', Nova Science Publishers, New York, 1999.

25 Bourbigot S, Bras M L, Dabrowski F, Gilman J W and Kashiwagi T, 'New development for using PA-6 nanocomposite in intumescent formulations' 10th annual conf. '*Recent Advances in Flame Retardancy of Polymeric Materials*', Lewin M (ed.), Proceedings of the 1999 Conference, Business Communication Co, Norwalk, USA, 1999.

26 Takekoshi T, Fouad F, Mercx F P M and De Moor J J M, US Patent 5 773 502. Issued to General Electric Company, 1998.

27 Okada K (Sekisui) Japan Patent 11-228 748, 1999.

28 Inoue H, Hosokawa T, Japan Patent Application (Showa Denko K K, Japan) Jpn. Kokai tokkyo koho JP 10 81 510 (98 81 510), 1998.

29 Kojima Y, Usuki A, Kawasumi M, Okada A, Kurauchi T and Kamigaito O, 'Fine structure of nylon-6–clay hybrid', *J. Polym. Sci. Part B: Polym. Phys.*, 1994, **32**, 625.

7

Recent developments in flame-retarding thermoplastics and thermosets

PAUL JOSEPH AND JOHN R. EBDON

The Polymer Centre
Lancaster University
Lancaster, UK

7.1   Introduction

Organic polymers degrade to give volatile combustible products when they are heated above certain critical temperatures, which in turn depend on their chemical structures. If the gaseous mixture resulting from the mixing of degradation volatiles with air is within the flammability limits, and the temperature is above the ignition temperature, then combustion begins. The combustion of a polymeric material is a highly complex process involving a series of interrelated and/or independent stages occurring in the condensed phase and the gaseous phase, and at the interfaces between the two phases.[1]

Fires involving organic polymers lead essentially to the same hazards as those fed by other fuels. The flammability and destruction of property are not the only problem. More important are fire fatalities due to the evolved smoke and toxic gases, exacerbated in some cases by poisonous fumes emitted from the synthetic organic polymers.[2,3] This has led to the introduction of stricter legislation and safety standards concerning flammability, and extensive research into the area of flame retardants for polymers has been the result.

Real fires involve not only substantially 'pure' polymers but also polymeric materials that contain considerable amounts of other constituents, such as artificially introduced fillers or plasticisers or naturally associated non-polymeric substances. Furthermore, polymers involved in a fire may be in the form of fabricated articles containing other materials; these products can take a variety of shapes, sizes and forms. Indeed, the whole of the surrounding system containing combustible articles in which a fire occurs is another aspect that needs to be considered. Therefore, in dealing with the combustion of organic polymers, in the broadest sense, it is necessary to consider the combustion behaviour of a wide range of polymers, and not to confine attention to those which are normally regarded as flammable.

The following sections review the literature that has appeared during the last ten years or so on the thermal degradation and flammability of commercially important thermoplastics and thermosets. The article concentrates in particular on the strategies adopted for improving the flame retardance of these polymers. The review does not, however, cover the patent literature on flame-retardant additives in commercial use; instead, priority has been given to describing research and development aimed at improving the inherent flame retardance of polymeric materials through modifications to chemical microstructures.

7.2   Thermoplastics versus thermosets

There are several ways of classifying polymers which are not mutually exclusive.[4] Among the criteria that can be used for the classification of polymers are the following:

(a)   origin – natural, synthetic and semi-synthetic;
(b)   physical properties – elastomers, plastics and fibres;
(c)   polymerisation mechanism – chain-growth and step-growth;
(d)   stereochemical configuration – syndiotactic, isotactic and atactic;
(e)   nature of the molecular chain – carbon, carbon–oxygen, carbon–nitrogen and wholly heteroatom chains.

When classifying polymers on the basis of physical properties, generally the values of the elastic modulus and the degree of elongation are chosen, and accordingly polymers fall into three main classes, namely, elastomers, plastics and fibres. Depending on the thermal behaviour, plastics are subdivided into thermoplastics and thermosets. Thermoplastic materials can undergo indefinite inelastic deformation at elevated temperatures that are above their melting or glass transition temperatures but which are not high enough to produce chemical decomposition, and they normally contain long individual chains that are not appreciably cross-linked. Thermosetting materials are, on the other hand, much less 'plastic' and on being heated (cured) undergo irreversible chemical changes which lead to substantially infusible products with strong three-dimensionally cross-linked chains; their molecular weights prior to curing are usually relatively low. Some commercially important plastics are listed in Table 7.1.

7.3   Factors affecting flammability and its reduction

7.3.1   General considerations

Successful strategies to reduce the flammability of a polymeric material involve interrupting the complex stages of the combustion process at one

Table 7.1 Some commercially important thermoplastics and thermosets

Polymer type	Example	Representative structural repeat unit
I. Thermoplastics		
Polyolefins	Polyethylene	$-CH_2-CH_2-$
	Polypropylene	$-CH_2-CH-$ with CH_3 branch
	Polybutadiene	$-CH_2-CH=CH-CH_2-$
Styrenics	Polystyrene	$-CH_2-CH-$ with phenyl group
Halogenated vinylics	Poly(vinyl chloride)	$-CH_2-CH-$ with Cl
	Polytetrafluoroethylene	$-CF_2-CF_2-$ (F, F / C-C / F, F)
Acrylics	Polyacrylonitrile	$-CH_2-CH-$ with CN
	Poly(methyl methacrylate)	$-CH_2-C-$ with CH_3, $C=O$, O, CH_3
Saturated polyesters	Poly(ethylene terephthalate)	$-O-CH_2-CH_2-O-C(=O)-$ ⬡ $-C(=O)-$
	Poly(butylene terephthalate)	$-O-CH_2-CH_2-CH_2-CH_2-O-C(=O)-$ ⬡ $-C(=O)-$
Polyamides	Nylon-6	$-NH(CH_2)_5CO-$
	Nylon-6,6	$-NH(CH_2)_6NHCO(CH_2)_4CO-$
	Nylon-11	$-NH(CH_2)_{10}CO-$
II. Thermosets		
Aminoresin	Urea-formaldehyde	
Phenolic resin	Phenol-formaldehyde	
Epoxy resin	Uncured – based on epichlorohydrin and bisphenol A	

Table 7.1 (cont.)

Polymer type	Example	Representative structural repeat unit
Polyurethane	General – based on linear polyols and diisocyanates	
Unsaturated polyesters	Uncured	

or more points to reduce the rate and/or change the mechanism of combustion at that point.[5] From a practical point of view, this is achieved either by the mechanical blending of a suitable flame-retardant compound with the polymeric substrate (i.e. by introducing an additive) or by the chemical incorporation of the retardant into the polymer molecule by simple copolymerisation or by chemical modification of the preformed polymer (i.e. using a reactive component).

Currently, synthetic polymers are usually made more flame retardant by incorporating additives. Such additives often have to be used at high loadings to achieve a significant effect, e.g. 30% by weight or more, which occasionally can have a more detrimental effect on the physical and mechanical properties of a polymer than that produced by reactive flame retardants. Nevertheless, additives are more generally used as they are often cheaper and more widely applicable.

Both additives and reactives can interrupt the burning cycle of a polymer in several ways: by altering the thermal decomposition mechanism of a polymer; by quenching the flame; or by reducing the heat transferred from the flame back to the decomposing polymer. Basically, there are two fundamental modes of action for flame-retardant compounds, namely vapour-phase inhibition and condensed-phase inhibition. If the radical intermediates of the combustion process that exist in the gas phase are intercepted by a flame retardant and converted to less reactive species, it is said to exert vapour-phase inhibition. On the other hand, if a flame retardant and/or its pyrolytic product(s) affect the solid-state degradation mechanism of a polymeric substrate in such way as to reduce the supply of flammable volatiles into the flame zone, thus interrupting the combustion cycle at this point, it is said to exert a condensed-phase action. The polymer breakdown may be accelerated by the flame retardant causing pronounced flow of the polymer, hence withdrawal from the sphere of influence of the flame. The flame retardant can also cause a layer of carbon to appear on the surface. This may occur through a dehydrating action of the flame retardant, generating unsaturation in the polymer. These unsaturated structures form a carbonaceous layer by cyclising and cross-linking.

Successful additives include (a) polyhalogenated hydrocarbons, which improve flame retardance by liberating halogen atoms that retard gas-phase chain oxidation reactions, (b) hydrated metal salts that decompose endothermically in a fire, thus reducing the overall heat of reaction, and which liberate water, which dilutes the flammable gases, and (c) phosphorus-based inorganic and organic additives, which promote the formation of an incombustible char, thus protecting the underlying, unburnt polymer.

The alternative method of flame-retarding a polymer, namely by chemical modification, has several potential advantages such as (a) that low levels of modification may suffice, (b) that the modifying groups are chemically attached and therefore are less likely to be lost during subsequent service, and (c) that the modification can more readily be molecularly dispersed throughout the polymer.

The selection of a flame retardant for a polymeric material is governed by the chemical and physical properties of the polymer (i.e. compatibility), the degradation characteristics of the polymer, and the chemical and physical properties of the flame-retardant composition. Ideally, the flame-retardant polymer system must have a high resistance to ignition and flame propagation, a low rate of heat release on combustion, and low smoke generation. It must also give little or no additional toxic vapour on burning and must retain its flame-retardant properties during normal usage. It must also be easily processed and acceptable in appearance with minimum cost of application.

It is common practice, especially from a commercial point of view, to use a combination of flame retardants for polymeric materials. In many cases, these flame-retardant mixtures can give an enhanced performance at low cost. The interaction of antimony, most commonly used as antimony oxide, with halogenated polymers or polymers containing halogenated additives, gives rise to a classic case of flame-retardant synergism.[6,7] The synergistic effects of phosphorus–nitrogen[8,9] and phosphorus–halogen[10,11] are also well-documented. Practical experience has led to the recognition of several useful combinations of flame-retardant ingredients, and these are frequently employed for flame-retarding commercially important plastics.

7.3.2 Relationship between flame retardance and structure

When subjected for a sufficient length of time to an external heat source, organic polymeric materials undergo thermal degradation, generating various products in varying concentrations over different temperature ranges. The chemical steps leading to the formation of volatiles may be homolytic or heterolytic, i.e. be radical or ionic. The three overall processes implicated in the thermal degradations of most thermoplastic polymers are as follows:

1   Random chain cleavage followed by chain unzipping is characterised by high monomer yields and a slow decrease in molecular weight of the polymer, e.g. poly(methyl methacrylate), poly(α-methyl styrene), polystyrene, polytetrafluoroethylene.

2   Random chain cleavage followed by further chain scission is characterised by very low monomer yields among the volatile degradation products and a rapid drop in molecular weight, e.g. polyethylene, polypropylene, poly(methyl acrylate), polychlorotrifluoroethylene.

3   An intra-chain chemical reaction followed by a cross-linking reaction and formation of a carbonaceous residue, or random chain cleavage. This generates a relatively high yield of volatiles from the intra-chain reaction, but produces little monomer, and produces no, or only a very slight, reduction in molecular weight during the initial stages of degradation, e.g. poly(vinyl chloride), poly(vinyl alcohol), polyacrylonitrile.

In some cases, several of these processes occur simultaneously, depending on sample size, heating rate, pyrolysis temperature, environment, and presence of any additives. Although polymer degradation schemes can be greatly altered by the presence of comonomers, side-chain substituents, and other chemical constitutional factors, the ultimate thermal stability is determined by the relative strength of the main-chain bonds. Many additives and comonomers employed as flame retardants are thermally labile; as a result the thermal stability of the polymer system is reduced. In order to reduce the observed effects of flame-retardant additives on the thermal stability of the polymeric materials, more thermally stable and hence inherently fire-resistant polymers are of increasing interest.

There are two general strategies which have been used to produce more thermally stable polymers. The first is to increase the strength of main-chain bonds. Resonance stabilisation can be achieved by utilising aromatic and heterocyclic ring structures with high resonance stabilisation energies.[12] There are several classes of polymers, such as polyphenylenes, poly(p-phenylene oxide)s, polybenzimidazoles, polypyrones, polybenzamides, that have relatively high thermal decomposition temperatures coupled with low levels of fuel production on degradation.

The second strategy is illustrated by the highly cross-linked three-dimensional network structure of several thermosets. The rupture of bonds does not initially generate combustible gases, and carbonisation is promoted. Ladder polymers, in which main chains are bonded together at each repeat unit by a cross-link, serve the same purpose.

7.3.3 Correlation between charring processes and flammability

The ability to form char is related to the flammability of a polymer.[13,14] The higher the amount of residual char after combustion, the lower the amount of combustible material available to perpetuate the flame and the greater the degree of flame retardance of the material. Therefore, one of the ways to achieve high degrees of flame retardancy or non-combustibility of polymeric materials is to increase the amount of char produced on combustion. This is illustrated by the fact that aromatic polymers, e.g. polycarbonate and poly(phenylene oxide), have lower flammabilities than purely aliphatic polymers. The greater thermal stability of cross-linked and aromatic structures in thermosets gives rise to a greater degree of condensation into aromatic chars, and therefore only relatively low levels of flammable gases are available to feed a flame.

It has been shown that the efficiency of conversion of carbon in the original polymer to carbon in the burnt residue is greater for less flammable polymers.[13] Quinn reported an effect of structural unsaturation upon char formation and hence flammability for copolyterephthalates and copolycarbonates.[15] A general decrease in flammability was noted with an increase in degree of unconjugated unsaturation and char-forming tendency in the polymer. In a comparative study of the flammabilities of polymers of different chemical structures relative to that of poly(ethylene oxide), it was shown that flammability increased with an increase in hydrogen and oxygen contents in the polymer chain.[16] It was also shown that an increase in the carbon to oxygen ratio gives rise to a less flammable material.

The literature dealing with flame retardancy pays considerable attention to char-forming mechanisms in polymeric materials.[17-19] Of particular importance to this area is the work of van Krevelen on the linear correlation between char and flammability parameters, on the 'group contribution to char formation' and on the 'char formation tendency' which is inherent in various groups in the polymer and can be estimated for aliphatic, aromatic and heterocyclic groups.[20,21]

The structural morphology and chemical nature of char residues from burning polymers can lead to invaluable information about the mechanistic aspects and mode of action of flame retardants.[22] Several authors have used scanning electron microscopy to investigate the complex surface features of char residues from the burning of flame-retarded polymeric systems.[23-26] Recently, X-ray photoelectron spectroscopy (XPS) has been used to study chemical bonding, elemental composition and depth profiles of surfaces of intumescent flame retardant systems.[27-30] Fourier transform infrared (FT-IR), solid-state nuclear magnetic resonance (NMR), and electron spin resonance (ESR) spectroscopies are also rapidly emerging as useful tools for the characterisation of solid residues from burning of polymers.[31–33]

7.4 Testing procedures and hazard assessments: general aspects

As a consequence of the complex nature and poor reproducibility of fire, there are many techniques for estimating the flammability characteristics of polymeric materials.[34,35] The most widely used laboratory test is the limiting oxygen index (LOI) technique,[36] a very convenient, precise and reproducible test developed by Simmons and Wolfhard[37] in 1957 and initially extended to polymers by Fenimore and Martin in 1966.[38] The LOI is a measure of the volume percentage of oxygen in a mixed oxygen and nitrogen gas stream that just supports candle-like combustion of a polymer sample. This value therefore enables the combustibility of a polymer to be expressed and compared with that of other materials. Also, the information concerning the type of mechanism involved with a particular flame retardant can be obtained by using a different oxidising medium, such as nitrous oxide, instead of oxygen (NOI test).[39,40] It is assumed that a flame retardant acting in the condensed phase works independently of the chemical oxidant, whereas a change in flammability is observed with a change in oxidant for a flame retardant acting in the vapour phase.[41] LOI values for some representative polymers are given in Table 7.2.

The high concentrations of oxygen used in making measurements of LOI are unrepresentative of a real fire, and generally there is a lack of correla-

Table 7.2 Limiting oxygen indices (LOI) of some polymers

Polymer	LOI	Polymer	LOI
Acetal (polyoxymethylene)	16	Typical polysulphone	33
Polyethylene	18	Typical polyarylate	34
Polypropylene	18	Typical liquid crystal polymer	35
Polystyrene	18	Poly(ether ether ketone)	35
Polyisoprene (natural rubber)	18	Poly(ether sulphone)	38
Poly(methyl methacrylate)	18	Poly(vinyl chloride)	42
ABS	19	Polyamide-imide	43
Poly(butylene terephthalate)	22	Poly(phenylene sulphide)	44
Nylon-6,6	24	Poly(vinylidene fluoride)	44
Poly(ethylene terephthalate)	25	Polybenzimidazole	48
Polychloroprene	26	Typical polyimide	50
Polycarbonate	27	Poly(vinylidene chloride)	60
Poly(phenylene oxide)	32	Polytetrafluoroethylene	95

tion between most of the small-scale tests and full-scale tests. Recently, several novel techniques for measuring a range of properties that correlate well with full-scale tests have been developed.[42] The most useful of these is undoubtedly the cone calorimeter, which enables pyrolysis profiles of polymeric materials to be obtained and important parameters, such as ignition times, overall and maximum rates of heat release, heats of combustion, mass losses and carbon dioxide, carbon monoxide and smoke concentrations, to be determined. The maximum rate of heat release is understood to be the most important measurable parameter concerned with fire hazards and fire scenarios. It controls the rate of burning, the rate of mass loss, and the ignition of the surrounding environment.[43] Attempts to reduce the rate of heat release are now beginning to dominate research into flame retardant and fire behaviour. Some typical (approximate) rates of heat release, culled from a variety of sources, are given in Table 7.3.

There are many different aspects to the burning of polymers which must be taken into account in assessing completely the hazards presented by any particular material. These include ease of ignition, rate of spread of flame, and the production of smoke and toxic and asphyxiant gases (Tables 7.4 and 7.5). In recent years, the latter hazards have become much more widely recognized. They arise because flames involving large amounts of polymer are fuel rich and result in incomplete combustion. This produces large amounts of carbon monoxide, which is the single largest cause of death in fires, and carbon-rich black smoke, which inhibits escape from the burning environment.

Table 7.3 Some representative rates of heat release obtained by cone calorimetry

Polymer	Rate of heat release[a]/$kW\,m^{-2}$
Polypropylene	680
Polystyrene	610
Poly(methyl methacrylate)	600
Polyethylene	590
Nylon-6,6	480
Poly(butylene terephthalate)	470
Poly(ethylene terephthalate)	270
Poly(ether ether ketone)	150
Polycarbonate	90
Polytetrafluoroethylene	70
Polyimide	30
Polybenzobisoxazole	10

[a] Measured after 300 s under an irradiance of $50\,kW\,m^{-2}$.

Table 7.4 Smoke emission on burning of some
polymers (NBS smoke chamber, flaming condition)

Polymer	Maximum smoke density (optical density, D)
ABS	800
Poly(vinyl chloride)	520
Polystyrene	475
Polysulphone	230
Polycarbonate	215
Typical polyamide-imide	169
Polyarylate	109
Polytetrafluoroethylene	95
Phenolic resin	75
Poly(ether sulphone)	35

Table 7.5 Toxic and asphyxiant gases from the combustion of polymers

Gas	Source
CO, CO_2	All organic polymers
HCN, NO, NO_2, NH_3	Wool, silk, nitrogen-containing polymers
SO_2, H_2S, COS, CS_2	Vulcanised rubbers, sulphur-containing polymers, wool
HCl, HF, HBr	PVC, PTFE, polymers-containing halogenated flame retardants
Alkanes, alkenes	Polyolefins and other organic polymers
Benzene	Polystyrene, PVC, aromatic polyesters
Phenol, aldehydes	Phenolic resins
Acrolein	Wood, paper
Formaldehyde	Polyacetals, formaldehyde-based resins
Formic and acetic acids	Cellulosics

7.5   Flame-retardant thermoplastics

7.5.1   Polyolefins: polyethylene, polypropylene and poly(1-butene)

Polyolefins are among the most important polymers in terms of production volume. By copolymerisation of ethylene and propylene with higher n-olefins, cyclic olefins, or polar monomers, product properties can be varied considerably, thus extending the range of possible applications.

Polyethylenes (PE) are manufactured in the largest tonnage of all thermoplastic materials. Several well-established families of polyethylenes are now on the market, each having a different molecular architecture and different behaviour, performance and applications, e.g. low-density polyethylene (LDPE), high-density polyethylene (HDPE), linear low-density polyethylene (LLDPE), very low-density polyethylene (VLDPE), high and ultra-high molecular weight polyethylenes. Other categories of polyethylene polymers include cross-linked, chlorinated or chloro-sulphonated polyethylenes and copolymers of ethylene. The main structural features that determine the properties of PE are the degrees of short- and long-chain branching, the average molecular weight and the polydispersity. The widespread use of PE is due to its excellent electrical insulation properties and chemical resistance, easy processability and low cost. The major applications of polyethylenes (LDPE, LLDPE and HDPE) have been as film for general packaging, in cable insulation, and in the building and agricultural industries. Polyethylene, especially HDPE, is an important injection moulding material for a wide range of products including toys, electrical fittings, containers and household goods.

Polypropylene (PP) also is a thermoplastic material of major importance, ranked third in terms of production volume after PE and poly(vinyl chloride) (PVC). Owing to its cost effectiveness, versatility and excellent environmental aspects, PP is the fastest growing commercial commodity plastic. The isotacticity of PP plays a major role in determining its properties. The non-polar nature of PP endows the material with excellent electrical insulation properties, similar to those of PE. Also, the resistance to most chemicals and solvents is exceptionally high. Around 30% of PP and its related copolymers are used as fibres and filaments, for instance, in carpets and geotextiles. Another large market for PP and related polymers is as film for packaging food and tobacco products.

Another important polyolefinic thermoplastic material is poly(1-butene) (PB). The major commercial products have high molecular weight, and are approximately 99% isotactic. PB exhibits the general properties of a polyolefin. However, the outstanding property of PB is its high creep resistance, which is advantageously exploited in the manufacture of pipes having a much reduced wall thickness compared to PE and PP pipes. The main use of PB pipes is in cold and hot water plumbing, as well as for the transportation of abrasive and corrosive materials in the mining, chemical and power generation industries. Atactic PB is widely used for making roof coverings and also for sealing compounds.

Generally, polyolefins are highly flammable, the principal mechanism of thermal degradation being homolytic chain scission followed by inter- and intra-molecular chain transfer resulting in the formation of volatile fragments. The long-chain fragments and the soot-like products, formed by cyclisation dehydrogenation, contribute to smoke development. Carbon dioxide and water are also formed during combustion. Polyolefins burn readily in air (LOI = 18) with melting and dripping, and produce little or no residual char.

Chlorinated polyethylene (CPE) is produced commercially from both LDPE and HDPE. In addition to increased flexibility, toughness and compatibility with a variety of other polymers, CPE has also an increased resistance to ignition.[44] Similarly, chemical modifications of PE, such as oxidative chlorophosphonylation[45] and radiation grafting of vinyl phosphonate oligomer,[46] result in a significant increase in the flame retardance of the polymer. The flame retardance in the case of phosphorus-modified PE is believed to arise, at least in part, from a condensed-phase mechanism.

Another strategy employed for chemical modification of PE, to improve its oxidative stability and resistance to ignition, is cross-linking the main chain by irradiation with γ rays[47] or with an electron beam.[48] Such cross-linked PEs are extensively used as insulating material in the cable industry and their superior flame retardance is attributed to the formation of a protective carbonaceous layer formed during the initial stages of combustion. Functional groups, mainly carbonyl and hydroxyl, can be introduced into LLDPE by exposure to γ rays in air. LLDPE, thus modified, has increased impact strength as well as improved compatibility towards conventional additive flame retardants such as aluminium trihydrate (ATH).[49] Another method used to modify PE is to form cross-links between polymer chains using low molecular-weight silane oligomers.[50] Silane-modified PEs are often used as an additive with unmodified PE flame retarded with metal hydroxides. Chlorosulphonated PE, by virtue of the chlorine atoms, is significantly more flame retarded than PE. Substantial improvements in LOI along with considerable reduction in smoke production is observed with chlorosulphonated PE containing tin compounds.[51] The improved fire performance of such systems is believed to arise from the formation of tin(II) chlorosulphonate and mixed-valence tin complexes during combustion.[52]

Chemical modification to impart flame retardance is less successful with PP and PB owing to substantial main-chain degradation encountered on treatment with common modifying reagents. Nevertheless, there are a few instances where chlorinated PP has been used as a flame retardant additive for polyolefins. In general for polyolefins, the use of additive flame retardants is a more common practice, mostly driven by commercial considerations. Halogen-containing additives are widely used, and they include chlorinated paraffins,[53] PVC or other chlorine- or bromine-containing aliphatic compounds such as hexabromocyclododecane (HBCD),[54] polybrominated dibenzofurans,[55] 3-(tetrabromopentadecyl)-2,4,6-tribromophenol,[56] and tris(2,3-dibromopropyl isocyanurate).[57] Bromine is more effective than chlorine as a constituent of flame retardants, and aliphatic compounds generally have a greater effect than aromatic compounds. Antimony oxide greatly increases the flame-retardant action of halogens, and antimony-halogen systems are widely used for reducing the flammability of polyolefins.[57,58] Another instance of metal–halogen synergism is the use of bismuth compounds in conjunction with chloroparaffins as flame-retardant additives for PP.[59,60] More conventional flame retardant additives such as ATH, Mg(OH)$_2$ and MgCO$_3$ are also still widely used for flame retarding polyolefins.[50,61–63]

The use of intumescent systems, largely consisting of ammonium polyphosphate and pentaerythritol, as efficient flame-retardant additives for PE, PP and PB, are well documented in the literature.[64–67] More recently, metallocene catalysed olefin copolymers and microencapsulated fire-retardant compounds were tried as novel ways of improving the fire performance of polyolefin polymers.[68,69]

7.5.2 Polystyrenes: PS, HIPS and ABS

Polystyrene (PS) is a hard, rigid, transparent thermoplastic polymer, which is substantially linear. Because of its relatively low cost, good mouldability, low moisture absorption, good dimensional stability and electrical insulation properties, it is widely used as an injection moulding and vacuum forming material. The low thermal conductivity has been made use of in polystyrene foam for thermal insulation. The principal limitations of the polymer are its brittleness, low resilience, relatively low thermo-oxidative stability, and flammability. More recently, polystyrene derivatives are being used in addition to PS and in replacing it.

Styrene forms, through copolymerisation with vinylic and diene monomers, a variety of copolymers and terpolymers. The important classes of such polymers are: styrene-acrylonitrile (SAN); styrene-butadiene (SB); styrene-divinyl benzene (S-DVB); acrylonitrile-butadiene-styrene (ABS) and rubber-modified or high impact polystyrene (HIPS). The ABS group possesses a very good combination of mechanical, thermal and electrical properties, as well as good chemical stability. Ease of processing, relatively low cost, processability, and ability to design grades to meet the requirements of a particular application are other important advantages of these polymers. The polymerisation of styrene in the presence of polybutadiene using radical initiators leads to HIPS, a rubber-modified PS. The impact strength of HIPS increases with rubber particle size and concentration, while gloss and rigidity decrease. The microstructure of polybutadiene has an important influence on properties of HIPS, and a 36% cis-1,4-polybutadiene provides optimal properties. Untreated polystyrenes are highly flammable (LOI = 19) because on heating they volatilise almost quantitatively, producing a higher proportion of monomers

(~50% or more) and lesser amounts of dimer, trimer, etc. These volatile products are formed rapidly, with a maximum rate at around 400 °C, and by about 450 °C the polymer is almost completely decomposed, leaving no residue.

Post-polymerisation modification of PS can be easily achieved through electrophilic aromatic substitution of the phenyl rings. By anchoring appropriate flame-retardant group(s) on the phenyl rings, a considerable degree of flame retardance can be imparted to styrenic polymers. Successful modification methods include boronation[70] silylation[71] and phosphorylation.[72] The resulting modified polymers are significantly more flame retardant than the unmodified polymer; the principal mode of flame-retardant action of these modifying groups involves a condensed-phase mechanism. Also, halogenation, either by free halogen or other halogen-containing reagents, or even processes such as irradiation, have been employed. In a recent report, the chemical resistance, thermal stability and fire performance of PS crosslinked with polyester resins of phthalic acid were investigated.[73]

However, functionalisation of linear PS using post-polymerisation methods often results in some degree of cross-linking, and therefore is not considered to be a satisfactory commercial method for achieving flame-retardancy in styrenic polymers. Because of this, considerable attention has been devoted to exploring the relatively straightforward method of chemical modification via copolymerisation of styrene with unsaturated compounds bearing flame-retardant groups. Most commonly used comonomers are phosphorus[74–76] and halogen[77–79] containing compounds. In addition to superior fire performance, flame-retardant synergistic effects were also noted in some of these modified polymers. In a recent report,[80] the effects of transition-metal complexes on the flame retardance of poly(styrene-co-4-vinyl pyridine) has been described. The LOIs of the modified polymers were significantly higher than those of the parent copolymers, and the production of considerable amounts of rigid, intumescent char suggests a predominantly condensed-phase mechanism of flame retardance.

Polystyrenes and related polymers like HIPS and ABS are more often flame retarded using additive since these are more cost effective, and are relatively easy to apply and process. The flame-retardant and synergistic effects of decabromodiphenylether–Sb_2O_3 have been the subject of a number of studies.[81–83] The principal flame-retardant mechanism in such systems involves the debromination of the flame retardant, decabromodiphenylether, to form less brominated diphenylethers, brominated polystyrene, and antimony bromides and antimony oxybromides. Triphenyl phosphate is often used as co-additive in flame-retardant formulations for HIPS.[84,85] The enhanced performance of these multicomponent flame-retardant formulations can be attributed to the effect of triphenyl phosphate in promoting char formation during combustion. The use of resorcinol bis(diphenyl phosphate),[86] chlorinated polyethylene[87] and fluorinated additives such as PTFE,[88] to improve the fire performance of HIPS has also been reported. PS and ABS are efficiently flame retarded with additives such as decabromodiphenylether–Sb_2O_3[88] and tetrabromobisphenol-A.[89] The synergistic effects of $Mg(OH)_2$ and $CaCO_3$ in flame-retarded ABS have been investigated.[90,91] The smoke-suppressing effects of iron compounds in ABS–PVC and ABC–PVC–PP blends was noted, in addition to enhanced flame retardation.[92,93] Blends of ABS with polycarbonate were successfully flame retarded using resorcinol bis(diphenyl phosphate),[94] triphenyl phosphate,[95] and phosphorus–bromine systems.[96–98]

7.5.3 Acrylics: poly(methyl methacrylate) and polyacrylonitrile

Acrylic polymers are obtained from derivatives of acrylic and methacrylic acids; the group includes also their copolymers with various vinylic and allylic monomers. Monomers commonly used in the production of these polymers are acrylonitrile, acrylic and methacrylic acids, and their amide and alkyl ester derivatives. The largest applications of acrylic polymers in terms of tonnages used are in moulded and fabricated plastic articles of many kinds made from poly(methyl methacrylate) (PMMA). The crystal clarity, light weight, outstanding weather resistance, formability and strength of PMMA have resulted in numerous applications in different technical fields and in many domestic products. Since PMMA is odourless, tasteless and non-toxic, it may be used in food-handling equipment.

Polyacrylonitrile (PAN), and copolymers with acrylonitrile in a predominant amount, are white powders having relatively high glass transition temperatures, T_g. However, they have a low thermal plasticity and cannot therefore be used as a moulding material. Their high crystalline melting points, T_m (~300 °C), limited solubility in certain solvents, and superior mechanical properties when used as fibres are due to the intermolecular forces between polymer chains. Staple acrylic fibres, being soft and resilient, are used as a substitute or diluent for wool, and fabrics made from them show good crease resistance and crease retention. PAN is also the most important raw material for the production of carbon fibres.

On heating, PMMA undergoes extensive chain unzipping or depolymerisation to produce a quantitative yield (>90%) of monomer and is, as a consequence, highly flammable (LOI = 18). The oxygen of the ester group assists complete combustion of the pyrolysis products and is the reason for the low smoke production in the burning polymer. The material melts and volatilises so that no residue remains. Acrylic fibres also burn readily

(LOI = 18) with melting and sputtering. The rate of burning and the amount of smoke produced depend on the acrylonitrile content of the fibre.

Generally, post-polymerisation chemical modification of PMMA with flame-retardant groups is far more difficult than in the case of PS, owing partly to the relatively less reactive ester-carbonyl groups, and partly to the fact that substitution reactions of backbone hydrogens with conventional modifying reagents invariably results in substantial chain degradation. However, chemical modification of PMMA by copolymerisation with a wide variety of comonomers bearing flame-retardant groups is relatively easy. Recent examples of such an approach include copolymerisation of methyl methacrylate with polymerisable cyclotriphosphazenes[78,99–101] and with a variety of phosphorus-containing unsaturated compounds.[74,75] The modified polymers were found to be significantly more flame retardant than PMMA, and predominantly a condensed-phase mechanism of flame retardation was found to be operative. In another study, phosphorus-containing groups, mainly phosphonate moieties, were incorporated in PMMA at specific positions on the polymer backbone, namely at chain ends and as pendent groups.[102] The degree of flame retardancy was found to depend on the topological dispositions of the modifying groups in addition to the extent of loading. Recently, enhanced char-forming tendency and increased flame retardancy were found in poly(methyl methacrylate-co-4-vinyl pyridine) polymers modified with transition metal complexes such as vanadium acetylacetonate, vanadyl chloride and ferric chloride.[80]

Probably, a much more widely used strategy to flame retard acrylics, in general, is the incorporation of flame-retardant compounds as additives. Such additives for PMMA include red phosphorus or its compounds, especially in combination with other inorganic nitrogen and halogen-containing flame retardants.[103] Other additive-type flame retardants include inorganic and organic tin halides and sulphur compounds. Wilkinson's salt of the type RhCl(PPh$_3$)$_3$ has also been used as a flame retardant for PMMA. RhCl(PPh$_3$)$_3$ reacts with PMMA at the carbonyl groups promoting crosslinking and eventually leading to char formation.[104] More recently, novel ecologically safe flame-retardant systems based on silica gels have been used to flame retard PMMA.[105]

There are several reports in the literature regarding the burning behaviour and the influence of various flame-retardant species on the flammability of fibre-forming homopolymer and copolymers of acrylonitrile.[106,107] A pressed, powdered, polymer sheet technique has been developed which allows a range of polymer compositions, in the presence and absence of flame retardants, to be assessed for LOI, burning rate and char residue.[106] It has been shown that the mechanism of thermal degradation of polymers of acrylonitrile is also dependent on the rate of heating. At low heating rates, cyclisation is the main reaction pathway whereas at high heating rates, commensurate with those encountered in fires, volatile-forming chain scission predominates. The common flame-retarding additives used for PAN and related polymers are red phosphorus,[108] brominated compounds,[109] and various aromatic phosphorus compounds in combination with metal oxides or hydroxides.[110]

7.5.4 Poly(vinyl chloride)

Poly(vinyl chloride) (PVC) is one of the most important large-volume thermoplastic polymers. In spite of its high volume production, PVC is one of the least thermally stable polymers in commercial use; for processing and subsequent end-uses, it requires special stabilisers. PVC is processed in rigid or plasticised products. Rigid PVC is used for making pipes, tubes, cladding, window and door frames, and roofing materials. PVC foams are used for thermal insulation and as separators for batteries. Flexible (plasticised) PVC is used for making hoses, cable insulation, wall-covering and flooring materials. Plastisols and organosols are used to obtain heavy-duty industrial clothing, handbags, shoes, gloves, and so forth. Good performance is shown by PVC composites containing mica and surface-treated wood fibres.

PVC, owing to its high chlorine content, will not undergo sustained combustion and is self-extinguishing (LOI = 45). Plasticised PVC may continue to burn with a smoky flame depending on the type and quantity of plasticiser used. Aliphatics, aromatics and condensed aromatics are further products of pyrolysis and combustion of PVC. The latter products, together with HCl, contribute to smoke emission during the burning of PVC. The burning behaviour of PVC foam, both rigid and plasticised, closely resembles that of solid PVC.

The chlorine atoms in PVC, especially the allylic and tertiary chlorines, are quite labile and, chemically, PVC is a highly reactive substrate.[111] However, substitution reactions of chlorine atoms with other nucleophiles, especially at higher temperatures, are generally accompanied by competing elimination reactions.[112] This could lead to dehydrohalogenation of the polymer chain thus affecting the thermal stability of the modified polymer. Furthermore, replacement of the inherently flame-retardant chlorines with other conventional flame-retardant groups has little or no advantageous effect on the flame retardance of the polymer. Therefore, attempts to improve the flame retardance of PVC by post-polymerisation modification are uncommon. Efforts in this area are often directed towards reducing smoke emissions and enhancing char yields through the incorporation of additives.

Chlorinated PVC (CPVC) has superior fire performance to PVC.[113] Substantial increases in LOI and char yield, with a decrease in the amount of smoke evolved, were observed with increases in the degree of chlorination.

CPVC has also been used as an additive with Fe(III) compounds for flame retarding ABS blends.[114] In another report, the thermal, flame and mechanical behaviour of ternary blends of PVC, poly (ethylene-co-vinyl acetate), and poly (styrene-co-acrylonitrile) blends were investigated.[115] These blends were found to have excellent thermal stability and mechanical properties compared to pure PVC and the binary blends. Measurements of smoke and volatile gas evolution revealed that the ternary blends are very efficient in flame-retardant applications.

The effects of Sb(III) and basic Fe(II) oxides on the smoke suppression of PVC have been described in a number of reports.[116–118] Ferrocene and its substituted derivatives have also been used as flame-retardant and smoke-suppressant additives in plasticised PVC.[119,120] Other additives include red phosphorus in combination with nitrogen and bromine compounds[121] and polyphosphate ester plasticizers.[122,123] Additive compounds, primarily used as smoke suppressants for PVC, also include sulphate glasses based on transition metals,[124] microzeolite–ammonium sulphamate[125] and organic Cr(III) complexes.[126]

7.5.5 Saturated polyesters

Aliphatic polyesters, although usually partly crystalline, generally have low melting points, poor mechanical properties and are susceptible to chemical attack, especially to hydrolysis. Their principal uses, therefore, are in blends or as reactive components in other materials, especially in polyurethanes. Aromatic polyesters such as poly(ethylene terephthalate) (PET), poly(butylene terephthalate) (PBT) and poly(cyclohexane terephthalate) (PCT) on the other hand, have high melting points (265 °C in the case of PET) and have important applications as fibres (PET) and as high performance thermoplastic moulding materials (PET, PBT and PCT).

Polyesters can be satisfactorily flame retarded by incorporation of a variety of conventional flame-retardant additives. For example, PET is conventionally flame retarded with low molecular weight or polymeric phosphorus esters, with brominated organics, such as oligobromostyrene, coupled with antimony trioxide, and with brominated phosphates (in which case antimony synergists are not required). PBT has additionally been flame retarded with decabromodiphenyl ether, and with brominated oligocarbonates. Less conventional additive strategies include the blending of polyesters with polymers having better flame retardance, for example PET with poly(sulphonyldiphenylene phenylphosphonate).[127] However, the nature of the general process by which polyesters are synthesised, namely step-reaction polymerisation, lends itself to the incorporation of flame-retardant groups as part of the polymer chain structure, i.e. to reactive flame-retardant strategies. In aliphatic polyesters, the incorporation of mono-, di- or trichloroacetic acid has been shown to lead to improved flame retardance when these polyesters are subsequently used as components in polyurethane coatings.[128-130] The functioning of the monoacids as chain stoppers during polyesterification is prevented by including appropriate amounts of the tri-functional alcohol, trimethylolpropane.

Phosphorus-containing flame-retardant groups also can be incorporated into polyesters during synthesis. Flame-retardant PETs containing phenylphosphate and phenylphosphine oxide units in the main chain have been reported,[131-133] as have polyesters containing flame-retardant units based on spirocyclic pentaerythritol diphosphate acid monochlorides (SPDPC).[134] In the latter, useful increases in T_g were achieved along with increases in flame retardance (oxygen index and char yield). Attempts to synthesise PET containing phenylphosphonate units by incorporating phenyl phosphonic acid in the polymerisation along with the other reactants have only limited success, however, since above 240 °C there is appreciable decomposition of the O–P–O bonds leading to low yields of polymer.[135] Phosphorus-containing PET copolymers have been prepared also by using DDP as a coreactant.[136] These copolymers form miscible blends with conventional PET and can be used to produce flame-retardant PETs with LOIs greater than 28. The use of 2-(6-oxido-6H-dibenz <c,e><1,2> oxaphosphorin-6-yl) dimethyl itaconate (**I**) as a reactive flame retardant in PET and poly(ethylene-2,6-naphthenate) (PEN) has also been reported.[137]

(**I**)

V0 ratings in UL 94 vertical burn tests on the modified polyesters were obtained with phosphorus contents as low as 0.75 wt% for PET and 0.5 wt% for PEN. The modified polymers also gave higher char yields than the unmodified equivalents and had greater thermal stabilities.

Polyarylates (polyesters based on aromatic acids, or acid derivatives, and aromatic diols) have also been reactively modified with phosphorus-containing groups. For example, bis[4-m-carboxyphenoxy)phenyl]phosphine oxide has been condensed with bisphenol A,[138] and 2-(6-oxido-6H-dibenz[c,e][1,2]oxaphosphorin-6-yl)-1,4-hydroxyethoxyphenylene (**II**) with

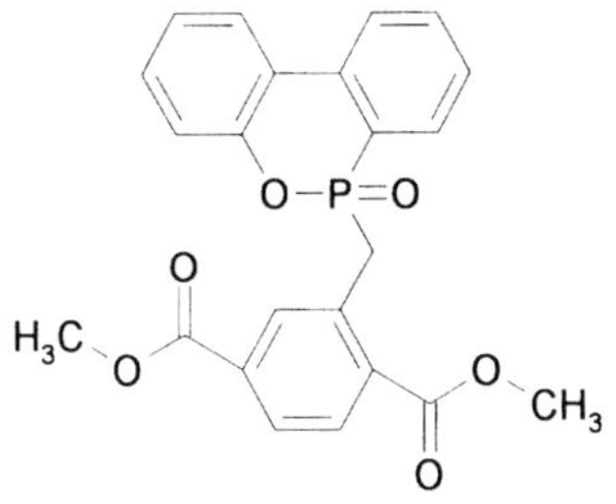

(**II**)

various aromatic acid chlorides.[139] The latter show considerable flame retardance (LOIs from 36 to 43 and char yields, under nitrogen, of 20–32% at 700 °C).

A strategy with potential for improving the flame retardance of a thermoplastic, but one which is very difficult to engineer successfully, is to incorporate reactive groups in the polymer chains that lead to thermal cross-linking below the onset temperature for thermal degradation but above the temperatures used for processing, i.e. rendering it, *in extremis*, a thermoset. As indicated earlier in this chapter, thermosets generally exhibit better flame retardance than thermoplastics owing to the extensive cross-linking in the former. An ingenious approach along these lines to flame retarding PET involves the incorporation of a benzocyclobutene-containing terephthalic acid derivative (XTA) (**III**) during polymerisation.[140] The cyclobutene moieties take part in cross-linking at around 350 °C. A modified PET containing 20 mol% XTA has an LOI of 35 (compared with 18 for unmodified PET) and a high char yield, although the melting point and onset temperature for thermal degradation are both slightly reduced.

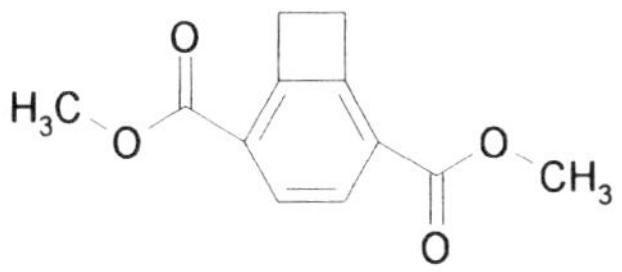

(**III**)

7.5.6 Polyamides

The aliphatic polyamides are widely used for making fibres, especially nylon-6,6, nylon-6,10 and nylon-11, and as thermoplastic moulding materials, especially nylon-6. Owing to their regular sequences of amide links, the nylons exhibit considerable interchain hydrogen bonding, and as a consequence are generally highly crystalline with high melting points. These features render it much easier to flame retard polyamides with particulate inorganic additives, such as magnesium hydroxide, ammonium polyphosphate and red phosphorus, and with polar nitrogen-containing organics, such as melamine and melamine cyanurate, than with, say, organic halides and phosphates. However, Dechlorane Plus™, the Diels-Alder adduct of hexachlorocyclopentadiene and octadiene, is used extensively to flame retard nylons at concentrations of up to 25% by weight with either antimony oxide or ferric oxide as a synergist. Other flame retardants that have been recommended for use with nylon include oligobromostyrene, oligo(dibromophenylene oxide) and low molecular weight brominated epoxy resins.

Ammonium polyphosphate has been shown to be particularly effective as a flame retardant in nylons, modifying degradation pathways and 'catalysing' the production of intumescent, protective chars.[141–145] The action of ammonium polyphosphate is also either largely unaffected or even enhanced by the presence of inert fillers[146,147] and is enhanced by the presence of oxidants such as manganese dioxide.[148] Oxidants, such as potassium nitrate, used alone are also effective flame retardants for polyamides[149] as are inorganic glass-forming substances such as ammonium pentaborate.[150] Melamine and melamine cyanurate also appear to promote char formation in aliphatic polyamides although at the expense of some reduction in thermal stability.[151–154]

A novel approach to flame-retarding aliphatic polyamides, claimed to be ecologically safe, involves the blending of the polyamide with poly(vinyl alcohol) or partly oxidised poly(vinyl alcohol). Poly(vinyl alcohol) readily dehydrates and forms a carbonaceous char when heated and, in mixtures with nylon-6,6, 'synergistic carbonisation' is observed.[155,156]

Reactive strategies, i.e. chemical modification of polyamides, appear not to have been explored to a large extent as a route to improved flame retardance in aliphatic polyamides, probably because chemical modification of aliphatic polyamides disrupts intermolecular hydrogen bonding and hence crystallinity, thus reducing melting points. This means that, although the reactive incorporation of bis(4-carboxyphenyl)phenylphosphine oxide (**IV**)

(**IV**)

at up to 20 mol% into nylon-6,6 gives flame-retardant polymers that are still partly crystalline; at 30 mol% incorporation, all crystallinity is lost.[157]

Similar results have been observed with incorporated diphenyl-methylphosphine oxide groups.[158] Aromatic polyamides modified with phosphine oxide groups have been synthesised also by condensation of bis(4-carboxyphenyl)phenylphosphine oxide but with various aromatic diamines.[159,160]

7.5.7 Polycarbonate and poly(phenylene oxide)

Polycarbonate (PC) is a mechanically tough and strong thermoplastic which, even though partly crystalline, is transparent. It is widely used for making impact-resistant mouldings and sheet glazing materials. Where transparency is not required, PC can be used blended with acrylonitrile –butadiene–styrene (ABS) terpolymers to give cheaper materials, still with good mechanical performance. PC and its blends can be flame retarded by the addition of conventional brominated flame retardants, such as decabro-modiphenyl ether, and organophosphates such as triphenylphosphate. However, use of such flame retardants leads to lower softening tempera-tures and some impairment of mechanical properties. The brominated flame retardants of choice for PC are polycarbonate oligomers of tetrabromo-bisphenol A. Particularly effective as flame retardants for PC and for blends of PC with ABS, PET and PBT are brominated organic phosphates, which give pronounced phosphorus–bromine synergy.[161–165]

Recently, a silicon-based flame retardant designed for use in PC blends destined for applications in electronic products has been described. It is a branched, oligo(methyl-phenyl siloxane) with methyl chain ends and is claimed to generate no toxic products during combustion.[166–168] Novel low-halogen flame retardants for PCs based on various alkyne-substituted com-pounds have also been reported.[169] These promote char formation by cross-linking at flame temperatures.

Reactive strategies for flame-retarding PC have also been advocated. For example, copolycarbonates based on mixtures of bisphenol A with bis(4-hydroxyphenyl)phenylphosphine oxide (**V**) have been reported to give char

(**V**)

yields on combustion that are significantly greater than those from conventional PCs (up to 30% char in air at 700 °C for a 50% copolymer) and to have lower rates of heat release. The phosphorus-containing PCs also have higher T_g's than the parent materials.[170]

Poly(phenylene oxide)s (PPOs) are most commonly used in blends with other thermoplastics such as high impact polystyrene (HIPS), in which the PPO component not only confers improved mechanical properties but also acts as a char-former. Such blends can be flame-retarded following additive strategies similar to those used with PC, with resorcinol bis(diphenyl phosphate) proving to be a particularly effective flame retardant.[171] A variety of reactively modified polyarylene ethers containing phosphorus, fluorine and/or heterocyclic groups have been described in the literature.[172,173] All tend to show improved thermal stability and/or flame retardance compared with unmodified counterparts.

7.6   Flame-retardant elastomers

7.6.1 Natural rubber and other polydienes

The common form of polyisoprene is *cis*-1,4-polyisoprene, which occurs in the latex of many plants and trees as natural rubber. The *trans* isomer can be isolated from some plants as *gutta purcha* or *balata*. Both of these forms, and their derivatives, may also be synthesised by the use of stereospecific catalysts. Despite the competition from highly developed synthetic polymers, natural rubber has retained a leading place among commodity and engineering elastomers. Natural rubber burns readily in air (LOI = 17).

For commercial purposes, natural rubber is generally vulcanised before use. The ingredients for vulcanisation may either increase or decrease the flammability of natural rubber, but generally increase the formation of smoke and other toxic products. Vulcanisation of natural rubber always involves the addition of fillers and currently extensive use is made of the flame-retardant properties of ATH and of carbon black (LOI = 56–63). Recently, additives such as bromo derivatives of cashew nut shell liquid[174] and polyphosphates[175] have been used to flame retard natural rubber.

Modified natural rubber products have wide commercial applications. Thermally degraded, low molecular weight rubber is used for potting compounds, binders for abrasive wheels and as casting moulds. Chlorinated rubber is extensively used as an adhesive for metal to rubber bonding and for chemically resistant paints. Rubber hydrochloride finds application in packaging owing to its low permeability to water vapour and good transparency. Chlorinated and hydrochlorinated rubbers, as would be expected, are more flame retardant than the unmodified forms and in this respect resemble the synthetic chlorine-containing polydiene, polychloroprene.

In an early report, chemical modification of poly-1,3-butadiene, to improve flame retardance, was achieved through copolymerisation reactions with halogen- and/or phosphorus-containing unsaturated compounds.[176] More recently, chemical modifications of 1,4-polydienes have been carried out by reactions on relatively low molecular weight polymers, including those on liquid natural rubber. Examples include phosphorus modification of epoxidised natural rubber with dialkyl or diaryl phosphates.[177,178] Phosphorus-modified polydienes were further cross-linked using methylnadic anhydride.[177] Formation of cross-linked, three dimensional structures led to improved thermal stability and superior fire performance. Char residues obtained from thermogravimetric analysis and burning of these polymers suggest a condensed-phase mechanism of flame retardance.

Bromination of polybutadienes also results in increased fire retardance owing to the gas-phase radical quenching reactions of bromine atoms during combustion.[179] Phosphorus reagents such as diethyl phosphonate and trichloromethyl phosphonyl dichloride can be added to relatively low molecular weight poly-1,2-butadienes under radical initiation; the modified polymers have been used as fire-retardant additives for natural rubber.[180,181] In another report, hydroxy telechelic polybutadiene has been modified by grafting with phosphonated thiols leading to macromolecular polyols containing phosphorus.[182] These polyols were used for preparing fire-retardant polyurethane networks.

7.6.2 Polyurethanes

The class of polymers known as the polyurethanes now encompasses a wide variety of materials ranging from surface coatings, both rigid and flexible, through elastomers, both thermoplastic and thermosetting (and curable by either conventional vulcanisation or at room temperature) to foams, which may be flexible, semi-flexible or rigid. However, the major applications for polyurethanes are in flexible (elastomeric) products. The chemical structures of polyurethanes also vary widely, although many are based on chain-extended oligomeric polyethers and/or polyesters (polyols). The one feature that all polyurethanes have in common is the urethane linkage which is formed by reaction of a hydroxyl group with an isocyanate group. However, the urethane linkage is not the only linking group in the structure; others include the ester and ether link present in polyurethane precursors, the urea link arising from reactions of amine groups with isocyanates, and allophanate, biuret and isocyanurate links arising from further reactions of isocyanate with urea, urethane and other isocyanate groups, respectively. Although the flammabilities of polyurethanes might be expected to be lower than those of many other polymers owing to their significant nitrogen content and the cross-linking usually present, in practice the fire performance of polyurethane-based materials is often poor owing to the high surface area of the product, particularly in foams. It is to the improvement of the flame-retardance of polyurethane foams, widely used in the furniture, building and automobile industries, that most efforts have, not surprisingly, been recently directed.

Many of the conventional additive types of flame retardant have been used, or at least advocated for use, in polyurethanes including halogenated materials such as chlorinated hydrocarbon waxes, phosphorus-containing additives such as tris(1,3-dichloro-2-propyl) phosphate, intumescents such as ammonium polyphosphate and melamine, and hydrated metal oxides and hydroxides such as alumina trihydrate and magnesium hydroxide. However, polyurethane synthesis lends itself particularly well to the reactive approach to flame retardance, and especially to the introduction of flame-retardant groups as part of the polyol structure. For example, polyurethane foams incorporating chloroendic diol (**VI**) or 2,3-dibromo-2-butene-1,4-diol as one of the reactive components exhibit good flame retardance.[183,184]

(**VI**)

Chlorine-containing polyester polyols have been synthesised by co-condensing chloroacetic acid or 2,4-dichlorobenzoic acid with 1,4-butanediol, trimethylol propane and adipic acid[185] and bromine-containing polyesters by similar reactions of 2,3-dibromopropanoic acid with 1,4-butanediol, trimethylol propane and adipic acid.[186] Polyurethanes based on these polyesters perform well as flame-retardant coatings.

Proprietary phosphorus-containing reactive flame retardants for polyurethanes have also been reported.[187,188] Polyols containing both phosphorus and nitrogen have been synthesised from tetrakis(hydroxymethyl)phosphonium chloride and have been shown to be suitable for producing flame-retardant rigid polyurethane foams.[189] Phosphorus has also been introduced into polyurethanes using phosphorus-containing isocyanates, such as bis(4-isocyanatophenoxy)phenyl phosphine oxide,[190] and by grafting phosphonated thiols onto hydroxy-telechelic oligobutadienes to give polyols containing between 3 and 5 wt% of phosphorus.[182] Phosphorus- and bromine-containing acrylic monomers have been employed to make UV-curable urethane–acrylate resins; maximum bromine–phosphorus synergism in flame retardance is seen at a Br:P atom ratio of two to one.[191]

Melamine has also been incorporated into reactive flame retardants for polyurethanes. For example, water-blown rigid polyurethane foams containing carbamylmethylated melamine polyols have been shown to be significantly more flame retardant than traditional foams containing aromatic amine or sucrose-based polyols.[192] Similar triazinic polyols have been synthesised by oxyalkylation of some simple melamine condensates with propylene oxide and shown also to be effective flame-retardant components in rigid polyurethane foams.[193]

Silicon has been introduced into polyurethanes by using hydroxy-terminated polydimethylsiloxane as one of the precursor polyols.[194] The materials are reported to show improved thermal stability and flame retardance over conventional, soft-segment block copolyurethanes. Segmented polyurethanes containing both phosphorus and silicon have been described.[195] Phosphorus is introduced through use of bis(hydroxypropyl) isobutylphosphine oxide as a chain-extender, and silicon through use of secondary amine-ended poly(dimethyl siloxane)s as the soft segments. Ferrocene-containing block copolyurethanes have also recently been reported; these too are more flame retardant than conventional materials.[196]

7.7   Flame-retardant thermosets

Thermosets are normally prepared as two-stage resins in which an initially soluble, low molecular weight oligomeric liquid, or low melting point solid (the first stage), is cross-linked (cured) during processing (the second stage) to give an insoluble, infusible and generally intractable and highly cross-linked product. Owing to the high cross-link density and the often aromatic and/or heteroatomic nature of the final product, thermosets generally display greater thermal stability and therefore better flame retardance than the average thermoplastic. Thermosets are especially amenable to further flame retardation by a reactive strategy, there usually being ample opportunity to incorporate flame-retardant groups during either the first or second stage of reaction.

7.7.1 Amino and phenolic resins

Cured amino and phenolic resins are above averagely flame retardant owing to the extensive cross-linking and to the high nitrogen contents, in the case of amino resins, and significant aromatic content, in the case of phenolic resins. Also, amino and phenolic resins tend to be used in materials containing high concentrations of non-flammable fillers and laminating agents, such as kaolin, glass powder and glass fibre. Amino resins, which themselves are used as components of flame retardants for other polymers, can be further flame-retarded by incorporating additive and reactive flame retardants. A recent publication, for example, has advocated the use of boric acid and borax to improve the flame retardance of urea–formaldehyde resins used in particle board manufacture.[197] Low molecular weight phenolic oligomers are also used in their own right as components of flame retardants, e.g. resorcinol bis(diphenyl phosphate), with triphenyl phosphate,[198] and in some recently described polycyclic phosphonates.[199] Addition to phenolic resins of, for example, boric acid,[200] and mixtures of metal hydroxides and halogenated organic phosphates,[201] can improve flame retardance, as can reactive modification with reagents such as monophenyl phosphoric acid and 2,4,6-tribromophenyl phosphoric acid.[202]

7.7.2 Unsaturated polyesters

Proprietary flame-retarded unsaturated polyesters are well known and several are based on a reactive strategy employing halogenated monomers such as tetra-bromophthalic anhydride, dibromoneopentyl glycol (**VII**), tetrabromobisphenol A or chlorendic anhydride (**VIII**), in their construction.

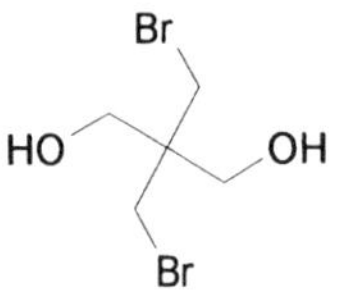

(**VII**)

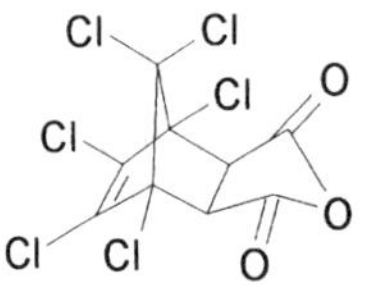

(**VIII**)

Reactive flame retardance in unsaturated polyesters can also be brought about by using dibromostyrene as part of the cross-linking monomer mixture. In such resins, the flame retardance and also the smoke suppression can be enhanced by the incorporation of zinc hydroxystannate as an additive.[203] There have been no reports of the significant use of phosphorus-containing monomers either in the first stage of polyester manufacture or in the curing reaction, probably because the use of such species would increase the cost considerably of these otherwise relatively inexpensive moulding and surface coating materials.

7.7.3 Epoxy resins

Cured epoxy resins are widely used as surface coating, encapsulating (potting) and adhesive materials, and as a resin matrix for glass and carbon fibre-reinforced composites. Flame-retardant epoxy resins based on reactive halogenated intermediates such as tetrabromobisphenol A are well-established. Flame retardance can also be improved by the use of several types of flame-retardant additive. Recent research, however, has concentrated on promoting increased char formation in cured epoxies through the use of phosphorus-containing monomers and/or curing agents. For example, dialkyl or diaryl phosphates can add to the epoxy groups of the 4,4′-diglycidyl ether of bisphenol A (DGEBA) and the resulting modified resins can then be cured in the usual way to give materials with better flame retardance than both unmodified resins and resins containing alkyl (or aryl) phosphates as additives.[204] Best results are observed with resins containing phenyl phosphate groups. Phosphate-containing epoxy resins have also been synthesised using the phosphorus-containing oxirane, bis-glycidyl phenyl phosphate, as a monomeric component and bis(4-aminophenyl)-phenyl phosphate as a curing agent.[205] Other phosphorus-containing reactive components recently incorporated in flame-retardant epoxy resins include: 2-(6-oxi-6H-dibenz<c,e><1,2>oxaphosphorin-6-yl)1,4-benzenediol (**IX**)[206] and the diglycidyl ether based upon it.[207]

(**IX**)

A range of phosphorus-containing curing agents has recently been disclosed.[208] The list includes more than twenty novel compounds, including di- and tri-amino cyclotriphosphazenes, di- and tri-hydroxy cyclotriphosphazenes and various phosphine oxides. The use of bis(*m*-aminophenyl)-

methylphosphine oxide (BAMPO) as a flame-retardant curing agent for epoxies is becoming more common and the mechanism of its action has recently been extensively explored.[209,210] It appears that at low concentrations, BAMPO mainly 'catalyses' the production of an intumescent char, but that at higher concentrations some BAMPO residues break down to give volatile phosphorus-containing species that act as a gas-phase inhibitor. Use of the aryl phosphinate anhydride, 9,10-dihydro-9-oxa-10-phosphaphenanthrene-10-oxide)-methyl succinic anhydride (**X**), as a curing agent in both epoxy resins and epoxy resin composites is reported also to give products with improved flame retardance.[211]

(**X**)

7.8   Inherently flame-retardant polymers

Several of the polymers mentioned already can be said to be inherently flame retardant through having flammabilities that are below the average. Such polymers tend to have relatively high thermal stabilities by virtue of significant aromatic content (e.g. the aromatic polyesters, polyethers and polyamides, the polyarylates and polycarbonates, and phenolic resins), or to decompose to give gas-phase inhibitors of combustion (e.g. PVC), or to contain significant quantities of nitrogen or similar heteroatoms (e.g. polyamides and aminoresins). In addition to these materials there are, however, a few other important groups including the poly(aryl ketone)s and ether ketones, poly(aryl sulphide)s and sulphones, polyimides, and the polybenzimidazoles, benzoxazoles and benzthiazoles, all of which show above average thermal and thermo-oxidative stability, both of which are important to flame retardance.

Although inherently flame retardant, polyimides (and ether imides) have been improved in this respect by the reactive incorporation of fluorine-containing,[212] and phosphorus-containing groups.[213] Polyimides have also been made more flame retardant by phosphorylation with phosphorus-containing reagents such as diethylchlorophosphate.[214] The resulting materials are slightly less thermally stable than the unmodified precursors but give high char yields and have LOIs in excess of 48. Flame-retardant randomly segmented copolymers with phosphorus-containing poly(aryl imide) hard segments and poly(dimethyl siloxane) soft segments have also been described.[215] Char yields in these copolymers were found to depend principally upon siloxane content.

7.9   Conclusions

The last decade has seen a significant move away from halogen-based flame retardants for thermoplastics and thermosets and towards those based on phosphorus, silicon, metal hydrates and other metal salts. It seems highly likely that this trend will continue given the current mounting concern over the release of halogenated species into the environment. It seems likely also that reactive flame retardants will be sought wherever practicable, especially for step-reaction polymers in which the incorporation of a reactive flame retardant can most easily be accomplished. Both of these developments are likely to see condensed-phase mechanisms of flame retardance, especially those involving intumescent formulations, assume even greater importance. It seems likely too, that the use of inherently flame-retardant polymers will increase, especially as new ways are found of processing such polymers, many of which are also rather intractable owing to high softening temperatures and/or melting points, or high degrees of cross-linking. Two interesting developments in this last area are of more versatile, low-temperature processing techniques for conventional phenolic resins and fibre-reinforced composites based on them,[216] and of polymers based on the ring-opening of bis-benzoxazines (**XI**), which have physical, mechanical and flame-retardant properties similar to those of phenolic resins but which are developed at lower cross-link densities.[217] Bis-benzoxazines can be made by reaction of amines, such as methylamine and aniline, with bisphenol-A: they cure with minimal shrinkage or even with slight expansion in volume.

(**XI**)

References

1 Cullis C F and Hirschler M M, *The Combustion of Organic Polymers*, Oxford, Clarendon Press, 1981, 93–125.

2 Levin B C, 'The development of a new small-scale smoke toxicity test method and its comparison with real-scale fire tests', *Toxicology Letters*, 1992, **64/65**(SISI), 257–64.

3 Purser D A, 'The evolution of toxic effluents in fires and the assessment of toxic hazards', *Toxicology Letters*, 1992, **64/65**(SISI), 247–55.

4 Cowie J M G, *Polymers: Chemistry and Physics of Modern Materials*, 2nd edn., New York, Blackie, 1991, 3–25.

5 Ebdon J R and Jones M S, 'Flame retardants (overview)', *Polymeric Materials Encyclopedia*, Salamone J C (ed.), Boca Raton, CRC Press, 1995, 2397–411.

6 Grassie N and Scott G, *Polymer Degradation and Stabilization*, Cambridge, Cambridge University Press, 1985.

7 Gann R G, Dipert R A and Drews M J, 'Flammability', *Encyclopedia of Polymer Science and Engineering*, **7**, 2nd edn., New York, Wiley-Interscience, 1987, 154–205.

8 Troitzsch J H, 'Methods for the fire protection of plastics and coatings by flame-retardant and intumescent systems', *Progress in Organic Coatings*, 1983, **11**(1), 41–69.

9 Kannan P and Kishore K, 'Novel flame retardant polyphosphoramide esters', *Polymer*, 1992, **33**(2), 418–22.

10 Green J, 'A Phosphorus-bromine flame-retardant for engineering thermoplastics – a review', *J. Fire Sci.*, 1994, **12**(4), 388–408.

11 Guo W, 'Flame-retardant modification of UV-curable resins with monomers containing bromine and phosphorus', *J. Polym. Sci., Part A Polym. Chem.*, 1992, **30**(5), 819–27.

12 Arnold Jr C, 'Stability of high-temperature polymers', *J. Polym. Sci., Macromol. Revs.*, 1979, **14**, 265–378.

13 Kuryla W C and Papa (eds.), *Flame Retardancy of Polymeric Materials*, **4**, New York, Dekker, 1978, 42.

14 Khanna Y P and Pearce E M, 'Flammability of polymers', *ACS Symposium Series*, 1985, **285**, 305–19.

15 Quinn C B, 'The flammability properties of copolyesters and copolycarbonates containing acetylenes', *J. Polym. Sci., Polym. Chem. Edit.*, 1997, **15**, 2587–94.

16 Stepek J and Daoust H, 'Flame retardants', in *Additives for Plastics, Polymer Properties and Application Series*, **5**, Berlin, Springer-Verlag, 1983, 201–16.

17 Cullis C F and Hirschler M M, 'Char formation from polyolefins: Correlation with low-temperature oxygen uptake and with flammability in the presence of metal-halogen systems', *Eur. Polym. J.*, 1984, **20**(1), 53–64.

18 Martel B, 'Charring process in thermoplastic polymers: effect of condensed phase oxidation on the formation of chars in pure polymers', *J. Appl. Polym. Sci.*, 1988, **35**(5), 1213–26.

19 Carty P and White S, 'Char formation in polymer blends', *Polymer*, 1994, **35**(2), 343–7.

20 Van Krevelen D W, 'Flammability and flame retardation in the case of organic high polymers and their relation to chemical structure', *Chem.-Ing.-Tech.*, 1975, **47**(19), 793–803.

21 Van Krevelen D W, 'Flame resistance of polymeric materials', *Polymer*, 1975, **16**(8), 615–20.

22 Green J, 'Char studies – flame-retarded polycarbonate/PET blend', *J. Fire Sci.*, 1994, **12**(6), 551–81.

23 Banks M, Ebdon J R and Johnson M, 'Influence of covalently bound phosphorus-containing groups on the flammability of some common addition homo- and co-polymers', *Proceedings from the Flame Retardants '94 Conference*, The British Plastic Federation, London, Interscience Communciations Limited, 1994, 183–91.

24 Bertelli G, Camino G, Marchetti E, Costa L and Locatelli R, 'Structural studies on chars from fire retardant intumescent systems', *Angew. Makromol. Chem.*, 1989, **169**, 137–42.

25 Bertelli G, Marchetti E, Camino G, Costa L and Locatelli R, 'Intumescent fire retardant systems: effects of fillers on char structure', *Angew. Makromol. Chem.*, 1989, **172**, 153–63.

26 Horrocks A R, Anand S C and Sanderson D, 'Complex char formation in flame retarded fibre-intumescent combinations: 1. Scanning electron microscopic studies', *Polymer*, 1996, **37**(15), 3197–206.

27 Wang J, 'The potential applications of X-ray photoelectron spectroscopy (XPS) to the flame retardance mechanism of polymers', *Makromol. Chem., Macromol. Symp.*, 1993, **74**, 101–10.

28 Bourbigot S, Le Bras M, Gengembre L and Delobel R, 'XPS study of an intumescent coating: application to the ammonium polyphosphate/pentaerythritol fire-retardant system', *Appl. Surf. Sci.*, 1994, **81**(3), 299–307.

29 Feng D M, Zhou Z M and Bo M P, 'An investigation of melamine phosphonite by XPS and thermal analysis techniques', *Polym. Degrad. Stab.*, 1995, **50**(1), 65–70.

30 Zhu W M, Weil E D and Mukhopadhyay S, 'Intumescent flame-retardant system of phosphates and 5,5,5',5',5'',5''-hexamethyltris(1,3,2-dioxaphosphorinanemethan)amine-2,2',2''-trioxide for polyolefins', *J. Appl. Polym. Sci.*, 1996, **62**(13), 2267–80.

31 Gilman J W, Lomakin S, Kashiwagi T, VanderHart D L and Nagy V, 'Characterization of flame retarded polymer combustion chars by solid-state ^{13}C and ^{29}Si NMR and EPR', *ACS Polym. Prep.*, 1997, **38**(1), 802–3.

32 Factor A, 'Fire and Polymer', Nelson G (ed.), *ACS Symposium Series*, 1990, **425**, 247.

33 Gilman J W, Lomakin S, Kashiwagi T, VanderHart D L and Nagy V, 'Characterization of flame retarded polymer combustion chars by solid-state C-13 and Si-29 NMR and EPR', *Fire Mat.*, 1998, **22**(2), 61–7.

34 Azeeva R M and Zaikov G E, 'Flammability of polymeric materials', *Advances in Polymer Science*, 1985, **70**, 171–229.

35 Sibulkin M, 'The dependence of flame propagation on surface heat transfer. II. Upward burning', *Combust. Sci. Tech.*, 1977, **17**(1–2), 39–49.

36 Brossas J, 'Fire retardance in polymers: an introductory lecture', *Polym. Degrad. Stab.*, 1989, **23**(4), 313–25.

37 Simmons R F and Wolfhard H G, 'Limiting oxygen concentrations for diffusion flames in air diluted with nitrogen', *Combust. Flame*, 1957, **1**, 155–61.

38 Fenimore C P and Martin F J, 'Flammability of polymers', *Combust. Flame*, 1966, **10**(2), 135–9.

39 Fenimore C P and Jones G W, 'Modes of inhibiting polymer flammability', *Combust. Flame*, 1966, **10**(3), 295–301.

40 Camino G and Costa L, 'Performance and mechanisms of fire retardants in polymers – a review', *Polym. Degrad. Stab.*, 1988, **20**(3–4), 271–94.

41 Kunz D H E, 'Flame retarding materials for advanced composites', *Makromol. Chem., Macromol. Symp.*, 1993, **74**, 155–64.

42 Tewarson A, 'Flammability parameters of materials: ignition, combustion, and fire propagation', *J. Fire Sci.*, 1994, **12**(4), 329–56.

43 Pagliari A, Cicchetti O, Bevilacqua A and Van Hees P, 'Intumescent flame retardants: new evidence of their higher fire safety', *Proceedings from the Flame retardants '92 Conference.* The Plastics and Rubber Institute, London, Elsevier Applied Science, 1992, 41–52.

44 Blanchard R R, 'Chlorinated polyethylene CPE', in *Handbook of Plastic Materials and Technology*, Rubi I I (ed.), New York, John Wiley and Sons, 1990, 67–75.

45 Banks M, Ebdon J R and Johnson M, 'Influence of covalently bound phosphorus-containing groups on the flammability of poly(vinyl alcohol), poly(ethylene-*co*-vinyl alcohol) and low-density polyethylene', *Polymer*, 1993, **34**(21), 4547–56.

46 Kaji K, Yoshizawa I, Kohara C, Komai K and Hatada M, 'Preparation of flame-retardant polyethylene foam of open-cell type by radiation grafting of vinyl phosphonate oligomer', *J. Appl. Polym. Sci.*, 1994, **51**(5), 841–53.

47 Irving M, Fodor Z, Body M, Baranovics P, Kelen T and Tudos F, 'The effect of processing steps on the oxidative stability of polyethylene tubing cross-linked by irradiation', *Angew. Makromol. Chem.*, 1995, **224**, 33–48.

48 Bae H J, Sohn H S and Choi D J, 'Development of high-voltage lead wires using electron-beam irradiation', *Radiation Physics and Chemistry*, 1995, **46**(4–6), Part 2, 959–62.

49 Wu W, Xiao W D and Xu X, 'A study on impact properties of γ-ray irradiated LLDPE filled with ATH', *Chemical Journal of Chinese Universities*, 1997, **18**(1), 140–4.

50 Yeh J T, Yang H M and Huang S S, 'Combustion of polyethylene filled with metallic hydroxide and cross linkable polyethylene', *Polym. Degrad. Stab.*, 1995, **50**(2), 229–34.

51 Donskoy A A, Shashikina M A, Azeeva R M and Ruban L V, 'Thermal stability and flammability of sulfochlorinated polyethylene compositions', *Int. J. Polym. Mater.*, 1994, **24**(1–4), 157–66.

52 Hornsby P R, Winter P and Cusack P A, 'Flame retardancy and smoke suppression of chlorosulfonated polyethylene containing inorganic tin compounds', *Polym. Degrad. Stab.*, 1994, **44**(2), 177–84.

53 Costa L and Camino G, 'Thermal-degradation of polymer fire retardant mixtures. 5: Polyethylene chloroparaffin mixtures', *Polym. Degrad. Stab.*, 1985, **12**(2), 105–16.

54 Vyazovkin S V, Bogdanova V V, Klimovtsova I A and Lesnikovich A I, 'Invariant kinetic-parameters of polymer thermolysis. 3: The influence of a fire-retardant additive on polypropylene thermolysis', *J. Appl. Polym. Sci.*, 1991, **42**(7), 2095–8.

55 Camino G, Costa L and Discortemiglia M P L, 'Overview of flame retardant mechanisms', *Polym. Degrad. Stab.*, 1991, **33**(2), 131–54.

56 Menon A R R, Pillai C K S, Sudha G D, Prasrad V S and Brahmakumar M,

'Processability characteristcs of polyethylene modified with 3-(tetrabromopen-tadecyl)-2,4,6-tribromophenol as a flame-retardant additive', *Polymer-plastics Tech. and Eng.*, 1995, **34**(3), 429–38.

57 Chiang W Y and Hu C H, 'The improvements in flame retardance and mechanical properties of polypropylene FR blends by acrylic acid graft copolymerization', *Eur. Polym. J.*, 1996, **32**(3), 385–90.

58 Vyazovkin S V, Bogdanova V V, Klimovtsova I A and Lesnikovich A I, 'Invariant kinetic-parameters of polymer thermolysis. 4: Influence of fire-retardant additives on polypropylene thermolysis', *J. Appl. Polym. Sci.*, 1992, **44**(12), 2157–60.

59 Costa L, Camono G and Discortemiglia M P L, 'Mechanism of condensed phase action in fire retardant bismuth compound-chloroparaffin-polypropylene mixtures. 1: The role of bismuth trichloride and oxychloride', *Polym. Degrad. Stab.*, 1986, **14**(2), 159–64.

60 Costa L, Camino G and Discortemiglia M P L, 'Mechanism of condensed phase action in fire retardant bismuth compound-chloroparaffin-polypropylene mixtures. 2: The thermal degradation behaviour', *Polym. Degrad. Stab.*, 1986, **14**(2), 165–177.

61 Jancar J, Kucera J and Vesely P, 'Peculiarities of mechanical response of heavily filled polypropylene composites. 2: Dynamic mechanical moduli', *J. Mater. Sci.*, 1991, **26**(18), 4883–7.

62 Rigolo M and Woodhams R T, 'Basic magnesium carbonate flame retardants for polypropylene', *Polym. Eng. and Sci.*, 1992, **32**(5), 327–34.

63 Chiu S H and Wang W K, 'The dynamic flammability and toxicity of magnesium hydroxide filled intumescent fire retardant polypropylene', *J. Appl. Polym. Sci.*, 1998, **67**(6), 989–95.

64 Delobel R, Ouassou N, Le bras M and Leroy J M, 'Fire retardance of polypropylene: action of diammonium pyrophosphate pentaerythritol intumescent mixture', *Polym. Degrad. Stab.*, 1989, **23**(40), 349–57.

65 Bourbigot S, Le bras M, Gengembre L and Delobel R, 'XPS study of an intumescent coating application to the ammonium polyphosphate pentaerythritol fire-retardant system', *Appl. Surf. Sci.*, 1994, **81**(3), 299–307.

66 Bourbigot S, Le bras M and Delobel R, 'Fire degradation of an intumescent flame-retardant polypropylene using the cone calorimeter', *J. Fire Sci.*, 1995, **13**(1), 3–22.

67 Zhu V M, Weil E D and Mukhopadhyay S, 'Intumescent flame-retardant system of phosphates and 5,5,5',5',5'',5''-hexamethyltris(1,2,3-dioxaphosphorinanemethan)amine 2,2',2''-trioxide for polyolefins', *J. Appl. Polym. Sci.*, 1996, **62**(13), 2267–80.

68 Huggard M, 'Flame retardant polyolefins: impact and flow enhancement using metallocene polymers', *J. Fire Sci.*, 1996, **14**(5), 393–408.

69 Zubkova N S, Butylkina N G, Chekanova S E, Tyuganova M A, Khalturinskii N A, Reshetnikov I S and Naganovskii Y K, 'Rheological and fireproofing characteristics of polyethylene modified with a microencapsulated fire retardant', *Fibre Chemistry*, 1998, **30**(1), 11–13.

70 Armitage P, Ebdon J R, Hunt B J, Jones M S and Thorpe F G, 'Chemical modification of polymers to improve flame retardance – I. The influence

of boron-containing groups', *Polym. Degrad. Stab.*, 1996, **54**(2–3), 387–93.

71 Ebdon J R, Hunt B J, Jones M S and Thorpe F G, 'Chemical modification of polymers to improve flame retardance – II. The influence of silicon-containing groups', *Polym. Degrad. Stab.*, 1996, **54**(2–3), 395–400.

72 Liu Y L, Hsiue G H, Chiu Y S, Jeng R U and Ma C, 'Synthesis and flame-retardant properties of phosphorus-containing polymers based on poly(4-hydroxystyrene)', *J. Appl. Polym. Sci.*, 1996, **59**(10), 1619–25.

73 Hiltz J A, 'Pyrolysis-gas chromatography mass-spectrometry identification of styrene cross-linked polyester and vinyl ester resins', *J. Anal. Appl. Pyrol.*, 1991, **22**(1–2), 113–28.

74 Banks M, Ebdon J R and Johnson M, 'The flame-retardant effects of diethyl vinyl phosphonate in copolymers with styrene, methyl methacrylate, acrylonitrile and acrylamide', *Polymer*, 1994, **35**(16), 3470–3.

75 Ebdon J R, Hunt B J and Joseph P, 'Flame retardance in styrenic and acrylic polymers with covalently-bound phosphorus containing groups', in *Proceedings of the 5th International Conference on Materials Science: Materials and Fire*, Alexandria Egypt, 1998, 105–17.

76 Allen D W, Anderton E C, Bradley C and Shiel L E, 'Fire-retardant polymers – Polymerization of 1-oxo-2,6,7-trioxa-1-phosphabicyclo[2,2,2]octa-4-yl methyl methacrylate, and its copolymerization with methyl methacrylate, styrene, and triallylcyanurate', *Polym. Degrad. Stab.*, 1995, **47**(1), 67–72.

77 Wang J L and Favstritsky N A, 'Flame-retardant brominated styrene based polymers. 9: Dibromostyrene based latexes', *J. Coating Tech.*, 1996, **68**(853), 41–8.

78 Bosscher G, Jekel A P and van de Grampel J C, 'Polymerization of an acetoxyvinyl substituted chlorocyclophosphazene', *J. Inorg. Organomet. Polym.*, 1997, **7**(1), 19–34.

79 Janovic Z, Ranogajec F and Kucisecdolenc J, 'Copolymerization and copolymers of N-(2,4,6-tribromophenyl) maleimide with styrene', *J. Macromol. Sci., Chem.*, 1991, **A28**(10), 1025–37.

80 Ebdon J R, Guisti L, Hunt B J and Jones M S, 'The effects of some transition-metal compounds on the flame retardance of poly(styrene-*co*-4-vinyl pyridine) and poly(methyl methacrylate-*co*-4-vinyl pyridine)', *Polym. Degrad. Stab.*, 1998, **60**(2–3), 401–7.

81 Luijk R, Govers H A J, Eijkel G B and Boon J J, 'Thermal-degradation characteristics of high-impact polystyrene decabromodiphenylether antimony oxide studied by derivative thermogravimetry and temperature resolved pyrolysis mass-spectrometry – formation of polybrominated dibenzofurans, antimony (oxy) bromides and brominated styrene oligomers', *J. Analyt. Appl. Pyrol.*, 1991, **20**, 303–19.

82 Luijk R, Wever H, Olie K, Govers H A J and Boon J J, 'The influence of the polymer matrix on the formation of polybrominated dibenzo-para-dioxins (PBDDS) and polybrominated dibenzofurans (PBDFS)', *Chemosphere*, 1991, **23**(8–10), 1173–83.

83 Luijk R, Govers H A J and Nelissen L, 'Formation of polybrominated dibenzofurans during extrusion of high-impact polystyrene decabromodiphenyl ether antimony (III) oxide', *Env. Sci. Tech.*, 1992, **26**(11), 2191–8.

84 Checchin M, Boscoletto A B, Camino G, Luda M and Costa L, 'Mechanism of

fire retardation in poly(2,6-dimethyl-1,4-phenyl ether) high-impact poly-styrene', *Makromol. Chem., Macromol. Symp.*, 1993, **74**, 311–4.

85 Boscoletto A B, Checchin M, Milan L, Camino G, Costa L and Luda M P, 'Characterization of fire-retardant poly(phenyl ether) high-impact polystyrene in the UL-94 test', *Makromol. Chem., Macromol. Symp.*, 1993, **74**, 35–9.

86 Murashko E A, Levchik G F, Levchik S V, Bright D A and Dashevsky S, 'Fire retardant action of resorcinol bis(diphenyl phosphate) in a PPO/HIPS blend', *J. Fire Sci.*, 1998, **16**(4), 233–49.

87 Utevski L, Scheinker M, Georlette P and Lach S, 'Flame retardancy in UL-94 V-0 and in UL-94 5VA high impact polystyrene', *J. Fire Sci.*, 1997, **15**(5), 375–89.

88 Roma P, Luda M P and Camino G, 'The use of fluorinated additives in fire retardation of polymers', *Makromol. Chem., Macromol. Symp.*, 1993, **74**, 299–302.

89 Luijk R and Govers H A J, 'The formation of polybrominated dibenzo-para-dioxins (PBDDS) and dibenzofurans (PBDFS) during pyrolysis of polymer blends containing brominated flame retardants', *Chemosphere*, 1992, **25**(3), 361–71.

90 Gutman E, Bobovitch A, Rubinchic I, Shefter S and Lach S, 'Thermal degradation of flame-retardant components in filled and unfilled ABS plastics', *Polym. Degrad. Stab.*, 1995, **49**(3), 399–402.

91 Markezich R L and Aschbacher D G, 'Chlorinated flame-retardant in combination with other flame retardants', *ACS Symposium Series*, 1995, **599**, 65–75.

92 Carty P and White S, 'A synergistic organoiron flame-retarding smoke-suppressing system for ABS', *Appl. Organomet. Chem.*, 1991, **5**(1), 51–6.

93 Carty P and White S, 'The effects of antimony (III) oxide and basic iron (III) oxide on the flammability and thermal-stability of a tertiary polymer blend', *Polym. Degrad. Stab.*, 1995, **47**(2), 305–10.

94 Murashko E A, Levchik G F, Levchik S V, Bright D A and Dashevsky S, 'Fire retardant action of resorcinol bis(diphenyl phosphate) in a PC/ABS blend. 1: Combustion performance and thermal decomposition behaviour', *J. Fire Sci.*, **16**(4), 278–96.

95 Jung H C, Kim W N, Lee C R, Suh K S and Kim S R, 'Properties of flame-retarding blends of polycarbonate and poly(acrylonitrile-butadiene-styrene)', *J. Polym. Eng.*, 1998, **18**(1–2), 115–30.

96 Green J, 'Phosphorus-bromine flame retardant synergy in polycarbonate/ABS blends', *Polym. Degrad. Stab.*, 1996, **54**(2–3), 189–93.

97 Antonov A V and Novokov S N, 'Study of the mechanism of fire retardants action in high impact-resistant ABS polycarbonate blend', *Vysokomol. Soed., Ser. A and B*, 1993, **35**(9), A1442–A1448.

98 Green J, 'Flame retarding polycarbonate ABS blends with a brominated phosphate', *J. Fire Sci.*, 1991, **9**(4), 285–95.

99 Selvaraj I I and Chandrasekhar V, 'Copolymerization of 2-(4′-vinyl-4-biphenylyloxy)pentachlorocyclotriphosphazene with acrylate and methacrylate monomers', *Polymer*, 1997, **38**(14), 3617–23.

100 Bosscher G and Vandergrampel J C, 'Synthesis and polymerization of gem-methyl(vinylbenzyl)tetrachlorocyclotriphosphazene', *J. Inorg. Organomet. Polym.*, 1995, **5**(3), 209–16.

101 Allen D W, Anderto E C, Bradley C and Schiel L E, 'Fire-retardant polymers: polymerization of 1-oxo-2,6,7-trioxa-1-phosphabicyclo[2,2,2]octa-4-yl methyl methacrylate, and its copolymerization with methyl methacrylate, styrene, and triallylcyanurate', *Polym. Degrad. Stab.*, 1995, **47**(1), 67–72.

102 Catala J M and Brossas J, 'Synthesis of fire-retardant polymers without halogens', *Progress in Organic Coatings*, 1993, **22**(1–4), 69–82.

103 Cullis C F, Hirschler M M and Tao Q M, 'Studies of the effects of phosphorus nitrogen bromine systems on the combustion of some thermoplastic polymers', *Eur. Polym. J.*, 1991, **27**(3), 281–89.

104 Sirdesai S J and Wilkie C A, 'Wilkinson salt – a flame-retardant for poly(methyl methacrylate)', *J. Appl. Polym. Sci.*, 1989, **37**(4), 863–6.

105 Lomakin S M, Zaikov G E and Artsis M I, 'New types of ecologically safe flame retardant systems for poly(methyl methacrylate)', *Int. J. Polym. Mater.*, 1996, **32**(1–4), 213–20.

106 Hall M E, Horrocks A R and Zhang J, 'The flammability of polyacrylonitrile and its copolymers', *Polym. Degrad. Stab.*, 1994, **44**(3), 379–86.

107 Zang J, Hall M E and Horrocks A R, 'The flammability of polyacrylonitrile and its copolymers. 1: The flammability assessment using pressed powdered polymer samples', *J. Fire Sci.*, 1993, **11**(5), 442–56.

108 Ballistreri A, Foti S, Montaudo G, Scamporrino E, Arnesano A and Calgari S, 'Thermal-decomposition of flame-retardant acrylonitrile polymers. 2: Effect of red phosphorus', *Makromol. Chem., Macromol. Chem. Phys.*, 1981, **182**(5), 1301–6.

109 Chou S and Wu C J, 'Effect of brominated flame retardants on the properties of acrylonitrile/vinyl acetate copolymer fibers', *Text Res. J.*, 1995, **65**(9), 533–9.

110 Tsai J S, 'The effect of flame retardants on the properties of acrylic and modacrylic fibers', *J. Mater. Sci.*, 1993, **28**(5), 1161–7.

111 Mijangos C, Martinez G and Millan J L, 'New approaches to the study of labile structures in poly(vinyl chloride) by phenolysis', *Eur. Polym. J.*, 1982, **18**(8), 731–4.

112 Naqvi M K and Joseph P, 'A study of the kinetics of acetoxylation of PVC under homogeneous conditions and the thermal stability of the modified polymer', *Polym. Commun.*, 1986, **27**, 8–11.

113 Chandler L A and Hirschler M M, 'Further chlorination of poly(vinyl chloride): effects on flammability and smoke production tendency', *Eur. Polym. J.*, 1987, **23**(9), 677–83.

114 Carty P and White S, 'Anomalous flammability behaviour of CPVC (chlorinated polyvinyl chloride) in blends with ABS (acrylonitrile-butadiene-styrene) containing flame-retarding/smoke-suppressing compounds', *Polymer*, 1997, **38**(5), 1111–9.

115 Lizymol P P and Thomas S, 'Thermal, flame and mechanical behaviour of ternary blends of poly(vinyl chloride), poly(ethylene-*co*-vinyl acetate) and poly(styrene-*co*-acrylonitrile)', *Thermochimica Acta*, 1994, **233**(2), 283–95.

116 Carty P, Metcalfe E and White S, 'A view of the role of iron containing compounds in char forming/smoke suppressing reactions during the thermal decomposition of semi-rigid poly(vinyl chloride) formulations', *Polymer*, 1992, **33**(13), 2704–8.

117 Carty P and White S, 'A review of the role of basic iron(III) oxide as char forming/smoke suppressing flame retarding additive in halogenated polymers and halogenated polymer blends', *Polymers and Polymer Composites*, 1998, **6**(1), 33–8.

118 Carty P and White S, 'The effects of antimony(III) oxide and basic iron(III) oxide on the flammability and thermal-stability of a ternary polymer blend', *Polym. Deg. and Stab.*, 1995, **47**(2), 305–10.

119 Carty P, Grant J and Metcalfe E, 'Flame-retardancy and smoke-suppression studies on ferrocene derivatives in PVC', *Appl. Organomet. Chem.*, 1996, **10**(2), 101–11.

120 Carty P, Metcalfe E and Saben T J, 'Thermal analysis of plasticized PVC containing flame-retardant smoke suppressant inorganic and organometallic iron compounds', *Fire Safety J.*, 1991, **17**(1), 45–56.

121 Cullis C F, Hirschler M M and Tao Q M, 'The effect of red phosphorus on the flamability and smoke-producing tendency of poly(vinyl chloride) and polystyrene', *Eur. Polym. J.*, 1986, **22**(2), 161–7.

122 Annakutty K S and Kishore K, 'Novel flame retardant plasticizer for poly(vinyl chloride)', *Eur. Polym. J.*, 1993, **29**(10), 1387–90.

123 Kannan P and Kishore K, 'Polyethylene stibinite phosphate esters: novel flame-retardant plasticizers for PVC', *Eur. Polym. J.*, 1997, **33**(10–12), 1799–809.

124 Reddy P V, Sridhar S and Ratra M C, 'Flame-retardant PVC compound for cable applications', *J. Appl. Polym. Sci.*, 1992, **46**(3), 483–8.

125 Stoeva S, Karaivanova M and Benev D, 'Poly(vinyl chloride) composition. 2: Study of the flammability and smoke-evolution of unplasticized poly (vinyl chloride) and fire-retardant additives', *J. Appl. Polym. Sci.*, 1992, **46**(1), 119–27.

126 Sharma S K, 'Metal-based organic complex as fire retardant and smoke suppressant for polyvinyl-chloride', *Research and Industry*, 1993, **38**(4), 268–72.

127 Wang Y Z, Zheng C Y and Wu D C, 'Properties of polyethylene terephthalate/polysulfonyldiphenylene phenylphosphonate flame-retardant systems', *Chemical Journal of Chinese Universities*, 1997, **18**(3), 472–6.

128 Park H S, Keun J H, Yeom K S, Kang D W and Im W B, 'Synthesis and physical properties of two-component polyurethane flame-retardant coatings using chlorine-containing aliphatic polyesters', *Polymer-Korea*, 1995, **19**(6), 891–902.

129 Park H, Keun J and Lee K, 'Synthesis and physical properties of two-component polyurethane flame-retardant coatings using chlorine-containing modified polyesters', *J. Polym. Sci., Part A, Polym. Chem.*, 1996, **34**(8), 1455–64.

130 Park H S, Ha K J, Keun J H and Kim T O, 'Preparation and physical properties of two-component polyurethane flame-retardant coatings using trichloro modified polyesters', *J. Appl. Polym. Sci.*, 1998, **70**(5), 913–20.

131 Lin R K, Way T F, Huang L C, Liou R J, Sheng C C, Hwang S K and Tang H I, 'A model study of the flame-retardant co-PET containing phenylphosphate units in the main-chain. 1', *Abstracts of Papers of the American Chemical Society*, 1995, **209**(2), 68.

132 Lin R K, Way T F, Liou R J, Huang L C, Fan C L, Sheng C C and Tang H I, 'A model study of the flame-retardant co-PET containing phenylphosphate units in the main-chain. 2', *Abstracts of Papers of the American Chemical Society*, 1995, **209**(2), 69.

133 Wan I Y, Keifer L A and McGrath J E, 'Phosphine oxide containing poly(ethylene terephthalate) copolymers', *Abstracts of papers of the American Chemical Society*, 1995, **209**(2), 118.

134 Ma Z L, Zhao W G, Liu Y F and Shi J R, 'Synthesis and properties of intumescent, phosphorus-containing, flame-retardant polyesters', *J. Appl. Polym. Sci.*, 1997, **63**(12), 1511–515.

135 Tang H I, Lin R K, Way T F, Liou R J, Huang L C, Lin J T and Sheng C C, 'A study of thermal stability of polyester containing phenyl phosphonate unit for flame retarded fiber', *Polym. Degrad. Stab.*, 1996, **54**(2–3), 373–7.

136 Chand S J and Chang F C, 'Characterizations of blends of phosphorus-containing copolyester with poly(ethylene terephthalate)', *Polym. Eng. and Sci.*, 1998, **38**(9), 1471–81.

137 Wang C S, Shieh J Y and Sun Y M, 'Synthesis and properties of phosphorus containing PET and PEN (1)', *J. Appl. Polym. Sci.*, 1998, **70**(10), 1959–64.

138 Delaviz Y, Gungor A, McGrath J E and Gibson H W, 'Phosphine oxide containing aromatic polyester', *Polymer*, 1992, **33**(24), 5346–7.

139 Wang C S, Lin C H and Chen C Y, 'Synthesis and properties of phosphorus-containing polyesters derived from 2-(6-oxido-6H-dibenz<c,e><1,2>oxaphosphorin-6-yl)-1,4-hydroxyethoxy phenylene', *J. Polym. Sci., Part A, Polym. Chem.*, 1998, **36**(17), 3051–61.

140 Pingel E, Markovski L J, Spilman G E, Foran B J, Tao J A and Martin D C, 'Thermally crosslinkable thermoplastic PET-co-XTA copolyesters', *Polymer*, 1999, **40**(1), 53–64.

141 Levchik S V, Costa L and Camino G, 'Effect of the fire-retardant, ammonium polyphosphate, on the thermal decomposition of aliphatic polyamides. 1. Polyamide-11 and polyamide-12', *Polym. Degrad. Stab.*, 1992, **36**(1), 31–41.

142 Levchik S V, Costa L and Camino G, 'Effect of the fire-retardant, ammonium polyphosphate, on the thermal decomposition of aliphatic polyamides. 2. Polyamide-6', *Polym. Degrad. Stab.*, 1992, **36**(3), 229–37.

143 Levchik S V, Costa L and Camino G, 'Effect of ammonium polyphosphate on combustion and thermal degradation of aliphatic polyamides', *Makromol. Chem., Macromol. Symp.*, 1993, **74**, 95–9.

144 Levchik S V, Costa L and Camino G, 'Effect of the fire-retardant ammonium polyphosphate on the thermal decomposition of aliphatic polyamides. 3. Polyamides 6,6 and 6,10', *Polym. Degrad. Stab.*, 1994, **43**(1), 43–54.

145 Siat C, Bourbigot S and LeBras M, 'Thermal behaviour of polyamide-6-based intumescent formulations – a kinetic study', *Polym. Degrad. Stab.*, 1997, **58**(3), 303–13.

146 Levchik S V, Levchik G F, Camino G and Costa L, 'Mechanism of action of phosphorus-based flame retardants in nylon-6. 2. Ammonium polyphosphate talc', *J. Fire Sci.*, 1995, **13**(1), 45–58.

147 Hornsby P R, Wang J, Rothon R, Jackson G, Wilkinson G and Cossick K, 'Thermal decomposition behaviour of polyamide fire-retarded compositions

containing magnesium hydroxide filler', *Polym. Degrad. Stab.*, 1996, **51**(3), 235–49.

148 Levchik G F, Levchik S V and Lesnikovich A I, 'Mechanisms of action in flame retardant reinforced nylon 6', *Polym. Degrad. Stab.*, 1996, **54**(2–3), 361–3.

149 Levchik S V, Levchik G F, Camino G, Costa L and Lesnikovich A I, 'Fire-retardant action of potassium nitrate in polyamide 6', *Angew, Makromol. Chem.*, 1997, **245**, 23–35.

150 Levchik S V, Levchik G F, Balabanovich A I, Camino G and Costa L, 'Mechanistic study of combustion performance and thermal decomposition behaviour of nylon 6 with added halogen-free fire retardants', *Polym. Degrad. Stab.*, 1995, **54**(2–3), 217–22.

151 Stern G and Horacek H, 'Melamine cyanurate, halogen and phosphorus free flame-retardant', *Int. J. Polym. Mats.*, 1994, **25**(3–4), 255–68.

152 Nagasawa Y, Hotta M and Ozawa K, 'Fast thermolysis/FT–IR studies of fire-retardant melamine-cyanurate and melamine-cyanurate containing polymer', *J. Analyt. Appl. Pyrolysis*, 1995, **33**, 253–67.

153 Shimasaki C, Watanabe N, Fukushima K, Rengakuji S, Nakamura Y, Ono S, Yoshimura T, Morita H, Takahura M, Shiroishi A, 'Effect of the fire-retardant, melamine, on the combustion and thermal decomposition of polyamide-6, polypropylene and low-density polyethylene', *Polym. Degrad. Stab.*, 1997, **58**(1–2), 171–80.

154 Casu A, Camino G, DeGiorgi M, Flath D, Morone V and Zenoni R, 'Fire-retardant mechanistic aspects of melamine cyanurate in polymide copolymer', *Polym. Degrad. Stab.*, 1997, **58**(3), 297–302.

155 Lomakin S M and Zaikov G E, 'New type of ecologically safe flame retardant based on polymer char former', *Polym. Degrad. Stab.*, 1996, **51**(3), 343–50.

156 Zaikov G E and Lomakin S M, 'Polymer flame retardancy: a new approach', *J. Appl. Polym. Sci.*, 1998, **68**(5), 715–25.

157 Wan I Y, McGrath J E and Kashiwagi T, 'Triarylphosphine oxide containing nylon-6,6 copolymers', *ACS Symposium Series*, 1995, **599**, 29–40.

158 Wan I Y and McGrath J E, 'Synthesis and characterization of diphenyl-methylphosphine oxide containing nylon-6,6 copolymers', *Abstracts of Papers of the American Chemical Society*, 995, **209**(2), 119.

159 Delaviz Y, Gungor A, McGrath J E and Gibson H W, 'Soluble phosphine oxide containing aromatic polyamides', *Polymer*, 1993, **34**(1), 210–13.

160 Zhang Y H, Tebby J C and Wheeler J W, 'Polyamides incorporating phosphine oxide groups. 1. Wholly aromatic polymers', *J. Polym. Sci., Part A, Polym. Chem.*, 1996, **34**(8), 1561–6.

161 Green J, 'Flame retarding polycarbonate ABS blends with a brominated phosphate', *J. Fire Sci.*, 1991, **9**(4), 285–95.

162 Green J, 'A phosphorus-bromine flame-retardant for engineering thermoplastics – A review', *J. Fire Sciences*, 1994, **12**(4), 388–408.

163 Green J, 'Phosphorus-bromine flame-retardant synergy in a polycarbonate polyethylene terephthalate blend', *J. Fire Sci.*, 1994, **12**(3), 257–67.

164 Green J, 'Char studies – Flame retarded polycarbonate/PET blend', *J. Fire Sci.*, 1994, **12**(6), 551–81.

165 Green J, 'Phosphorus-bromine flame retardant synergy in polycarbonate/ABS blends', *Polym. Degrad. Stab.*, 1996, **54**(2/3), 189–93.

166 Iji M and Serizawa S, 'New flame-retarding polycarbonate resin with silicone derivative for electronics product', *NEC Research and Dev.*, 1998, **39**(2), 82–7.

167 Moore S, 'Additives – Silicone flame retardant boosts properties of polycarbonate', *Modern Plastics*, 1998, **9**, 35–6.

168 Masatoshi I and Serizawa S, 'Silicone derivatives as new flame retardants for aromatic thermoplastics used in electronic devices', *Polymers for Advanced Technologies*, 1998, **9**(10/11), 593–600.

169 Morgan A B and Tour J M, 'Synthesis and testing of nonhalogenated alkyne-containing flame-retarding polymer additives', *Macromolecules*, 1998, **31**(9), 2857–65.

170 Knauss D M, McGrath J E and Kashiwagi T, 'Copolycarbonates and poly(arylates) derived from hydrolytically stable phosphine oxide comonomers', *ACS Symp. Ser.*, 1995, **599**, 41–55.

171 Murashko E A, Levchik G F, Levchik S V, Bright D A and Dashevsky S, 'Fire retarded action of resorcinol bis(diphenyl phosphate) in a PPO/HIPS blend', *J. Fire Sci.*, 1998, **16**(4), 233–49.

172 Riley D J, Gungor A, Srinivasan S A, Sankarapandian M, Tchatchoua C, Muggli M W, Ward T C, McGrath J E and Kashiwagi T, 'Synthesis and characterization of flame resistant poly(arylene ether)s', *Polym. Eng. and Sci.*, 1997, **37**(9), 1501–11.

173 Srinivasan S, Kagumba L, Riley D J and McGrath J E, 'Synthesis of hydrolytically stable phosphorus-containing polymers via polycondensation and their utilization in polymer blends', *Macromol. Symp.*, 1997, **122**, 95–100.

174 Bauer R G, Kavkoch R W and O'Connor J M, 'Fire retardant emulsion polymers', *J. Fire Retard. Chem.*, 1976, **3**(2), 99–110.

175 Menon A R R, 'Flame-retardant characteristics of natural rubber modified with a bromo derivative of phosphorylated cashew nut shell liquid', *J. Fire Sci.*, 1997, **15**(1), 3–13.

176 Gosh S N and Maiti S, 'A polymeric flame-retardant additive for rubbers', *J. Polym. Mater.*, 1994, **11**(1), 49–56.

177 Derouet D, Morvan F and Brosse J-C, 'Flame resistance and thermal stability of 1,4-polydienes modified by dialkyl (or diaryl) phosphates', *J. Nat. Rubb. Res.*, 1996, **11**(1), 9–25.

178 Derouet D, Radhakrishnan N, Brosse J-C and Boccaccio G, 'Phosphorus modification of epoxidised liquid natural rubber to improve flame resistance of vulcanized rubbers', *J Appl. Polym. Sci.*, 1994, **52**(9), 1309–16.

179 Camino G, Guaita M and Priola A, 'Study of flame retardance in brominated liquid polybutadienes', *Polym. Degrad. Stab.*, 1985, **12**(3), 241–7.

180 Brosse J C, Koh M P and Derouet D, 'Modification chimique du polybutadiene-1,2 par le phosphonate d'ethyle', *Eur. Polym. J.*, 1983, **19**(12), 1159–65.

181 Derouet D, Brosse J C and Pinazzi C P, 'Modification chimique des polyalcadienes par les composes du type CX_3POZ_2-I: Modification des polybutadienes-1,2 par le dichlorure de trichloromethylphosphonyle', *Eur. Polym. J.*, 1981, **17**(7), 763–72.

182 Boutevin B, Hervaud Y and Mouledous G, 'Grafting phosphonated thiol on hydroxy telechelic polybutadiene', *Polym. Bull.*, 1998, **41**(2), 145–51.

183 Stull D P, Fogerty R and Chen S M, 'Chlorendic diol-antimony trioxide syner-gism – Use and performance of a novel rigid urethane foam retardant', *J. Cellular Plastics*, 1986, **22**(2), 147–66.

184 Brzozowski B K, Pietruszka N, Gajewski J and Stankiewicz R, '2,3-Dibromo-2-butene-1,4-diol (DBBD) – A new ecological reactive fire retardant for self-extinguishing polyurethane foams. Part 1. Synthesis and solubility in polyols', *Polimery*, 1998, **43**(4), 252–5.

185 Park H S, Keun J H, Yeom K S, Kang D W and Im W B, 'Syntheses and phys-ical properties of two-component polyurethane flame-retardant coatings using chlorine-containing modified polyesters', *J. Polym. Sci., Part A, Polym. Chem.*, 1996, **34**(8), 1455–64.

186 Park H S, Hahm H S and Park E K, 'Preparation characteristics of two-component polyurethane flame retardant coatings using 2,3-dibromo modified polyesters', *J. Appl. Polym. Sci.*, 1996, **61**(3), 421–9.

187 Lee F T, Nicholson P and Green J, 'New reactive phosphorus flame-retardant for rigid polyurethane application. 1', *J. Fire Ret. Chem.*, 1982, **9**(3), 194–205.

188 Lee F T, Green J and Gibilisco R D, 'New reactive phosphorus flame-retardant for rigid polyurethane application. 2', *J. Fire Ret. Chem.*, 1982, **9**(4), 232–44.

189 Sivriev C and Zabski L, 'Flame retarded rigid polyurethane foams by chemical modification with phosphorus-containing and nitrogen-containing polyols', *Eur. Polym. J.*, 1994, **30**(4), 509–14.

190 Liu Y L, Hsiue G H, Lan C W and Chiu Y S, 'Flame-retardant polyurethanes from phosphorus-containing isocyanates', *J. Polym. Sci., Part A, Polym. Chem.*, 1997, **35**(9), 1769–80.

191 Guo W J, 'Flame-retardant modification of UV-curable resins with monomers containing bromine and phosphorus', *J. Polym. Sci., Part A, Polym. Chem.*, 1992, **30**(5), 819–27.

192 Singh B, Proskoff H, Lekovich H, Sendijarevic A, Sendijarevic V, Klempner D and Frisch K C, 'Carbamylmethylated melamine polyols in rigid water-blown urethane foams', *Cellular Polymers*, 1992, **11**(6), 475–89.

193 Ionescu M, Mihalache I, Zugravu V and Mihai S, 'Inherently flame-retardant rigid polyurethane foams based on new triazinic polyether polyols', *Cellular Polymers*, 1994, **13**(1), 57–68.

194 Benrashid R and Nelson G L, 'Flammability improvement of polyurethanes by incorporation of a silicone moiety into the stucture of block-copolymers', *ACS Symposium Series*, 1995, **599**, 217–35.

195 Ji Q, Muggli M, Wang F, Ward T C, Burns G, Sorathia U and McGrath J E, 'Synthesis and characterization of modified segmented polyurethanes displaying improved fire resistance', *ACS Polym. Prepr.*, 1997, **38**(1), 219–20.

196 Najafi-Mohajeri N and Nelson G L, 'New flame retardant ferrocene modified polyurethane block copolymers', *ACS Polym. Prepr.*, 1998, **39**(2), 380–1.

197 Yalinkilic M K, Imamura Y, Takahashi M and Demirici Z, 'Effect of boron addi-tion to adhesive and/or surface coating on fire-retardant properties of parti-cleboard', *Wood and Fibre Science*, 1998, **30**(4), 348–59.

198 Costa L, DiMontelera L R, Camino G, Weil E D and Pearce E M, 'Flame-retardant properties of phenol-formaldehyde-type resins and triphenyl phosphate in styrene-acrylonitrile copolymers', *J. Appl. Polym. Sci.*, 1998, **68**(7), 1067–76.

199 Mandal H and Hay A S, 'Polycyclic phosphonate resins: thermally crosslinkable intermediates for flame-resistant materials', *J. Polym. Sci., Part A, Polym. Chem.*, 1988, **36**(11), 1911–8.

200 Mamleev V S, Bekturov E A and Gibov K M, 'Dynamics of intumescence of fire-retardant materials', *J. Appl. Polym. Sci.*, 1998, **70**(8), 1523–42.

201 Jang J, Chung H, Kim M and Kim Y, 'Improvement of flame retardancy in phenolics and paper sludge/phenolic composites', *J. Appl. Polym. Sci.*, 1998, **69**(10), 2043–50.

202 Antony R and Pillai C K S, 'Synthesis and thermal characterization of chemically-modified phenolic resins', *J. Appl. Polym. Sci.*, 1994, **54**(4), 429–38.

203 Cusack P A, Heer M S and Monk A W, 'Zinc hydroxystannate – A combined flame-retardant and smoke suppressant for halogenated polyesters', *Polym. Degrad. Stab.*, 1991, **32**(2), 177–90.

204 Derouet D, Morvan F and Brosse J C, 'Chemical modification of epoxy resins by dialkyl (or aryl) phosphates: evaluation of fire behaviour and thermal stability', *J. Appl. Polym. Sci.*, 1996, **62**(11), 1855–68.

205 Liu Y L, Hsiue G H and Chiu Y S, 'Synthesis, characterization, thermal, and flame retardant properties of phosphate-based epoxy resins', *J. Polym. Sci., Part A, Polym. Chem.*, 1997, **35**(3), 565–74.

206 Cho C S, Chen L W and Chiu Y S, 'Novel flame retardant epoxy resis – 1: Synthesis, characterization, and properties of aryl phosphinate epoxy ether cured with diamine', *Polym. Bull.*, 1998, **41**(1), 45–52.

207 Wang C S and Shieh J Y, 'Synthesis and properties of epoxy resins containing 2-(6-oxid-6H-dibenz⟨c,e⟩⟨1,2⟩oxaphosphorin-6-yl)1,4-benzenediol', *Polymer*, 1998, **39**(23), 5819–26.

208 Buckingham M R, Lindsay A J, Stevenson D E, Muller G, Morel E, Coates B and Henry Y, 'Synthesis and formulation of novel phosphorylated flame retardant curatives for thermoset resins', *Polym. Degrad. Stab.*, 1996, **54**(2–3), 311–5.

209 Levchik S V, Camino G, Luda M P, Costa L, Muller G, Costes B and Henry Y, 'Epoxy resins cured with aminophenylmethylphosphine oxide. 1. Combustion performance', *Polymers for Advanced Technologies*, 1996, **7**(11), 823–30.

210 Levchik S V, Camino G, Luda M P, Costa L, Muller G and Costes B, 'Epoxy resins cured with aminophenylmethylphosphine oxide. II. Mechanism of thermal decomposition', *Polym. Degrad. Stab.*, 1998, **60**(1), 169–83.

211 Cho C S, Fu S C, Chen L W and Wu T R, 'Aryl phosphinate anhydride curing for flame retardant epoxy networks', *Polym. Internat.*, 1998, **47**(2), 203–9.

212 Rogers M E, Brink M H, McGrath J E and Brenan A, 'Semicrystalline and amorphous fluorine-containing polyimides', *Polymer*, 1993, **34**(4), 849–55.

213 Tan B, Tchatchoua C N, Dong L and McGrath J E, 'Synthesis and characterization of arylene ether imide reactive oligomers and polymers containing diarylalkylphosphine oxide groups', *Polymers for Advanced Technologies*, 1998, **9**(1), 84–93.

214 Liu Y L, Hsiue G H, Lan C W, Kuo J K, Jeng R J and Chiu Y S, 'Synthesis, thermal properties, and flame retardancy of phosphorus containing polyimides', *J. Appl. Polym. Sci.*, 1997, **63**(7), 875–82.

215 Wescott J M, Yoon T H, Rodrigues D, Kiefer L A, Wilkes G L and McGrath J E, 'Synthesis and characterization of triphenylphosphine oxide-containing poly(aryl imide)-poly(dimethyl siloxane) randomly segmented copolymers', *J. Macromol. Sci. – Pure and Applied Chem.*, 1994, **A31**(8), 1071–85.

216 Brown J R and St John N A, 'Fire-retardant low-temperature-cured phenolic resins and composites', *Trends in Polym. Sci.*, 1996, **4**(12), 416–20.

217 Ishida H and Allen D J, 'Physical and mechanical characterization of near-zero shrinkage polybenzoxazines', *J. Polym. Sci., Part B, Polym. Phys.*, 1996, **34**(6), 1019–30.

Applications of halogen flame retardants

PIERRE GEORLETTE

Technical Service Manager FR SBU
Dead Sea Bromine Group
Beer Sheva
ISRAEL

8.1 Introduction

Flame-retardants, by far the largest group of plastics additives, are playing a major role in the plastics industry by improving life safety. In 1997, some $US 2.2 billion of flame retardants were consumed accounting for 27% of the plastics additives market (Fig. 8.1).

The rapid development of plastics applications in the electronic, building and automotive industries is very demanding with regard to properties and cost. Between 150 and 200 flame retardants have been designed to cover most of the requirements of the market. They are mainly based on halogen (bromine and chlorine), phosphorus, inorganics and melamine compounds. Among them brominated flame retardants are known for their very efficient role in saving lives and goods due to their optimal combination of properties. They are the main players in the market with around 39% of its share (Fig. 8.2).

Chemical names, structures, physical properties and major applications of the most important halogenated fire retardants are given in Table 8.1.

Since the early nineteen nineties, 'Green' parties in some European countries have been investing considerable effort to limit as much as possible the uses of halogenated flame retardants, particularly those based on diphenyl oxide (DPO) claiming that they may be a source of toxic fumes under fire conditions or during incineration. Consequently some producers of electronic goods have started to offer products with non-halogen flame retardants or even without flame-retardant which still satisfy the lower standards of flame retardancy presently applied in Europe. Since then in some European countries an increase in the number of fires has been observed.[1]

Conscious of the danger of such a trend, fire experts, fire brigades and the major producers of halogenated flame retardants have started to inform the relevant authorities and the consumers more systematically about the safety of using commercial halogenated flame retardants offered in the

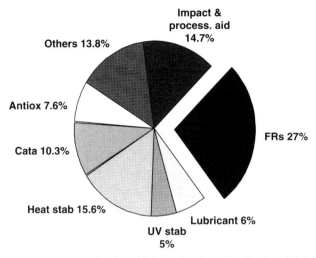

8.1 Worldwide plastic additives. Market distribution ($8.5 bn 1997)

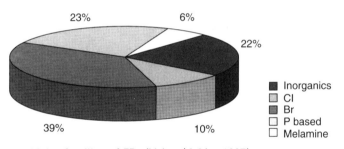

8.2 Major families of FRs (Value $2.2 bn 1997)

market. Though not as well-known, improvement of fire safety by use of halogenated flame retardants is also an important factor in protecting the environment as it reduces the production of considerable quantities of toxic smoke.[2]

8.2   Mechanism of flame retardancy with halogenated compounds

A detailed account of the various theoretical explanations for the efficiency of flame retardants is presented in Chapter 2. A general outline only is presented here.

A major factor in the combustion of plastics is the presence of highly active OH$^\bullet$ and H$^\bullet$ free radicals which play an important role in the chain reactions leading to decomposition and burning. The hydrogen halides

Table 8.1 Part I: properties and main applications of halogenated flame retardants

Chemical name	Chemical structure	Tradename	Halogen content %	Melting/ soft. range °C	TGA[a] 5% weight loss °C	Main applications
Hexabromocyclododecane (HBCD)		FR-1206 (DSBG) CD-75P (GLCC) HBCD (Albemarle)	74.7 (Br)	175–197	230	EPS-XPS[b] & textiles
Tribromophenol allyl ether		PHE-65 (GLCC)	64.2 (Br)	74–76	154	EPS-XPS
Tetrabromobisphenol A bis (allyl ether)		BE-51 (GLCC)	51.2 (Br)	115–120	238	XPS

[a] Thermogravimetric analysis under air, 10 °C per minute.
[b] Expanded and extruded polystyrene foams.

Table 8.1 (cont.) Part II

Chemical name	Chemical structure	Tradename	Halogen content %	Melting/ soft. range °C	TGA[a] 5% weight loss °C	Main applications
Brominated indan (Br indan)	$m+n=7.8$	FR-1808 (DSBG)	73 (Br)	240–255	325	HIPS, ABS, PE, polyamides
Brominated epoxy oligomers (BEOs)		F-2016 (DSBG)	49–51 (Br)	105–115	340	ABS & PC/ABS alloys
Modified brominated epoxy oligomers (MBEOs)		F-3000 series (DSBG)	56 (Br)	113–127	360	HIPS, ABS & phenolic laminates

[a] Thermogravimetric analysis under air, 10°C per minute.

Table 8.1 (cont.) Part III

Chemical name	Chemical structure	Tradename	Halogen content %	Melting/ soft. range °C	TGA[a] 5% weight loss °C	Main applications
Decabromodiphenyl oxide (DECA)		FR-1210 (DSBG) DE-83 (GLCC) Saytex 102 (Albemarle)	83 (Br)	305 min.	362	HIPS, PE, PP, PBT, polyamide 6 epoxy & textiles
Proprietary		Saytex 8010 (Albemarle)	82 (Br)	380	na[b]	HIPS, ABS, PE, PBT & polyamides
Octabromodiphenyl oxide (OCTA)		FR-1208 (DSBG) DE-79 (GLCC)	79 (Br)	70–150	304	HIPS & ABS
Tetrabromobisphenol A (TBBA)		FR-1524 (DSBG) BA-59P (GLCC) RB-100 (Albemarle)	58.5 (Br)	181	305	Epoxy laminates, ABS & FR intermediates

[a] Thermogravimetric analysis under air, 10°C per minute.
[b] Non available.

Table 8.1 (cont.) Part IV

Chemical name	Chemical structure	Tradename	Halogen content %	Melting/ soft. range °C	TGA[a] 5% weight loss °C	Main applications
Ethylenebistetrabromo-phthalimide		BT-93 (Albemarle)	67.2 (Br)	450–455	442	HIPS, PE, PBT
Bis (tribromophenoxy) ethane		FF-680 (GLCC)	70 (Br)	223–228	290	ABS
Tris(tribromophenyl) cyanurate		SR-245 (DKS) & FR-245 (DSBG)	67 (Br)	230	385	HIPS & ABS
Chlorinated paraffin		Chlorez 760 et al	74 (Cl)	160	na[b]	PVC, PE & HIPS

[a] Thermogravimetric analysis under air, 10°C per minute.
[b] Not available.

Table 8.1 (cont.) Part V

Chemical name	Chemical structure	Tradename	Halogen content %	Melting/ soft. range °C	TGA[a] 5% weight loss °C	Main applications
Brominated polystyrene		Pyrocheck 68 PB (Ferro) HP-7010 (Albemarle)	67 (Br)	240–260	360	PBT & polyamides
Phenoxy-terminated carbonate oligomer of TBBA		BC-52 & 58 (GLCC) FG 7/8000 series (Teijin)	51 & 58 resp. (Br)	210–230 & 230–260 resp.	430	PBT & PC
Poly (pentabromobenzyl acrylate)		FR-1025 (DSBG)	70 (Br)	190–220	345	PBT, polyamides & HIPS

[a] Thermogravimetric analysis under air, 10°C per minute.

Table 8.1 (cont.) Part VI

Chemical name	Chemical structure	Tradename	Halogen content %	Melting/ soft. range °C	TGA[a] 5% weight loss °C	Main applications
High MW Brominated epoxy		F-2000 series (DSBG)	51–54 (Br)	130–155	344	PBT, polyamides & PC alloys
Dodecachloropentacyclo octadeca-7,15 diene		Dechlorane plus (Oxy)	65 (Cl)	350	320	Polyamides & polyolefins
Poly-(dibromostyrene)		PDBS-80 (GLCC)	59 (Br)	210–230	381	Polyamides & PBT/PET

[a] Thermogravimetric analysis under air, 10°C per minute.

Table 8.1 (cont.) Part VII

Chemical name	Chemical structure	Tradename	Halogen content %	Melting/ soft. range °C	TGA[a] 5% weight loss °C	Main applications
Poly-dibromophenylene oxide		PO-64P (GLCC)	62 (Br)	210–240	400	Polyamides
Bis(2,3-dibromopropyl ether) of TBBA		PE-68 (GLCC) FG-3100 (Teijin) SR-720 (DKS) FR-720 (DSBG) HP-800 (Albemarle)	68 (Br)	90–105	310	PP & HIPS
Tris(tribromoneopentyl) phosphate		CR-900 (Daihachi) FR-370 (DSBG)	70 (Br) 3 (P)	181	309	PP & HIPS

[a] Thermogravimetric analysis under air, 10°C per minute.

Table 8.1 (cont.) Part VIII

Chemical name	Chemical structure	Tradename	Halogen content %	Melting/ soft. range °C	TGA[a] 5% weight loss °C	Main applications
Tris-2,3-dibromopropyl-iso cyanurate		TAIC-6B (Asahi Glass)	65 (Br)	107	316	PP
Ethylene bis-dibromonorbornane dicarboximide		BN-451 (Albemarle)	45 (Br)	294	307	PP
Stabilised hexabromocyclododecane		FR-1206 HT(DSBG) SP-75 (GLCC)	56–72 (Br)	150–197	>250	PP & × PS HIPS

[a] Thermogravimetric analysis under air, 10°C per minute.

Table 8.1 (cont.) Part IX

Chemical name	Chemical structure	Tradename	Halogen content %	Melting/ soft. range °C	TGA[a] 5% weight loss °C	Main applications
Dibromoneopentyl glycol		FR-522 (DSBG)	61 (Br)	109.5	225	Unsaturated polyester & PUR
Tribromoneopentyl alcohol		FR-513 (DSBG)	73.8 (Br)	65	180	PUR & FR intermediate
Tetrabromophthalic anhydride		PHT4 (GLCC) RB-49 (Albemarle)	68.2 (Br)	270–276	250	Unsaturated polyester & FR intermediate

[a] Thermogravimetric analysis under air, 10°C per minute.

Table 8.1 (cont.) Part X

Chemical name	Chemical structure	Tradename	Halogen content %	Melting/ soft. range °C	TGA[a] 5% weight loss °C	Main applications
Tetrabromophthalate diol		PHT4-DIOL (GLCC) RB-79 (Albemarle)	46 (Br)	Liquid	188	PUR
Pentabromodiphenyl oxide		DE-71 (GLCC)	70.8 (Br)	Liquid	243	PUR, phenolic laminates & rubbers
Chlorendic anhydride		HET Acid (Oxy)	54.7 (Cl)	208–210	na[b]	Unsaturated polyester
Halogenated polyetherpolyol	Proprietary	Ixol (Solvay)	32 (Br) 6.5 (Cl) 1.1 (P)	Liquid	na[b]	PUR
Ammonium bromide	NH_4Br	FR-11 (DSBG)	81.6 (Br)	452 (Sublimes)	na[b]	Wood treatment

[a] Thermogravimetric analysis under air, 10°C per minute.
[b] Not available.

released by decomposition of halogenated flame retardants are believed to react with these radicals to produce radicals which are much less active and thus inhibit the propagation of fire. Fractions of the decomposing polymer probably also react with the flame retardant and these entities further assist in de-activation of the OH$^{\bullet}$ and H$^{\bullet}$ free radicals.

The activity of halogenated flame retardants occurs mainly in the gas phase but they are thought to inhibit further burning in the solid or liquid phase where the heavy halogenated (usually brominated) molecular fragments help exclude oxygen from the burning material and possibly encourage char formation.

Halogenated flame retardants are frequently used in combination with synergists, particularly antimony trioxide, which is not itself a flame retardant. The synergist significantly enhances the efficiency of the halogenated flame retardants probably via the formation of volatile antimony halides[3] which not only terminate the chain reaction of combustion but react with the solid phase reducing its flammability.

8.3   Guidelines for choice of halogenated flame retardants

An optimal use of flame retardants is highly recommended and a set of simple rules can help the user to choose the best flame-retardant system for each given application.

8.3.1 Thermal properties

The thermal properties of flame retardants are important for several reasons. Such properties include decomposition temperatures, thermal stability and their role in the thermal ageing of plastics.

The flame retardant must have a decomposition temperature sufficiently lower than that of the polymer in order to ensure its efficiency. Figure 8.3 illustrates this by an example involving polypropylene: it compares the better efficiency of a flame retardant based on aliphatic bromines with that based on aromatic bromines which are more thermally stable.

In another example, flame retardancy of brominated indan (FR-1808 DSBG) has been compared with that of brominated polystyrene (Table 8.1) in nylon 6,6. Brominated indan, a new environmentally friendly (DPO-free) flame-retardant additive, has been recently introduced in order to address market needs for high melt flow moulding with thin walls, thermal stability, cost efficiency and good impact properties (see molecular structure and properties in Table 8.1). It is particularly useful in 'styrenics' and engineering thermoplastics. Its flame-retardant efficiency has been compared with that of brominated polystyrene in nylon 6,6. Thermal stability of both flame

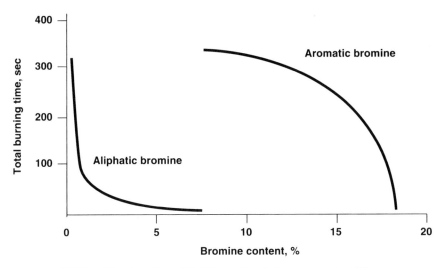

8.3 Total flaming time (UL 94 – vertical), flame retarded PP compounds (1.6 mm)

Table 8.2 Flash-ignition temperature of nylon 6,6 and thermal stability of flame retardants

FR type	Brominated indan (FR-1808 – DSBG)	Brominated polystyrene
Flash ignition temperature of nylon 6,6 (°C)	490	
Flame retardant thermal stability: 10% weight loss by TGA under N_2, 10 °C/min	349	400

retardants and flash-ignition temperature of nylon 6,6[4] shown in Table 8.2 indicate that FR-1808 will decompose 50 °C lower than brominated polystyrene and about 150 °C before the flash-ignition temperature of nylon 6,6. As can be seen in Table 8.3, the flame retardancy achieved with brominated indan, measured by both the Limited Oxygen Index (LOI) and according to the UL 94 standard is significantly better than with brominated polystyrene. Brominated indan also contributes to good melt flow properties.

The thermal stability of the flame retardant should allow it to remain stable throughout the processing of the final compound. Flame-retardant producers usually provide recommended processing temperatures for their additives. Figures 8.4 to 8.6 show temperature limitations normally

Table 8.3 Effect of flame retardant on nylon 6,6 properties

Flame Retardant	Ref No FR	FR-1808 Br indan	Brominated polystyrene
Composition, %			
Nylon 6,6	100	72	70.6
Flame retardant	—	20.6	22
Antimony trioxide	—	7	7
Anti-drip agent (PTFE)	—	0.4	0.4
Bromine content, %	—	15	15
Flammability			
UL 94 (1.6 mm):			
Maximum flaming time, sec	47	1	17
Total flaming time, sec	81	3	69
Numbers of flaming drips	5	0	0
Rating	V-2	V-0	V-1
UL 94 (0.8 mm):			
Maximum flaming time, sec		1	8
Total flaming time, sec		1	13
Numbers of flaming drips	5	0	0
Rating	NR	V-0	V-0
L.O.I., %	22	39	27
Properties			
Spiral flow at 300 °C, cm	99	118	81
Tensile strength at break, MPa	88	82	71
Tensile modulus, MPa	4200	4000	3400
Elongation at break, %	9.6	3.5	4.4
Izod notched impact, J/m	37	29	25
HDT (1.8 Mpa), °C	77	78	80
CTI, V	600	300	450

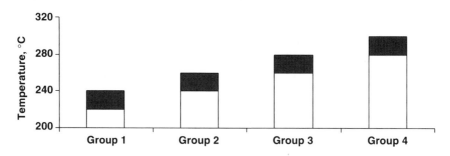

Group1 : HBCD stabilised, TBBA
Group 2: OCTA, Tris(tribromoneopentyl) phosphate
Group 3: Br Indan, Tris(Tribromophenyl) cyanurate, MBEO
Group 4: DECA, BEO

8.4 Maximum processing temperatures in styrenics

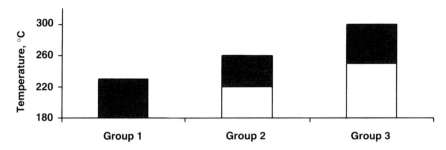

Group 1: HBCD stabilised
Group 2: Tris(tribromoneopentyl) phosphate, TBBA bis(2,3 dibromopropyl ether)
Group 3: DECA

8.5 Maximum processing temperatures in polyolefins

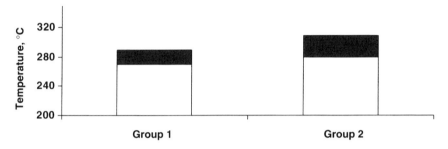

Group 1: Poly (pentabromobenzyl acrylate), Phenoxy-terminated carbonate oligomer of TBBA
Group 2: DECA, Br Indan, High MW Br epoxy, Brominated polystyrene

8.6 Maximum processing temperatures in engineering thermoplastics

applicable to some flame retardants in relation to families of thermoplastic resins.

Dynamic and static thermogravimetric analytical studies may be used to screen and guide the choice of flame retardants (Fig. 8.7).

The thermal stability of flame retardants plays an important role during the thermal ageing test simulating long-term behaviour of finished parts with high working temperatures. Several brominated flame retardants have been shown to improve thermal ageing behaviour of plastics. Poly(pentabromobenzyl acrylate) (FR-1025 DSBG) is a polymeric flame retardant particularly suitable for use with polyamides and PBT with or without fibre reinforcement (see molecular structure and properties in Table 8.1). Poly(pentabromobenzyl acrylate) exhibits inherent advantages over other halogenated FR additives currently offered for the same applications, as a result of its polymeric nature, high bromine content and

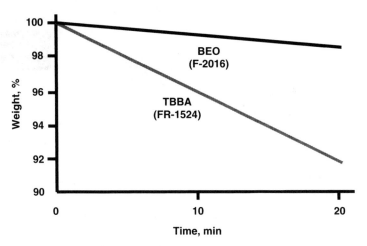

8.7 Isothermal TGA at 240 °C under N$_2$

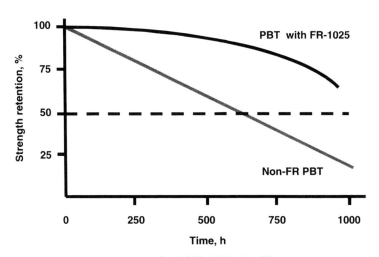

8.8 Thermal ageing at 190 °C of GFR PBT with FR-1025

excellent thermal stability. In addition, the processability of polymers containing it is very good. After a 1000 h thermal ageing treatment at 190 °C of glass reinforced PBT flame retarded by poly(pentabromobenzyl acrylate), tensile properties are maintained above 50% of their initial value, while non-flame retarded PBT would fail during this test (Fig. 8.8). Similar effects of thermal stabilisation of PP by brominated flame retardants are observed after thermal ageing at 150 °C for 1 month.

8.3.2 Purity

The level of purity required for flame retardants depends upon the final application. For instance, production of multilayer epoxy laminates demands superior electric insulation and the purity of the tetrabromo-bisphenol A (Table 8.1) used in this production is typically higher than 99% and its maximum bromide content is strictly controlled. Another example is the production of clear UV stable and flame-retarded unsaturated polyester (UPE) for application in the building industry; the flame retardant of choice is dibromoneopentyl glycol (Table 8.1) with a typical purity above 98% which guarantees low discoloration of sheets during long sun and light exposure. High purity flame retardants with low metal contents are also required for the production of wire and cables with good electrical insulation.[5,6]

8.3.3 Melting or softening range

Applications in the electronic and business machine industries are very demanding in terms of weight reduction, thinner walls and miniaturisation. To cope with this challenge, new flame-retardant systems have been developed which are melt blendable. Such flame-retardant systems provide improvement in flow during injection moulding (Figs. 8.9 and 8.10) allowing reduction in production cost (less weight and shorter moulding cycle). Moreover, easily melt blendable polymeric flame retardants are more efficient and contribute to improvement of mechanical properties (Table 8.4).

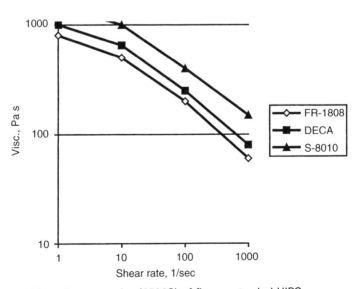

8.9 Viscosity properties (250 °C) of flame retarded HIPS

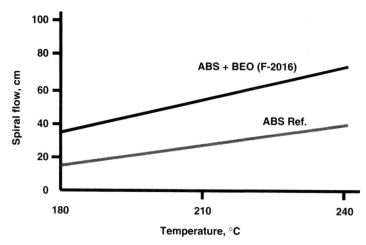

8.10 Spiral flow of FR ABS

Table 8.4 Flame retarded and glass-reinforced PBT

FR Type FR softening range, °C	None	FR-1025 190–220	Br polystyrene 240–260
Composition, %			
30% GR-PBT	100	84.5	79.5
FR-1025	—	10	15
Antimony trioxide	—	5	5
PTFE (antidripping)	—	0.5	0.5
Bromine content, %	0	7	10
Flammability			
UL 94 (0.8 mm)	Burning	V-0	V-0
Limiting oxygen index, %	22	32	28
Properties			
MFI (250 °C – 2.16 Kg), g/10 min	13	24	20
Tensile strength, MPa	98	107	95
Elongation at break, %	2.6	2.1	2.0
Tensile modulus, MPa	11 000	11 000	8200
Izod notched impact (J/m)	87	80	57
HDT (1.8 MPa), °C	192	199	200

8.3.4 Bromine content

The bromine content is generally closely related to flame-retardant efficiency and therefore should be as high as possible in order to reduce the loading required and hence limit changes in properties of the substrate. In Table 8.5, a comparison in high impact polystyrene (HIPS) is made between ethylene bis-tetrabromophthalimide (67% of bromine) and

Table 8.5 Bromine content and properties of UL 94 V-0 HIPS

FR type	Ethylene bis tetra-bromophthalimide	Decabromodiphenyl oxide
Bromine content in the FR, %	67	83
Tensile strength, MPa	13	16
Elongation at break, %	6	20
Notched IZOD, J/m	42	109

Table 8.6 Effect of flame retardant on HIPS properties

Flame Retardant	Ref No FR	BEO (F-2016 DSBG)	MBEO (F-3014 DSBG)
Composition, %			
HIPS	100	77.6	80.6
Flame retardant	—	16	13.8
Antimony trioxide	—	6.4	5.6
Bromine content, %	—	8	8
Flammability			
UL 94 (3.2/2.5 mm)[a]:			
Rating	NR	V-0	V-0
Properties			
MFI (200 °C – 5 Kg)	5	18	15
Tensile strength at break, MPa	21	21	22
Tensile modulus, MPa	1900	2000	2000
Elongation at break, %	52	16	41
Izod notched impact, J/m	143	72	93
HDT (1.8 MPa), °C	77	75	73

[a] V-0 on 1.6 mm is achieved with circa 12–14% of bromine.

decabromodiphenyloxide (83% bromine) both of which are filler-like brominated flame retardants. Ethylene bis-tetrabromophthalimide (Table 8.1) is recommended for UV stable applications but the higher loading needed for good flame retardancy has a detrimental effect on mechanical properties. Table 8.6 gives another comparison between brominated epoxy oligomers and modified brominated epoxy oligomers in HIPS. These flame retardants (molecular structure and properties in Table 8.1) are very similar, melt blendable and recommended when light stability is required. The modified brominated epoxy has a greater bromine content and its use in HIPS allows significant improvement in impact properties.

8.3.5 Compatibility with polymer

Good compatibility with the chosen polymeric system is a way to combine better efficiency in flame retardation with improvement of mechanical properties and avoidance of surface migration. Reactive flame retardants for this purpose must be soluble in the polymerisation medium, for example tetrabromobisphenol A in a solution of epoxy in methyl ethyl ketone or similar. Tribromoneopentyl alcohol and dibromoneopentyl glycol also have good solubility in systems used for the production of polyurethane foam, for instance in polyol and in trichloropropyl phosphate (TCPP) (Table 8.7). Formulations and properties of rigid polyurethane foam flame retarded by these compounds are given in Table 8.8.

Table 8.7 Solubility data in g/100 g at 23 °C

FR Type	Tribromoneopentyl alcohol (FR-513 DSBG)	Dibromoneopentyl glycol (FR-522 DSBG)
Type of solvent		
CMHR polyol	194 (48.5% Br)	35 (15% Br)
CME polyol	233 (51.1% Br)	32 (14% Br)
TCPP	54 (39.8% Br)	25 (12% Br)

Table 8.8 Properties of FR rigid PUR

FR Type	Tribromoneopentyl alcohol (FR-513 DSBG)		Dibromoneopentyl glycol (FR-522 DSBG)
Composition, ppw			
Polyol	80		75
Surfactant		2	
Catalyst		1.5	
HCFC 141B		25	
Water		1	
FR	18		22
DEEP		11	
Isocyanate	127		138
Bromine content, %		5	
Properties			
Cream time, sec	19		19
Gel time, sec	140		68
Tack free time, sec	203		108
Density, Kg/m³	31		34
Heat resistance, °C (DIN 53424)	143		135
Flame retardancy (DIN 4102)	B-2		B-2

Table 8.9 Effect of flame retardant on ABS properties

TYPE OF FR	Ref No FR	BEO (F-2016 DSBG)	Bis(tribromo-phenoxy)ethane (FF-680 GLCC)
Composition, %			
ABS general purpose	100	72.2	72.9
Brominated FR		21	18.6
Antimony trioxide		6.8	8.5
Bromine content, %	0	11	14
Properties:			
Tensile strength, MPa	41	45	39
Tensile modulus, MPa	2300	3100	2500
Izod impact, J/m (notched)	220	35[a]	48
HDT (1.8 MPa), °C (annealed 24 h 65 °C)	83	86	73
UL 94 (1.6 mm)	NR	V-0	V-0
Blooming 1 month RT		none	heavy

[a] Impact of 150 to 200 J/m achievable with high impact ABS.

Polymeric flame-retardant additives exhibit inherent advantages over others regarding compatibility with polymeric systems. This is illustrated by the two following examples. In Table 8.9, a polymeric brominated epoxy flame retardant (BEO) is compared with a typical non-polymeric flame retardant recommended for ABS application. With the BEO, a UL 94 class V-0 (1.6 mm) is achieved at a level of only 11% of bromine while with the non-polymeric flame retardants, some 14% is needed. Moreover, other thermomechanical properties are also improved and no surface blooming is observed. In the second example shown in Table 8.10, mechanical properties of glass-reinforced PBT are improved with increased loading of poly(pentabromobenzyl acrylate) which acts as a compatibiliser between the polymeric matrix and the glass reinforcement.

8.3.6 Conditions of use

The choice of the flame retardant must take into account the conditions of use such as maximum operating temperature, UV or sun exposure, level of electrical insulation properties or chemical environment of the final product. Brominated polystyrene flame retardants with high purity are preferred when high values of tracking index are needed in nylon applications (Table 8.3). For other brominated flame retardants, tracking index value may be improved by adding 10 to 20% of surface-treated magnesium

Table 8.10 Flame-retarded and glass-reinforced PBT properties

FR Type	None	FR-1025 DSBG	
Composition %			
30% GR-PBT	100	85	79
FR-1025	—	10	14
Antimony trioxide	—	5	7
Bromine content, %	0	7	9.7
Flammability			
UL 94 (0.8 mm)	Burning	V-0	V-0
Limiting oxygen index, %	22	32	38
Properties			
Tensile strength, MPa	98	107	116
Elongation at break, %	2.6	2.1	2.0
Tensile modulus, MPa	11 000	11 000	12 000
Izod notched impact (J/m)	87	80	86
HDT (1.8 Mpa), °C	192	199	200

Table 8.11 Methods of testing and standards of fire retardancy

Applications	United States of America	Europe
Computers Business machines & others	UL 94 (V-0)	IEC 65 (HB) & UL 94 Glow wire test
Electronics Building	ASTM E 84 = UL 723	BS 476 Part 7 DIN 4102 – (B1 Brandschacht – B2) NF P 92–501-(Epiradiateur)
Automotive Textiles & furniture Wire & cables	FMVSS 302 California test IEEE 383	FMVSS 302 BS 5852 VDE 0472

(General method of testing, LOI – Cone calorimeter)

hydroxide[7] (FR-20 310-DSBG). Brominated epoxy oligomers and tris(tribromophenyl) cyanurate (FR-245) are optimal choices for UV stable ABS applications. Properties of the recently introduced FR-245 are shown in Table 8.1 and comparative data on UV stability are given in Fig. 8.11.

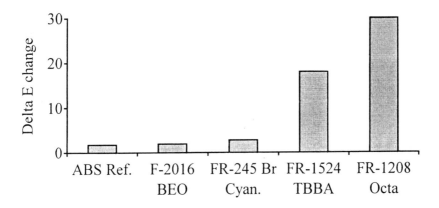

Compounds with 10% Br and 4.8% antimony trioxide
0.25% of UV absorber and 0.25% HALS
4% if TiO_2

8.11 UV Stability of FR ABS, ASTM 4459-86 (300 h, Xenon Arc)

8.3.7 Toxicity

The toxicological properties of flame retardants are important in order to ensure safety during production, application and end use.

Safety during production and application requires a series of tests dealing with skin and eye irritation, mutagenicity and lethal dose levels. Since the early 1990s, 'Green' parties in Germany and in several Scandinavian countries have invested considerable efforts to try to ban halogenated flame retardants claiming that they may be a source of toxic halogenated dibenzodioxins and dibenzofurans. The spreading of many rumours[8,9] and actions via multiple channels to introduce restrictions on brominated flame retardants have resulted in a major decrease in the use of flame-retarded TV housings in the European market.[1] As a consequence of several years of such practice, televisions have become a primary cause of domestic fires in Germany,[10] while in Sweden the number of TV fires per million sets more than doubled during the time period 1990–95.[11] Recently, a leading European fire expert has raised his voice to draw the attention of authorities to the weakness of the European standard of flame retardancy for TV sets.[12]

Regarding end use and also recycling and incineration aspects, a number of studies have been conducted on smoke and products of decomposition such as a recent study by Stevens and Mann.[13] This study concludes that in spite of the known effects of adding halogenated flame retardants to consumer products, the benefits in saving human lives outweighs the risks in using them. A test program, conducted by the National Institute of Standards and Technology (formerly NBS) to measure the effects of flame-retardant chemicals on the fire hazard of plastics has shown that the total quantities of toxic combustion products released by the flame-retarded plastics (expressed as carbon monoxide equivalents) was one-third that of non-flame-retarded plastics.[2] It must be emphasised that the flame-retardants involved included brominated types.

Work done in 1996 at the TAMARA pilot incinerator of Forschungszentrum Karlsruhe (the Karlsruhe Research Center), a well-reputed testing centre in Germany, has shown that, even under extreme conditions, electrical and electronic (E&E) plastic waste can safely undergo co-combustion with municipal solid waste.[14] The work carried out at Karlsruhe has given further proof that municipal solid waste combustion is an *ecologically acceptable* and *economically sound* disposal route for typical amounts of electric and electronic waste containing brominated flame retardants.

Some of the main conclusions of this work are:

1   The high amounts of E&E plastic waste material added *had a positive influence* on the combustion process (that is combustion was cleaner and more complete) and the 'burn-out' of the bottom ashes was improved.

2   The levels of dioxin and furan remained within the range typically found in such incineration plants – the addition of the extra E&E waste did *not* increase these levels.

The positive contribution of brominated flame retardants as life savers is a proven fact[15,16] and European manufacturers of electronic goods should not jeopardise fire safety in an attempt to gain market acceptability.

8.3.8   Standards

One of the last and probably most important obstacles to pass before being allowed to offer flame-retarded products in the market is that of the standards for fire retardancy. These are numerous and different standards apply depending on the country and the use to which the product is put. An excellent summary of FR standards is given by Troitzsch in *International Plastics Flammability Handbook 1990*[4] (recently being re-edited and updated). A brief summary of some of the most common testing methods and standards is given in Table 8.11.

Since 1993, Europe has been trying to harmonise fire testing for all construction materials except flooring by introducing a new standard of flame retardancy called the Single Burning Item (SBI) test.[17]

8.3.9 Cost of production

With regard to costing, not only the price of the flame retardant but also its efficiency and its specific gravity must be taken into account, since use of a more cost effective flame retardant often results in a final product which is cheaper and has better mechanical properties.

8.4   New trends in halogenated flame retardants

The rapid development in computerised systems for the electronics, communication and automotive industries is creating demand for component plastic properties at low costs. High flow during moulding, good compromise between impact and stiffness as well as tolerance to high temperature when in use are some of the requirements for the plastic parts used in their production. Due to miniaturisation and the consequent increase in operating temperature, more stringent flame retardancy properties are needed. Moreover under the pressure from environmentalists, the European market has been looking for diphenyl oxide-free flame-retardant systems. As a result, a new generation of multipurpose environmentally-friendly brominated flame retardants has been recently introduced that offers additional benefits which widen the usage of the host polymeric systems. Flame retardants with appropriate softening temperatures such as brominated indan,[18] tris(tribromophenyl) cyanurate[19] and tris(tribromoneopentyl) phosphate[20] provide processing aid effects and better flow properties. Reduced cycle times during injection moulding are possible with these flame retardants and they enable production of parts with thinner walls. FR-1808 (brominated indan DSBG), according to recent testing, is not considered to pose any risk to the health of the general population or to the environment. FR-1808 and its compounding in HIPS has been tested for the presence of dibenzo-para-dioxins/dibenzofurans (PBDD/PBDF) prior to and after incineration. The results show that in all cases the level of PBDD/PBDF is in full compliance with the German Ordinance and EPA requirements. Saytex 8010 (Table 8.1), a proprietary flame retardant produced by Albemarle,[21] does not contain any detectable quantities of brominated dioxins or furans and also meets the German Dioxin Ordinance. It displays excellent thermal stability, bloom resistance and high bromine content and it provides an intermediate level of UV stability.

Very interesting developments have also occured with chlorinated flame retardants offering cost-efficient systems with minimal environmental impact (DPO free) and optimal flame retardancy.[22] This is achieved by blending chlorinated wax (Chlorez ICC Industries) with brominated indan (FR-1808 DSBG) and with a partial replacement of antimony trioxide by zinc sulphide. In other work, synergism between chlorine and bromine has been investigated in ABS using, for instance, blends of Dechlorane Plus (see Table 8.1) with brominated epoxy oligomers.[23,24] This combination is also recommended for its UV stability.

Polymeric brominated flame retardants have been shown to increase the heat distortion temperature in polypropylene-based compounds. Improved thermal stability in several plastics has been observed when these flame retardants are used.[25] Recently FMC (flame retardant business acquired by Great Lakes) also investigated the enhanced synergism between bromine and phosphorus when they are in the same compound.[26] Applications have been tested in styrenics and in polypropylene. Tris(tribromoneopentyl) phosphate is of interest in the production of flame retarded polypropylene fibre as no antimony trioxide is needed in order to generate required levels of flame retardancy.

Brominated acrylate polymer is an excellent coupling agent between plastic and fibre or filler reinforcement. This results in better retention of tensile properties and freedom from blooming in the end-product. New flame retardant systems for polypropylene combining brominated compounds, antimony trioxide and magnesium hydroxide offer optimal combination of cost and properties.[27]

8.5   Conclusions

There is a clear requirement for flame retardants in many plastic applications but several factors must be taken into account in the choice of flame retardant. Principally, these are:

- The thermal properties must be matched to the polymer so that the flame retardant decomposes when the polymer burns but remains intact during polymer processing and is stable at the operating temperature of the final product.
- High chemical purity may be required for electrical/electronic and building applications.
- Suitable melting range and compatibility can influence physical properties and migration in the polymer.
- Higher bromine content reduces flame retardant loading leading to improved physical properties.
- Freedom from toxic hazards from manufacture through to end-use or disposal needs to be assured.
- The chosen flame retardant may depend upon the standard to be met.
- Cost-efficiency with respect to the end-use is more important than cost of the flame retardant *per se*.

A new generation of flame retardants is being introduced to address the rapid and demanding development in computerised systems for the electronic, communication and automotive industries. Due to miniaturisation and the consequent increase in operating temperature, more stringent flame retardancy is needed. These new flame-retardant systems also take into account the increasing pressure of environmentalists in the European market which is looking for diphenyl oxide-free flame-retardant systems.

Acknowledgements

The author is very grateful to J. Simons, A. Teuerstein, R. Minke, M. Peled, N. Kornberg, Y. Bar Yaakov, L. Utevski, G. Reznick, Y. Scheinert, I. Finberg, M. Shenker, J. Reyes, A. Heijboer, S. Bron and I. Antonir for their work and help which render this publication possible.

References

1 Spates B, 'Fire fighting', *World Plastics & Rubber Technology*, Cornhill Publications London, 1999.
2 Babrauskas V, *Flame Hazard Comparison of Flame-retarded and Non-flame-retarded Products*, NBS special publication, US Government Printing Office Washington, 1988.
3 Touval I, 'Antimony flame retarder synergists', *Fire Retardant Chemical Association Conference*, Baltimore, FRCA, March 1996.
4 Troitzsch J, *International Plastics Flammability Handbook 2nd edition*, Munich, Hanser Publishers, 1990, 18.
5 Terry D G, 'Future trends involving the use of fire retardants chemicals in wire & cable applications', *Fire Retardant Chemical Association Conference*, Baltimore, FRCA, 1996.
6 Green J, 'The Flame retardation of polyolefins', *Flame-Retardant Polymeric Materials*, Vol. 3, Plenum Publishing, New York, 1982.
7 Georlette P, 'Optimization of flame retarded thermoplastics for engineering applications', *Flame Retardants 98*, London, Interscience Communications, 1998.
8 *Modern Plastics International*, 1997, April, p 16.
9 *Modern Plastics International*, 1996 November, p 20.
10 'Stiftung Warentest 10/98' at homepage http://www.firesafety.org
11 De Poortere M, Schonbach C and Simonson M, 'The Fire Safety of TV Set Enclosure Materials, A Survey of European Statistics' (To be published). *Polym. Int.*, 2000.
12 Troitzsch J, 'Trends in Flame Safety', 2nd UBA *Workshop on the use of flame retardants*, Berlin, 1995.
13 Stevens G C and Mann A H, 'Risks and benefits in the use of flame retardants in consumer products', Department of Trade and Industry (DTI ref: URN 98/1026), University of Surrey, Guildford, 1999.
14 *Co-Combustion of Electrical and Electronic Plastics Waste with Municipal Solid Waste at the TAMARA pilot incinerator of Forschungszentrum Karlruhe*, The

Forschungszentrum Karlsruhe (http://www.firesafety.org), Association of Plastics Manufacturers in Europe (APME) Report 1996.

15 See Homepage of the BSEF about Fire Safety at http://www.firesafety.org

16 Troitzsch J, *Fire Safety of TV-sets and PC-monitors*, prepared for the European Brominated Flame Retardant Industry Panel (EBFRIP) and the European Flame Retardants Association (EFRA), Wiesbaden, Germany. 1998.

17 Smith D A and Shaw K, Evolution of the single burning item test – in *Flame Retardants 98*, Interscience, London 1998, 1–14.

18 Reyes J, 'FR-1808, A novel flame retardant for environmentally friendly applications', 6th annual conf Recent Advances in Flame Retardancy of Polymeric Materials, Business Communications Co, Norwalk, USA, 1995.

19 Reyes J, 'A new brominated cyanurate as flame retardant for application in styrenics', 7th annual conf Recent Advances in Flame Retardancy of Polymeric Materials, Business Communications Co, Norwalk, USA, 1996.

20 Squires G E, 'Flame retardant polypropylene – a new approach that enhances form, function and processing', *Flame Retardants Conference*, Interscience Communications Ltd, London, 1996 in Flame Retardants.

21 Landry S D and Reed J S, 'Recyclability of Saytex 8010 flame retardant in high impact polystyrene', *Fire Retardant Chemicals Association Conference*, Lake Buena Vista, 1995.

22 Lee V W and Siddhamalli S K, 'Injection moldable FR-HIPS formulations based on blends of chlorinated wax and non diphenyl oxide (DPO)-brominated flame retardant in conjunction with metal oxide/metal sulfide synergists', *Journal of Vinyl & Additive Technology*, **4**, (2), June 1998.

23 Yun S and Kim H, 'Bromine-chlorine synergy to flame retard ABS resins', *Fire Retardant Chemical Association Conference*, San Francisco, FRCA, 1997.

24 Markezich R L, 'Use of chlorine/bromine synergism to flame retard polymers', 6th annual conf Recent Advances in Flame Retardancy of Polymeric Materials, Business Communications Co, Norwalk, USA, 1995.

25 Heijboer A, Utevski L, Bar Yaakov Y, Finberg I and Georlette P, 'Beneficial effects of brominated flame retardants in polymeric systems', *Advances in Plastics Technology Conference, Katowice*, 1997.

26 Green J, 'Mechanism for flame retardancy and smoke suppression-a review', *Fire Retardant Chemical Association Conference*, Baltimore, FRCA, 1996.

27 Montezin F, Lopez Cuesta J-M, Crespy A and Georlette P, 'Flame retardant and mechanical properties of a copolymer PP/PE containing brominated compounds/antimony trioxide blends and magnesium hydroxide or talc', *Fire and Materials*, **21**, 245–52, 1997.

9

Natural polymers, wood and lignocellulosic materials

RYSZARD KOZLOWSKI AND
MARIA WLADYKA – PRZYBYLAK
Institute of Natural Fibres, Poznan, Poland

9.1 Combustion of natural polymers, wood and lignocellulosic materials

Wood and lignocellulosic materials belong to the natural biocomposites of plant origin, containing cellulose, hemicelluloses, lignin and other compounds. Their chemical structures and compositions depend on the plant nature such as: tree, annual, biannual and perennial plants rich in cellulose (bast plants like flax, hemp, kenaf, jute, roselle (karkadeh), and others like sisal, grass-like *Miscanthus*, grain straw, reed, bagasse, bamboo, curava, etc. They have found use in a great number of applications which include textiles, geotextiles, furnishings and composites. The basic chemical compositions of the most important lignocellulosic raw materials are given in Table 9.1.[1–6]

Natural polymers, wood and lignocellulosic materials, when exposed to fire or any other high-intensity heat source, are subject to thermal decomposition and combustion depending on conditions. Combustion is accompanied by heat release and chemiluminescence. However, the latter does not always occur during combustion. To make combustion of natural polymers possible, particular conditions have to be met, such as direct contact with air and a physical, chemical or microbiological stimulus associated with heat release. The intensity of heat stimulus, oxygen concentration and circulation of gas in the area where combustion can occur influence the time to ignition of natural polymers and wood as does the intensity of the combustion process. Relevant factors include rate of flame spread, heat release rate, mass loss and carbonisation rates. Thermal decomposition proceeding in anaerobic conditions or in an atmosphere poor in oxygen is called pyrolysis.

There are three stages of heat action on natural polymers and wood.[7–9]

Table 9.1 Basic chemical composition of lignocellulosic raw materials

Material	Approximate contents of main chemical compounds [%]		
	Cellulose	Lignin	Pentosans
Coniferous wood[a]	40–45	26–34	7–14
Deciduous wood[a]	38–49	23–30	19–26
Cotton	90–95	–	–
Kapok	65–70	5–15	2–10
Flax fibres	64–71	2–15	2–5
Flax shives	36–47	24–30	21–30
Hemp fibres	60–67	3–14	5–10
Hemp shives	40–52	22–30	17–25
Kenaf[a]	31–39	15–19	22–23
Roselle fibres[d]	70–72	12–13	1–3
Jute	55–65	10–15	15–20
Ramie	60–70	1–10	5–12
Abaca (Manila)	55–65	7–10	16–19
Sisal[b]	63	7.5	22
Date-palm[b]	58	15.3	20
Pineapple[b]	69.5	7.5	21.8
Bagasse[a]	32–44	19–24	27–32
Esparto[a]	33–38	17–19	27–32
Elephant grass (Miscantus)	35–40	10–15	10–20
Bamboo[c]	33–45	20–25	30
Reed[a] Phragmites communis	44.75	22.8	20
Grain Straws	27–37	12–21	20–34

[a] SOURCE: Reference[3]
[b] SOURCE: Reference[4]
[c] SOURCE: Reference[5]
[d] SOURCE: Reference[6]

9.1.1 Preliminary (flameless) stage

This stage includes dehydration and release of liquid and volatile products and heating of natural polymers and wood to their decomposition temperatures.

Heating of natural polymers and wood to 105 °C results in the removal of moisture. Reactions occurring at this stage proceed slowly and are mostly endothermic. After exceeding the above temperature, a slow thermal decomposition of the components of natural polymers begins and at 150–200 °C gas products of the decomposition start to be released, bonds between components of natural polymers and wood weaken and the latter turns yellow.

At 160 °C decomposition of lignin begins. Lignin under thermal degradation yields phenols from cleavage of ether and carbon–carbon linkages, resulting in more char than in the case of cellulose. Most of the fixed carbon in charcoal originates from lignin.[10] At 180 °C hydrolysis of low molecular weight polysaccharides (hemicelluloses) begins.[11] Thermal stability of hemicelluloses is lower than that of cellulose and they release more incombustible gases and fewer tarry substances.[12] According to Saunders and Allcorn,[7] gases released most frequently contain about 70% of incombustible CO_2 and about 30% of combustible CO. Depending on availability of oxygen, subsequent reactions can be exothermic or endothermic.

In the temperature range of 200–260 °C exothermic reactions begin. They are characterised by increased emission of gaseous products of decomposition, release of tarry substances and appearance of local ignition areas of hydrocarbons with low boiling points. Spontaneous ignition does not occur at these temperatures, but under favourable conditions ignition can start from a pilot flame. Cellulose decomposes in the temperature range between 260 and 350 °C, and it is primarily responsible for the formation of flammable volatile compounds.[10,12] The thermal degradation of cellulose can be accelerated in the presence of water, acids, and oxygen.[12] Natural polymers and wood turn brown (holocellulose fractions become darker) and pyrophoric carbon is formed. However, below 260 °C the reaction rate still remains low.[7]

Detail of the products of the thermal decompositions of cellulose, lignin and hemicelluloses has been given by LeVan and Winandy,[10] Shafizadeh,[12] Hirata,[13] and others.

9.1.2 Main (flame) stage

This stage includes ignition of thermal decomposition products, flame spread by combustible gases and increase in heat release and mass loss rates. This is the active process of decomposition.

The above stage occurs in the temperature range from 260 to 450 °C. After reaching the thermal decomposition temperature of wood, which is about 275–280 °C, uncontrolled release of considerable quantities of heat begins and increased amounts of liquid and gaseous products (280–300 °C), including methanol, ethanoic acid and its homologues, are formed. The amount of evolving carbon monoxide and dioxide decreases, mechanical slackening of wood structure proceeds and ignition occurs. Mass loss of wood reaches about 39%. Tarry matter begins to appear at 290 °C. The release of gases still increases and rapid formation of charcoal takes place. The reactions are highly exothermic with peak temperatures over the range 280–320 °C. Secondary reactions of pyrolysis become the main processes and this fact results in an increased ability of the mixture of gases formed to ignite if there is sufficient access of oxygen. Combustion proceeds in the gas phase at a small distance from the surface rather than on the wood surface itself. From this moment, wood can burn even after the removal of heat stimulus. The ignition of wood occurs at 300–400 °C, and the exact ignition temperature of wood depends on its origin. In the case of wood obtained from coniferous trees, it ranges from 350 to 365 °C, whereas in wood of deciduous trees it ranges from 300 to 310 °C. This difference is the result of the higher lignin content in wood derived from coniferous trees.[7,14]

9.1.3 Final (flameless) stage

This stage includes slow burning of the residue and ashing of the remaining matter.

This final stage occurs above 500 °C and while the formation of combustible compounds is small, the formation of charcoal increases. When the combustion is complete, the stoichiometric balance for wood and carbohydrates is given by the following equations:[15]

Cellulose: $C_6H_{10}O_5 + 6O_2 \rightarrow 6CO_2 + 5H_2O$ [9.1]

Hemicellulose: $C_5H_8O_4 + 5O_2 \rightarrow 5CO_2 + 4H_2O$ [9.2]

Wood: $C_{1.7}H_{2.5}O_{1.05} + 1.8O_2 \rightarrow 1.7CO_2 + 1.25H_2O$ [9.3]

The course of the process of combustion of wood, lignocellulosic materials and natural polymers and the continuity and intensity of accompanying phenomena depend on a number of factors, such as the density of the material, the content of easy-to-ignite compounds such as resins and essential oils, moisture content, thermal conductivity and heat capacity, air circulation velocity, ratio of reacting surface to mass of burning material, geometry of combustion system, intensity and direction of incoming heat flux and rate of heat transfer to the material. Of particular importance to the process of wood combustion, especially in the case of a large cross-section, are low thermal conductivity and formation of charcoal on the surface of the burnt material. The charcoal formed is characterised by a thermal conductivity several times lower than that of wood. It therefore makes an insulating layer which renders heat transfer more difficult, thus preventing the temperature from increasing to the level at which pyrolytic decomposition of wood occurs. This results in fire retardancy of wooden constructions of large cross-sections. The protective action of the charcoal layer is only temporary, because in the case of increased heat transfer to the system, a recurrence of intensive combustion can occur. Wooden constructions maintain their ability to withstand fire conditions, in contradistinction to steel constructions which frequently collapse suddenly.[8,16,17] Wood competes with steel in some construction applications and fire-retarded wood may be safer than steel. By observations and understanding of the processes of thermal decomposition and combustion, it has been possible to develop very efficient fire-retarding systems for natural polymers and wood.

9.2   Chemistry of fire retardancy

9.2.1 Historical background

According to historical sources, vinegar and alum were used by two ancient civilisations, the Chinese and the Egyptian, as combustion limiting agents for natural polymers and wood. The Chinese applied a solution of vinegar and alum to wood and then covered it with a layer of clay which delayed and reduced fire spread. In Egypt, 3000 years ago, reed and grass were soaked in sea-water before being used for roofing purposes. Mineral salts, which crystallised during drying, acted as fire retardants. In ancient Roman times, a fire retardant was described by Aulus Gellius. During the siege of Piraeus by Sulla in 86 BC, wood withstood fire because it had been previously soaked with alum.[18] In more modern times, the use of fire retardants was described by several, including Sabatini in 1658, Wild in 1735, Fagot in 1740, Brugnatelli in 1821 and others.[84]

The first approach to identify the scientific principles of fire retardancy was made in 1821 by Gay-Lussac, who presented a number of recipes for making fire retardants for cellulosic materials by using ammonium phosphates and borax. These recipes could be successfully applied today.[18] Lyons[19] in his book on fire retardants wrote that 'references on fire retardant studies date back at least 200 years and were initially directed to textiles'. Comprehensive reviews of fire retardants were also published by Brahmhall,[20] Juneja,[21] Lewin[22,36] and others.

9.2.2 Fire retardants

The consumption of fire retardants is closely related to the development of fire safety codes. According to Green,[23] world consumption of fire retardants in 1992 was estimated at (358 million kg) 788 million lbs and the value of them was US$ 1169 million. The above amount includes bromine compounds 254, organophoshorus 224, antimony oxide 134, chlorine compounds 91, others 85 (in million lbs). The application profile of fire retardants was as follows: 65% for plastics, 25% for rubber (including aluminium trihydrate ATH in carpet underlay), 5% for textiles, 3% for coatings/adhesives and only 2% for wood and paper. From the above data it is apparent that the fire retardant market for wood and lignocellulosics is relatively narrow and highly dispersed over many final products. Moreover, fire retardants for wood are frequently manufactured by small companies.

Fire-retardant treatments for wood can be divided into two general classes: (1) those impregnated into the wood or incorporated into wood composite products, and (2) those applied as paint surface coatings (non-intumescent or intumescent).[9,24] Coatings are easier to apply and more economical. They are used mostly to protect materials which are already elements of a construction. Their drawbacks are the formation of cracks, susceptibility to abrasion and wear which result in the loss of fire retardant efficiency. Impregnation usually consists of maximising penetration of fire retardant into wood cells under pressure and it is costly.[9] It is particularly useful for fire retardancy of new elements before their assembly. As a result of impregnation, fire retardants are located in the interior of wood, and it follows that even after destruction of the surface, the fire retardant still remains in place.

There are two basic groups of fire retardants: additive and reactive.[22,25] Additive compounds are those whose interaction with a substrate is only physical in its nature, whereas reactive compounds interact chemically with cellulose, hemicellulose or lignin.[22,25]

The applied additive compounds include: mono- and diammonium phosphate, halogenated phosphate esters, phosphonates, inorganic compounds such as antimony oxide and halogens, ammonium salts (ammonium bromide, ammonium fluoroborate, ammonium polyphosphate and ammonium chloride); amino resins (compounds used for their manufacture are dicyandiamide, phosphoric acid, formaldehyde, melamine and urea); hydrated alumina; stannic oxide hydrate; zinc chloride and boron compounds (boric acid, borax, zinc borate, triammonium borate, ethyl and methyl borates).[16,22,26] The above compounds reduce the combustion potential of wood, but they can unfavourably affect such properties as strength, hygroscopicity, stability, toxicity, adhesive and mechanical properties, and receptivity to paint-coatings.[27,28] They are used mainly in the form of impregnants for wood in relatively large doses (10–20% by weight).[29] One of the most widely used fire retardants is monoammonium phosphate. Its fire retardant activity consists in the acceleration of the formation of a charred mass from cellulosic material and suppression of flammable volatiles.[16,30,31] However, it causes an increase in smoke emission and a rise in CO and decrease in CO_2 concentrations. Compounds of phosphorus and nitrogen are frequently applied jointly because of the synergy of their interaction.[32] A drawback of many inorganic fire retardants based on salts is their susceptibility to leaching by water. Some of them can absorb moisture and the penetration of water into the bulk of wood promotes not only its decay (e.g. due to fungi), but also leads to the destruction of metal joints and building elements.[33] Fire retardants based on boron compounds give a long-lasting protection due to their deep penetration into wood. Boric acid acts as the inhibitor of flame and smoke formation. Moreover, boron compounds have fungicidal[34] and insecticidal[16] properties. An ideal fire retardant should be characterised by the presence of a synergy effect which enables the stabilisation of the cellulose structure. It is recommended that phosphates are applied to the interior and boron compounds to the exterior of wood in order to increase the amount of charred mass on its surface.[35] The presence of nitrogen compounds, especially in the form of urea, increases the efficiency of phosphates. Moreover, during pyrolysis nitrogen-containing compounds liberate nitrogen gas, which can dilute combustible volatile products released from wood.[16,25]

Some of the proposed fire retardants which are resistant to water leaching contain a combination of chemical compounds capable of forming water-insoluble complexes. These are systems based upon amino resins and monomers which undergo polymerisation inside wood.[9,22] Examples of reactive compounds are chlorendic anhydride (1,4,5,6,7,7-hexachloro-5-norbornene-2,3-dicarboxylic anhydride), tetrabromophthalic anhydride and derivatives of polyhydric alcohols (halogeno-phosphorus polyols, chlorinated bisphenols and chlorinated neopentyl glycols).[36–38] Most of the reactive fire retardants are based on bromine, fluorine and chlorine compounds such as, for example SF_3Br, CH_2BrCl, CF_2BrCl, CF_2Br_2, $CF_2Br–CF_2Br$, although they are very toxic.[16,41] It is well-known that halogens such as chlorine and bromine are efficient inhibitors of the formation of free radicals, but a large content of them (15–30% by weight) is necessary to reach efficient level of fire retardancy.[9] However, for reasons of environmental considerations the use of halogen-containing fire retardants (especially those containing bromine), chlorinated paraffins and substances containing antimony oxide is debatable.[39,40]

One of the most effective methods of protecting wood and lignocellulosics from fire is the use of fire retardant coatings, particularly intumescent ones. When heated, they form a thick, porous carbonaceous layer. This provides an ideal insulation of the protected surface against an excessive increase in temperature and oxygen availability, thus preventing thermal decomposition which plays a decisive role in retarding the combustion process. Under the influence of heat, intumescents expand to 50 times their volume, and, in some cases, to even 200 times.[42–46] Intumescent systems such as paints, lacquers, mastics and linings must contain ingredients which, when heated to high temperatures, will form large amounts of non-flammable residues. These residues, under the influence of emitted gases should take the form of foam with good insulating properties. This foam should be sufficiently durable and adhere to the surface of the substrate in order to resist the currents of extremely hot gases, as well as other forces appearing under fire conditions.[42,45]

Intumescent coatings have many advantages among which are a low mass requirement as well as a relatively thin coating in order to secure effective fire protection over a given period. They exhibit the unique property of intumescence when exposed to flame and of creating a barrier against flame and oxygen access to the protected surface. To make intumescent coatings effective, the following ingredients are necessary: carbonising compounds (e.g. polyhydric alcohols, polyphenols, carbohydrates or resins), dehydrating agents (e.g. Lewis acids, phosphoric acid, mono- and diammonium phosphates, polyphosphates of ammonium, urea or melamine, ammonium sulphate), foam-forming agents (e.g. dicyandiamide, melamine, urea or guanidine) and modifying agents to obtain the maximum amount of carbonaceous mass as well as film-forming substances.[46] A proper selection of film-forming substances is extremely important. Amino-formaldehyde, polyvinyl and acrylic resins are preferred. Currently there is a trend towards using epoxy resins in intumescent systems.[42,45]

The classification of fire retardants according to their chemical composition is presented in Fig. 9.1.

9.2.3 Mechanism of fire retardancy

Most of the fire retardants for wood and lignocellulosics reduce the formation of combustible volatile matter as well as retarding oxidation of those materials, thus efficiently limiting flame access and combustion and lowering the temperature. Mechanisms of fire retardancy depend on the character of the action of chemical compounds present in fire retardants and on the environmental conditions of fire, i.e. properties of the environment in which active pyrolysis proceeds. Moreover, the chemical compounds playing the role of fire retardants hinder the formation of 1,6-anhydroglucopyranose which is highly flammable and is the main basic volatile fraction evolving during thermal degradation of cellulose.[9]

A comprehensive review of different theories of wood fire retardancy was published by Browne.[18] According to the main theories, the role of fire retardant consists in (1) reducing the flow of heat to prevent from further combustion, (2) quenching the flame, or (3) modifying the thermal degradation process.[47] Based on the mechanisms described by Browne, le Van[9] has discussed several theories concerning the inhibition of pyrolysis and combustion of wood, namely barrier theories, thermal theories, dilution by noncombustible gases theories, increased char/reduced volatiles theories and reduced heat content of volatiles theories. Eickner[48] has discussed theories of wood fire retardancy caused by chemical treatment. In the light of these considerations, such chemical compounds should:

1   Create a liquid or glassy layer, forming a barrier or insulation that prevents the exchange of air and flammable products at the combustion zone.

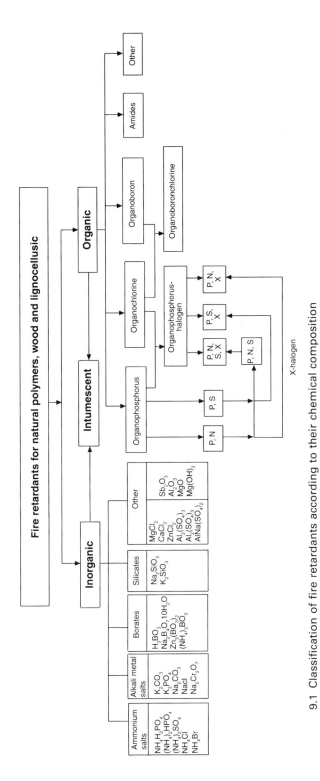

9.1 Classification of fire retardants according to their chemical composition

2 Form coatings, glazes or foams insulating the surface and restraining the pyrolysis by reducing/preventing access of air.
3 Increase the thermal conductivity allowing heat to be dissipated faster than it is supplied by the ignition source.
4 Create endothermic chemical and physical changes that absorb heat at a level that prevents ignition conditions.
5 Form non-flammable gases which dilute the flammable gases and form less flammable gaseous mixtures, reducing the combustible matter.
6 Change the mechanism of pyrolysis by terminating free radical reactions, lower the temperature for pyrolysis and promote greater char formation which reduces the production of flammable products such as tar and gases.[16]

Mechanisms of protective activity as well as drawbacks and advantages of fire retardants are presented in Table 9.2. In this Table, fire retardant systems and means of fire retardancy are divided into five main groups depending on the mechanism of their activity.[2,24,49–51] On the basis of current knowledge of mechanisms presented in Table 9.2, the experience of many years of studies and the results of the most recent research demonstrate that an ideal fire retardant should have the following properties:[2,49–52]

- High fire protection effectiveness (an amount not exceeding 10% of total weight of protected material).
- Chemical stability at normal service conditions.
- No effect on mechanical strength and aesthetics of protected materials.
- No emission of toxic and corrosive substances during their use.
- No increase of emission of toxic products of thermal decomposition and combustion.
- Easy application.
- Insecticidal and fungicidal effectiveness.
- Resistance to water and UV degradation.
- Relatively low price.

Many fire retardants are available in the market, but only a few of them come close to meeting the above requirements. It is therefore worthwhile to develop and apply fire retardants which are the closest to this ideal and are also environmentally friendly.[53] The perception of such efforts by consumers should in turn lead to a larger market share by companies which are distinguished by their enterprise and experience in the field of environmental protection.

Table 9.2 The mechanism of protection of fire retardants

Mechanism of protective activity	Compounds and agents	Remarks
Generation of protective gaseous coatings and non-flammable gases at raised temperatures. Action inhibiting combustion by interruption of free radicals; a cyclic reaction in flammable gases due to the generation of free radicals. Ability of building a polymer into the structure that is converted into fire-retardant form.	Ammonium compounds, halogen compounds (Br, Cl, F), sulphates, organic halogen derivatives for instance derivatives of 4-bromophthalic acid, HET acid (chlorendic anhydride). Synergistic combinations of halogenated metal oxide (antimony trioxide).	**Advantages:** effective protection against flame propagation, easily applicable. **Disadvantages:** easily leached, generally (except organic derivatives) poor stability, emission of great amounts of toxic gases during combustion.
Inhibiting temperature increase of flammable materials e.g. lignocellulose particles due to high melting heat, decomposition and the ability to convert at flame temperature into non-flammable liquid forms which cut-off oxygen supply. Catalysts of flame reaction for the creation of protective carbon layer and insulating foams.	Boron compounds (boric acid, borax, zinc borate, methyl and ethyl borate), polyphosphate, tungstic acid, phosphate derivatives (mineral binders) e.g. aluminium phosphates, molybdenum oxides.	**Advantages:** very effective by increasing material ignition temperature. Decrease of emitted heat energy. **Disadvantages:** in the case of phosphates, they have a great influence on the material properties and cause unfavourable catalysis of the combustion reaction to favour the emission of toxic CO.

Table 9.2 (cont.)

Mechanism of protective activity	Compounds and agents	Remarks
Activity that comprises the properties of two aforementioned groups.	Ammonium phosphates, ammonium polyphosphates, ammonium borates, phosphate additives with urea, melamine, biuret, dicyandiamide, etc. ammonium dimolybdenate and molybdenate in composition with hydrated aluminium oxide, proteins.	**Advantages:** extremely advantageous and multiple activities. **Disadvantages:** easily leached out, emission of a great amount of toxic gases, especially in case of phosphate systems.
Lowering of the thermal conductivity coefficient, resistance to fire.	Mineral fibres (asbestos, glass fibre, mineral wool), granulated glass, kaolin, diatomite, mica and its derivatives (vermiculite), also by covering surfaces with silver, carbon fibres, synthetic fibres from aluminium oxide and blends with titanium dioxide, synthetic fibres based on polymers including 'ladder' polymers.	**Advantages:** significant improvement of insulating power. Decrease of toxic gas emissions. **Disadvantages:** They are not cost effective for solid wood protection. They do not dissipate heat generated during fire.
Insulation of lignocellulose material against penetration of heat energy. Screening and reflecting of heat radiation.	Mirrors of Al and Ag foil, silicate boards, intumescent coatings including the latest solution: flexible intumescent interlayers – fire blockers (for special application e.g. aircraft interiors, ancient buildings).	**Advantages:** ready to use, relatively low cost of protection, complete inhibition of fire propagation over the surface and insulation of flammable materials from the fire zone. **Disadvantages:** protection of surface only. In case of surface damage there is a fire hazard. In the case of wood, they cover the natural design and grain.

9.3   Fire retardancy of lignocellulosic boards and panels

There are two methods of increasing the fire retardancy of lignocellulosic boards and panels:

- Application of different types of fire retardants during the production process.
- Application of fire retardants, especially intumescent coatings, at the finishing stage.

Urea-formaldehyde and urea-melamine resins, which are most frequently used for board gluing, have some fire retardant properties. These properties are, however, quite poor, therefore the flammability of boards based on the above resins does not differ to a significant extent from that of wood.[2]

The ready-made product, resistant to fire, can be obtained by the following methods:[54]

- Impregnation of lignocellulosic particles or fibres with fire retardants before the production process.
- Treatment with fire retardants in liquid or solid form during the production process.
- Addition of mineral particles to the lignocellosic particles.
- Application of non-flammable binders.
- Insulation of lignocellulosic material to prevent penetration of the heat flux using intumescent coatings and fire barriers (e.g. aluminium foil).

These methods can be used alternatively or jointly. In general, the addition of non-combustible components is aimed at covering and separating the lignocellulosic particles and this is accompanied by migration of fire retardants into the interior of the flammable particles, thus resulting in fire protection of the latter.[54] The most commonly used fire-retardant additives are the following chemical compounds: boric acid; ammonium phosphates and borates, ammonium sulphate and chloride; zinc chloride and borate; phosphoric acid; dicyanodiamide; sodium borate and antimony oxide.[22,54–56] They are usually added in the form of powder in amounts ranging from 5 to 10% in relation to dry mass (8% on average). The particle size of a fire retardant strongly influences its efficiency and the amount to be added. Fire retardants are most frequently introduced at the forming stage into partially glued lignocellulosic particles. Organic fire retardants based mainly on organobromine compounds can also be used. Their application is, however, limited by the high toxicity of products of thermal decomposition and combustion of materials containing such compounds.[54,57,58]

The addition of mineral fillers in the form of tiny particles results in a separation of the flammable material (i.e. lignocellulosic particles present in the board) as well as in a decrease in thermal conductivity and, as a consequence, a board acquires fire-resistant properties. The most common inorganic materials used as mineral fillers are rock wool, glass fibres, vermiculite, perlite and even protein waste from the leather industry which is highly effective.[33,54]

Mineral binders, by covering lignocellulosic particles and penetrating into them, cause their fire retardancy. It is well-known that boards based on gypsum and cement are fire-resistant to a greater or lesser extent.[59-61] Cement-board products are impact resistant and fire resistant. Cement-board roofing sheets are the ideal because they provide universal appeal in addressing regional roofing concerns.[60] They are not used as covering and furniture boards, however, and usually are characterised by a high density (over 800 kg/m^3). Other alternative mineral binders, instead of synthetic resins, used for production of fire-resistant boards, are calcium silicate and magnesite.[54]

A product of polycondensation of urea borates and urea phosphates with silicates is another member of the above family of binders.[63] It can be obtained by injecting trimethylborate into phenolic- and isocyanate-bonded flakeboards during pressing with the object of improving fire resistance. It is worth adding that the deposition of borates improves fire resistance, but reduces board strength.[62] A recent development is a three-layer non-flammable composite particleboard based on lignocellulosic particles and mineral filler, expanded vermiculite with the urea-formaldehyde resin as a binder. This type of board can be manufactured using different lignocellulosic raw materials, such as wood particles, annual bast plant waste (flax, hemp, jute, kenaf, etc.) and shives.[2,54,63-65] The technological flow chart of three-layer composite board production based on flax shives and vermiculite is presented in Fig. 9.2.

Fire characteristics of three-layer composite boards have been determined and compared with those of typical shive particleboards of similar density and the same thickness of 20 mm, in accordance with ISO 5660 by cone calorimetry.[2,65] The composite board did not ignite at the heat flux of 30 kW/m^2, although exposure time to the heat flux was extended to 20 minutes, i.e. by 100% compared to the standard exposure time (10 min). A typical particleboard ignited after 89.3 seconds in the same conditions. When exposed to a heat flux of 50 kW/m^2, the composite particleboard ignited after over 14 minutes and all parameters, except for smoke release, were significantly reduced in comparison with the typical shive particleboard. Results are shown in Figs. 9.3 and 9.4. Three-layer lignocellulosic-mineral composite boards can be used as a construction material after finishing with methods typical of particleboards (e.g. natural veneer, paint,

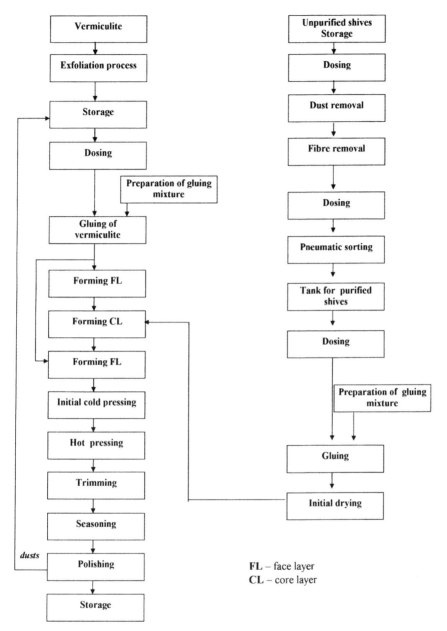

9.2 The scheme of production technology of composite, three-layer particleboards based on flax or hemp shives and vermiculite

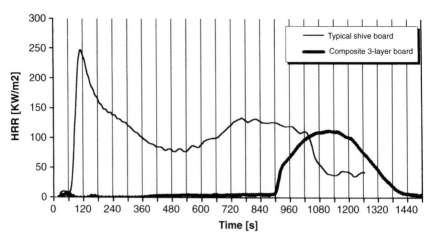

9.3 Heat release rate (HRR) from composite board in comparison with typical shive board in accordance with ISO 5660 at heat flux of 50 kW/m²

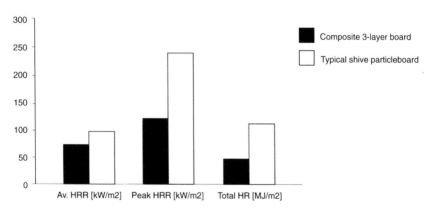

9.4 Heat release parameters in accordance with ISO 5660 (heat flux 50 kW/m²) for typical shive board and composite board based on shives and vermiculite

foils) for furniture and decorative elements. Mechanical properties of the above boards are similar to those of typical lignocellulosic ones but they are characterised by much lower water absorption and obviously by fire retardancy.

Wood wool and other lignocellulosic materials, e.g. grain straw, can be used for the manufacture of insulating boards by implementation of fire retardant and mineral binders (e.g. cement, gypsum). Reed and hemp insulating boards are an example of annual plant utilisation for the manu-facture of boards.[2,66–68] They make a cheap and interesting alternative to insulating materials made of polystyrene or polyurethane foams. The problem of fire retardancy of insulation boards was solved by the addition of vermiculite.

In the case of plywood a surface protection is used, e.g. intumescent systems, impregnation of ready-made plywood by means of the vacuum-pressure method (high pressures are applied) or manufacture of plywood from veneers pre-impregnated with fire retardants using special glues to overcome difficulties with strength and adhesion.[69–71] Östman and Tsan-daris[72] has tested fourteen fire retardant-treated and thirteen untreated wood-based products chosen from five sets of building products. All the products were commercially available from different European countries and believed to be a representative sample of typical fire-retarded wood products. Test results were obtained from measurements carried out on a cone calorimeter and from a full-scale room fire test (See Table 9.3.).

9.4   Test methods

Methods used for the determination of the efficiency of fire retardants for wood and natural polymers include:

- Thermal analysis – TG, DTA and DSC give basic information on the mechanism of pyrolysis and combustion as well as data on the effect of fire retardants and different materials. However, the relation between thermal analysis and real fires is as yet unknown.[9]
- Flame spread measurements – samples are exposed to an ignition source and spread of flame, ignition, extinguishment of flame and heat for sustained burning are measured.
- Oxygen index testing consists of the measurement of the minimum concentration of oxygen in oxygen–nitrogen mixtures, which is necessary to sustain flame burning of a tested sample. Readily flammable materials are characterised by low values of oxygen index, whereas slow-burning ones by high values. The method can be used for the investigation of the mechanism of fire retardant activity in the gas phase.[9]
- Heat release rate during fire is the most fundamental fire property of a material. The higher heat release rate, the greater is the fire.[73,74] Fire retardants can reduce the heat release rate and prolong the time of heat release. Heat release rate, mass loss rate, effective heat of combustion, total heat released, specific extinction area, time to sustained ignition, and CO/CO_2 formation can be measured in the bench scale by means

Table 9.3 Cone calorimeter data frame at 50 kW m^{-2} and horizontal orientation, and room fire test data by Östman[72] and Tsandaris .

Products	Thickness (mm)	Density (kg m^{-3})	Time to ignition (s)	Cone calorimeter				Room Fire Test	
				THR (MJ m^{-2})	RHR$_{60}$	RHR$_{300}$ (kW m^{-2})	RHR$_{max}$	Time to flashover min:s	
								Measured	Predicted
Beech plywood	5	440	31	19.1	123	99	278	–	2:30
Beech plywood	15	640	26	61.0	99	79	199	–	3:55
Birch plywood	12	600	28	77.1	159	121	253	2:17	2:33
Ordinary birch plywood	12	600	30	73.3	163	127	340	2:30	2:26
Wood panel, spruce	11	450	20	52.8	98	83	132	2:11	2:22
Particleboard	10	670	34	78.5	213	141	243	2:37	2:24
Particleboard	12	720	33	84.3	179	134	205	–	2:54
Particleboard	22	660	33	144.9	163	128	179	–	2:44
Hardboard	6	920	38	60.0	182	192	421	–	2:52
Medium-density wood fibreboard	9	750	22	73.5	125	109	299	–	3:25
Medium-density wood fibreboard	12	655	31	72.6	139	109	147	2:11	3:05
Insulating wood fibreboard	13	250	12	33.1	128	102	179	0:59	1:16
Insulating wood fibreboard	13	260	11	33.8	100	108	186	–	1:18
FR beech plywood, F	15	590	NI	NI	NI	NI	NI	–	>20:00
FR beech plywood, F	15	600	30	49.8	67	58	167	–	5:03
FR plywood, Injecta, Fin	9	620	469	3.0	49	29	54	>20:00	22:09
FR plywood, Noncom, DK	13	700	629	7.5	77	50	84	–	14:55
FR particleboard, S	12	750	700	17.2	67	57	58	>20:00	14:33
FR particleboard, Pyroex, D	12	750	NI	NI	NI	NI	NI	–	>20:00
FR particleboard, type B1, D	13	710	48	57.1	87	63	119	–	6:26
FR particleboard, F	16	630	21	33.7	55	35	110	10:30	8:54
FR particleboard, F	22	690	48	59.5	78	59	116	–	6:43
FR hardboard, F	22	710	97	37.9	52	44	62	–	11:33
FR hardboard, GB	6	1040	30	20.7	75	64	124	–	10:17
FR[a] medium-density wood fibreboard, S	9	810	37	40.4	22	41	253	–	12:32
FR[a] medium-density wood fibreboard, S	12	850	515	38.4	60	90	165	–	9:31
FR[a] insulating wood fibreboard, S	15	300	21	41.7	78	70	172	–	1:51

THR – Total heat release; RHR – Rate of heat release; NI – No ignition; FR – Fire retardant treated; [a] – Fire retardant treatment on one side.

of a *cone calorimeter*. The latter has recently become one of the most commonly used instruments for comparing the combustibility of different materials[74].

- Toxicity – most of the methods for smoke analysis give only the total amount of gases formed or enable measuring of single gases only. The combination of a cone calorimetry with FTIR spectroscopy which can be used for continuous measurement of different gas components is a recent development which is gaining credibility in this area.[75]

9.5 Recent progress in fire retardants for wood and lignocellulosics

Studies of new flame retardant systems for the protection of natural polymers, wood and lignocellulosics against fire continue very intensively in many research centres in the world.

In general, this research is directed to three areas:

- Development of fire retardants with reduced leachability and lower environmental impact (no toxic properties, higher efficiency of usage).
- Chemical and biochemical modification of natural polymers, wood and lignocellulosics.
- Use of more efficient intumescent systems and fire blockers.[24,50]

Wood fire retardancy with limited leachability can be achieved by means of a two-stage impregnation (first a fire retardant treatment and then a water-proofing treatment) or by chemical modification of wood. This is called flame retardancy created '*in statu nascendi*' by chemical reaction with components of lignocellulosics such as cellulose, hemicelluloses and lignin. It is based on the chemical reaction between hydroxyl groups of cellulose or lignin with reactive fire retardant. In the case of wood and other lignocellulosic raw materials, lignin and its hydroxyl groups are much more accessible to chemical reagents than crystalline cellulose.

Cellulose as a polyhydric compound can be modified by the formation of chemical derivatives through typical reactions of alcohols – substitution of hydroxyl groups (all or a part of them). However, the physical structure of natural cellulose fibres, especially in its crystalline regions, results in reduced chemical reactivity of cellulose. The crystalline structure is the reason for impenetrability of the molecule by chemical reagents: for this reason, when considering a reaction system, the most important point to determine is how best to disrupt the crystalline regions to make them as accessible to reagents as are the amorphous regions. For this reason pretreatment processes are often applied which result in cellulose activation.[22,49,76–78] The activation of cellulose can be achieved by various processes involving treatment with water or aqueous solutions and is associated with swelling of cellulose. Sodium hydroxide is still important and widely used for the process of mercerisation of flax and other cellulosic fibre yarns. During the mercerisation process the crystalline form of cellulose changes from cellulose I to cellulose II and a complete or partial removal of hemicelluloses, pectins and lignin occurs. As a consequence the structure becomes more accessible to chemical impregnation. A very promising treatment is the application of enzymes, such as lactase, in order to obtain a higher yield of the reaction of cellulose hydroxyl group substitution.[49,79]

From the chemical point of view, reactions aimed at modifying cellulose can be classified as follows:[49]

• Addition, with water and alkalis.
• Esterification, inorganic and organic.
• Etherification.
• Oxidation.
• Other reactions such as halogenation, grafting, cold plasma and corona treatment.

When considering the strength of chemical bonds between cellulose and modifying compounds, the strongest are those of the ether type.

Another promising development seems to be the application of genetic engineering to obtain transgenic lignocellulosic materials with incorporated fire retardants which are partially substituted for hydroxyl groups. This application is, however, still at the stage of fundamental research.

Currently a very important problem is the emission of toxic products from thermal decomposition and combustion as well as smoke release. According to the literature, 55–70% of fatal accidents during fires are caused by gaseous toxic products of decomposition and combustion of materials and not by direct action of high temperature and flames.[2,80] An equally important problem is the elimination of toxic substances which enter the environment in the form of sewage, gases, fumes, dust and solid waste. This danger can be lessened by moving in the direction of developing and introducing zero-waste technologies, i.e. full-recycling production and application technologies.[2] Another problem is the physiological effect of protected wood and lignocellulosic materials in direct contact with the human body. Opinions on this matter have changed rapidly in the last ten years, especially in Europe and North America and the carcinogenic activity of some components of fire retardants has cast a shadow on their use.[82,83]

References

1 Kozlowski R, Helwig M, 'Lignocellulosic Polymer Composites', Science and Technology of Polymers and Advanced Materials. Edited by Prasad P N, Mark J E, and others. *Plenum Press* New York and London, 679–98.

2   Kozlowski R, Helwig M, 'Fire Retardancy of Lignocellulosic Composites'. 9th annual conf *Recent Advances in Flame Retardancy of Polymeric Materials*, Business Communications Co, Norwalk, USA, 1998.

3   Rowell R M, 'Chemical modification of non-wood lignocellulosics', in Hon D N-S (ed.) *Chemical Modification of Lignocellulosic Materials*, Chapter 9, 1996, 229–45.

4   Pandey S N, Ghosh S K, 'The chemical nature of date-palm (*Phoenix dactylifera – L*) leaf fibre', *J Text Inst*, 1995, **86**(3), 487–89.

5   Liese W, 'Production and utilization of bamboo and related species in the year 2000', Procedings of the 5th int conf on Forest Products, vol. 2, Nancy, publisher, 1992, 743–51.

6   Eweida M H T, El-Hariri D M, Fayed M H, 'Influence of planting dates, stem diameter and retting methods on the chemical changes in roselle fibres', *Al-Azar Agricultural Research Bulletin*, 3, 1974. Cairo, Egypt.

7   Saunders R G, Allcorn R T, 'Course of Thermal Decomposition', Fire Performance of Timber TRADA, Timber Research and Development Association, High Wycombe UK, Chapter 4.1,9.1,14, 1972.

8   Deppe H J, 'On the fire performance of wood-based panel products', *J Inst of Wood Sci*, 1973, **6**(4), 20–7.

9   LeVan S L, 'Chemistry of fire retardancy', in Rowell, R (ed.) *The chemistry of solid wood*. Washington DC, American Chemical Society, 1984, 531–74.

10 LeVan S, Winandy J, 'Thermal degradation', in Schiewind A P (ed.) *Concise Encyclopedia of Wood & Wood-Based Material*. Elmsford, New York: Pergamon Press, 1989, 271–73.

11 Hurst N W, Jones T A, 'A review of products evolved from heated coal, wood and PVC', *Fire Mater*, 1985, **9**(1), 1–9.

12 Shafizadeh F, 'The chemistry of pyrolysis and combustion', in Rowell R (ed.) *The Chemistry of Solid Wood*, Washington DC: American Chemical Society, 1984, 489–529.

13 Hirata T, 'Changes in degree of polymerization and weight of cellulose untreated and treated with inorganic salts during pyrolysis', (Forest and Forest Prod. Res. Inst. 304), *J Japan Wood Res Soc Bull*, 1979, 77–124.

14 Janssens M, 'Rate of heat release of wood products', *Fire Safety J*, 1991, **17**(3), 217–38.

15 Tran H C, 'Rates of heat and smoke release of wood in an Ohio State University Calorimeter', *Fire Mater*, 1988, **12**, 143–51.

16 Leão A L, 'Fire retardants in lignocellulosics composites', *Lignocellulosics-plastics Composites*, Universidade de Sao Paulo, Brazil, 1997, 111–61.

17 Tuve R L, 'Principles of fire protection chemistry', *National Protection Association NFPA*, Quincy, Madison, 1976.

18 Browne F L, 'Theories of the combustion of wood and its control' Forest Service Report No 2136 *Forest Products Laboratory*, Madison Wis. 1958, reviewed and affirmed, 1963.

19 Lyons J W, '*The Chemistry and Uses of Fire Retardants*', New York, Wiley-Interscience, 1970.

20 Brahmall G, '*Wood fire behavior and fire retardant treatment*', *A review of the Literature*, Canadian Wood Council, Ottawa, Canada, 1966.

21 Juneja S C, 'Advances in Fire Retardants', *Progress in Fire Retardancy Series*, Bhatnagar V M (ed.) Westport USA Technomic Publ, 1973, **3**(2), 31–53.

22 Lewin M, Atlas S M, Pearce E M, '*Flame-Retardant Polymeric Materials*', vol 1, New York and London, Plenum Press, 1975.

23 Green J, 'An overview of the fire retardant chemicals industry, past-present-future', *Fire Mater*, 1995, **19**, 197–204.

24 Kozłowski R, Helwig M, 'Progress in flame retardancy and flammability testing', 1st int conf *Progress in Flame Retardancy and Flammability Testing*, Poznan, Poland, Institute of Natural Fibres, 1995.

25 Chamberlain D L, 'Mechanisms of fire retardancy in polymers', in Kuryla W C, Papa A J, (eds.) *Flame Retardancy of Polymeric Materials*, vol 4, New York, Dekker, Chapter 2, 1978.

26 Sumi K, Tsuchiya Y, Fung D P C, '*The Effect of Inorganic Salts on the Flame Spread and Smoke Producing Characteristics of Wood*', Nat Res Council of Canada, Eastern Forest Products Lab, Ottawa, Dept. of the Environment, No 81, 1971.

27 Holmes C A, *Improvement by fire retardant treatments*, American Wood Preserves Association, Washington, USA, 1974, 95–102.

28 Holmes C A, 'Effect of fire-retardant treatments on performance properties of wood', *ACS Symposium Series 43*, Washington, DC, *American Chemical Society*, 1977.

29 Nussbaum R, 'The effect of low concentration fire retardant impregnations on wood charring rate and char yield', *J Fire Sci*, 1988, **6**, 290–307.

30 Ishihara S, 'Fire endurance of wood composites with gradiented phosphorylation of phosphoric acid', in Better Wood Products, IURPO Meeting, Nancy, 1992.

31 Grexa O, Kosik M, 'Flammability of wood treated with diammonium phosphate and toxity of its fire effluents', *in: Better Wood Products, IURPO Meeting*, Nancy, France, 1992.

32 Lewin M, 'Some aspects of synergism and catalysis in FR of polymeric materials – an overview', 9th annual conf *Recent Advances in Flame Retardancy of Polymeric Materials*, Business Communications Co, Norwalk, USA, 1998.

33 Sarvanta L, *Fire retardant wood, polymer and textile materials*, Technical Research Centre of Finland, ESPOO, 1996.

34 Cockcroft R, Levy J F, 'Bibliography on the use of boron compounds', in 'The preservation of wood', *J Inst Wood Sci*, 1973, **6**(3), 28–37.

35 Hirrata T, Kawamoto S, 'A conception for development of fire-retardant wood materials from research on thermal analysis of wood materials treated with water-insoluble retardants': *XIX IUPRO World Congress Division 5*, Montreal, 1990.

36 Lewin M, 'Flame retarding of wood by chemical modification with bromate–bromide solutions', *J Fire Sci*, 1997, **15**, 29–51.

37 Rowell R M, Susott R A, DeGroot W F, Shafizadeh F, 'Bonding fire retardants to wood. Part I. Thermal behavior of chemicals bonding agents', *Wood and Fiber Sci*, 1984, **16**(2), 214–23.

38 Bhatnagar V M, 'Generalization and new approaches in fire retardants', Advances in Fire Retardants – Part 2, 3, ed. *Technomic*, Westport, Conn. U.S.A., 1973, 1–17.

39 Caney D, 'Flame retardants: no burnout yet. Chemical Marketing Reporter' Apr 1992, p. SR9. In: PROMT. Abstract no 92:297329 in Flame Retardant for Plastics.

40 Gann R G, 'Flame retardants–overview' in: *Kirk-Othmer Encyclopedia of Chemical Technology*, 4th ed, Vol. 10, New York, John Wiley, 1994, 930–6.

41 Petrella R V, Sellers G D, 'Flame inhibition by bromine compounds', *Fire Technology*, 1970, **6**(2), 93–101.

42 Kozłowski R, Wesołek D, Władyka-Przybylak M, 'An intumescent, transparent coating for protection of wood and wood-based materials, 2nd *Beijing International Symposium/Exhibition on Flame Retardants*, Beijing, 1993, 387–92.

43 Kozlowski R, Wladyka Przybylak M, Wesolek D, 'An intumescent, transparent, fire retardant and protective coating for wood-derivative materials', *27th International Particleboard/Composite Materials Symposium*, Pullman, USA, 1993.

44 Wladyka-Przybylak M, Kozlowski R, 'The thermal characteristics of different intumescent coatings', *Fire Mater*, 1999, **23**.

45 Wladyka-Przybylak M, 'Fire retardant efficiency of intumescent coatings on particular modifiers', Doctoral Thesis, *Institute of Natural Fibres*, 60–630 Poznań, Wojska Polskiego str 71b, Poland, 1997.

46 Vandersall H L, Intumescent coating systems, their development and chemistry. *J Fire Flamm*, 1971, 2 April: 97–140.

47 Sweet M S, 'Fire performance of wood: test methods and fire retardant treatments', in *Recent Advances in Flame Retardancy of Polymeric Materials*, M Lewin (editor) Vol. 4, Business Communications Co, Norwalk, USA, 1993.

48 Eickner H W, 'Wood and wood products', in *Fire Safety Aspects of Polymeric Materials Vol 1*, State of the Art Nat Mat Adv Board Publ NMAB 318-1, National Academy of Science, Washington DC, 1997, 13–18.

49 Kozłowski R, Helwig M, 'Progress in fire retardants for lignocellulosics materials', 5th Arab int conf on Materials science *Materials and Fire*, Alexandria, 1998.

50 Kozłowski R, Helwig M, Wesołek D, 'Fire retardants and their special application in historic buildings', 2nd int symp on *Fire Protection of Ancient Monuments*, Institute Natural Fibres, Poznan, Poland, 1995, 145–52.

51 Kozlowski R, Helwig M, Rulewicz J, 'Application of fire retardant systems to museums', 23rd int conf on *Museum Security*, Berlin, 1997.

52 Kozlowski R, Helwig M, Rulewicz J, 'Fire protection of historic buildings and sites in Poland', World Meeting on *Integral City Protection against Fire and other Hazards*, Toledo, Spain, 1997.

53 Holme I, 'Flammability: the environment and the eco movement', *Flammability '93*, Textile Institute Manchester, 1993.

54 Kozlowski R, Helwig M, Przepiera A, 'Light-weight, environmentally friendly fire retardant composite boards for panelling and construction', in Moslemi A A (ed.) *Inorganic-Bonded Wood and Fiber Composite Materials*, **4**, University of Idaho, Idaho, 1995, 6–11.

55 Moslemi A A, *'Particleboard vol. 1–2'*, Carbondale, USA, Southern Illinois University Press, 1974.

56 Maloney T M, *'Modern Particleboard and Dry-Process Fibreboard Manufacturing'*, San Francisco, Miller Freeman, 1977.

57 Kozlowski R, Wesolek D, Muzyczek M, 'toxic products of thermal deterioration and combustion of wood-based and textile materials', 1st *Beijing International Symposium on Flame Retardants*, International Academic Publishers, Beijing, 1989, 689–92.

58 Kozlowski R, Wesolek D, Wladyka-Przybylak M, 'Combustibility and toxicity of board materials used for interior fittings and decorations', *Polym Degrad Stab*, Special Issue, **64**, No 3, Elsevier, 1999, 595–600.

59 Kozlowski R, Mieleniak M, Helwig M, 'Fire resistant lignocellulosics-mineral composite particleboards', *Polym Degrad Stab*, Special Issue, **64**, No 3, Elsevier, 1999, 511–16.

60 Schwarz H, Wentworth R, Eilmus G, 'The new age of inorganic-bonded wood compositions in North America', *28th int Particleboard/Composite Materials Symposium*, Pulman, Washington, 1994, 143–52.

61 Natus G, Schäfer K, 'New developments of gypsum fiberboard in Europe', 28th *inter Particleboard/Composite Materials Symposium*, Pulman, Washington State University, 1994, 153–7.

62 Geimer R, Leao A, Armbruster D, 'Property enhancement of wood composites using gas injection', *28th int Particleboard/Composite Materials Symposium*, Pulman, Washington State University, 1994, 243–59.

63 Kozlowski R, Helwig M, Przepiera A, 'Flame Retardant composite particleboards based on by-products of fibrous plants and other materials', *Pacific Rim Bio-Based Composites Symposium*, Rotorua, New Zealand, FRI Bulletin No 117, 1992, 320–5.

64 Kozlowski R, Mieleniak B, Przepiera A, 'Composite, non-flammable and non-toxic particleboards based on lignocellulosics and mineral particles', 4th *European Regional Workshop on Flax*, Rouen, 1996, 361–70.

65 Kozlowski R, Mieleniak B, Helwig M, 'Fire resistant lignocellulosic-mineral composite particleboards', *FRPM'97 Conference*, Lille, 1997, 110–11.

66 Kozlowski R, Mieleniak B, Przepiera A, 'Hemp straw and hemp shives as raw materials for the production of particleboards and insulating board', *The Commercial & Industrial Hemp Symposium*, Vancouver, 1997.

67 Kozlowski R, Mieleniak B, Przepiera A, 'Composite fire retardant particleboards made of lignocellulosic wastes, int conf on *Sustainable Agriculture Food, Energy and Industry*, FAO, Rome Italy and Federal Agricultural Research Centre (FAL) Braunschweig, Germany, June 1997.

68 Kozlowski R, Mieleniak B, Przepiera A, 'The production and properties of hemp insulating boards', inter conf on *Sustainable Agriculture Food, Energy and Industry*, FAO, Rome Italy and Federal Agricultural Research Centre (FAL) Braunschweig, Germany, June 1997.

69 Still R M, LeVan S L, Shuffleton J D, 'Degradation of fire-retardant-treated plywood: current theories and approaches', int symp on *Roofing Technology*, Montreal, 1991, 517–22.

70 Ross R, Cooper J, Wang Z, 'In-place evalution of fire-retardant-treated plywood', *8th International Nondestructive Testing of Wood Symposium*, Vancouver, Pullman, Washington State University, 1992, 247–52.

71 Winandy J, LeVan S L, Ross R, 'Thermal degradation of fire-retardant treated plywood: development and evaluation of test protocol', *Res. Pap. FPL-RP-501*. Department of Agriculture, Forest Service, Forest Product Laboratory, Madison, USA, 1991.

72 Östman B, Tsantaridis L, 'Heat release and classification of fire retardant wood products', *Fire Mater*, 1995, **19**, 253–8.

73 Babrauskas V, Peacock R D, 'Heat release rate: the single most important variable in fire hazard', *Fire Safety* 1992, **18**, 255–72.

74 Babrauskas V, Grayson S J, *'Heat Release in Fires'*, E & FN SPON London Chapman & Hall, 1996.

75 Innes J D, Talandis J, Faulkner, 'Combustion toxicity analysis from the cone calorimeter', int symp *Progress in Flame Retardancy and Flammability Testing* Inst of Natural Fibres, Poznań, Poland, 1995, 91–6.

76 Lai Y Z, 'Reactivity and accessibility of cellulose, hemicelluloses, and lignins', in Hon D N-S (ed.) *Chemical Modification of Lignocellulosics Materials*, New York, Dekker, USA, 1996, 35–96.

77 Hon D N-S, 'Chemical modification of cellulose', Hon D N-S (ed.) *Chemical Modification of Lignocellulosics Materials*, New York, Dekker, USA, 1996, 97–128.

78 Sharama H S, Fraser T W, McCall D, 'Fine Structure of Chemically Modified Flax Fibre', *J Text Inst*, 1995, **86**(4), 359–548.

79 Felby C, Olesen A B, Olesen P O, 'Quantification of surface lignin on thermo-mechanical fibres from beech, spruce and wheat using laccase based bioassay and polyelectrotyle adsorption analysis', submitted to *Wood and Fiber Science*, 1997.

80 Kimmerle G, 'Aspects and methodology for evaluation of toxicology parameters during fire exposure', *J Comb Toxic*, 1980, **I**, 42.

81 Kozłowski R, Kaźmierczak R, Helwig M, 'Fires of wooden buildings in Poland and trends of preventive activities', int meeting of *Fire Research and Test Centres*, Avila, Spain, 1986, 649–66.

82 Ray D R, 'Upholstered furniture flammability: regulatory options for small open flame & smoking material ignited fires', *US Consumer Product Safety Commission*, Washington, DC, 1997.

83 Hoebel J F, 'US Consumer Product Safety Commission activities related to polymer flamability', 6th annual conf *Recent Advances in Flame Retardancy of Polymeric Materials*, Business Communications Co, Norwalk, USA, 1998.

84 Sandholzer M W, 'Flameproofing of textiles.' National Bureau of Standards, Circular C455. 1946.

10
Intumescent materials

G CAMINO AND S LOMAKIN

Dipartimento di Chimica Inorganica, Chimica Fisica e Chimica del
Materiali dell'Università, Torino, Italy

10.1 Introduction

Intumescence can be described as fire-retardant technology which causes
an otherwise flammable material to foam, forming an insulating barrier
when exposed to heat. A common characteristic of intumescent materials
is that heat exposure initiates a chemical process that makes the material
intumesce. It is the intumesced and porous part of the material which gives
the insulation effect.

In recent years, the use of some traditional halogenated flame-retardants
has been limited on account of the possible formation of extremely toxic
halogenated dioxins or dibenzofurans.[1,2] Chlorinated dibenzo-*p*-dioxins and
related compounds (commonly known simply as dioxins) are contaminants
present in a variety of environmental media. This class of compounds has
caused great concern to the general public as well as intense interest in the
scientific community. Much of the public concern revolves around the characterisation of these compounds as being among the most potent man-made
toxicants ever studied. Indeed, these compounds are extremely potent in
producing a variety of effects in experimental animals based on traditional
toxicology studies at levels hundreds or thousands of times lower than most
chemicals of environmental interest. In addition, human studies demonstrate that exposure to dioxin and related compounds is associated with
subtle biochemical and biological changes whose clinical significance is as
yet unknown and with chloracne, a serious skin condition associated with
these and similar organic chemicals. Laboratory studies suggest the probability that exposure to dioxin-like compounds may be associated with
other serious health effects including cancer. Recent laboratory studies
have provided new insights into the mechanisms involved in the impact of
dioxins on various cells and tissues and, ultimately, on toxicity. Dioxins have
been demonstrated to be potent modulators of cellular growth and differ-
entiation, particularly in epithelial tissues.

The flame retardant chemical industry has historically been driven by regulations and standards. The normal fire-, smoke-, and toxicity-related standards have been joined by environmental standards caused by the alleged environmental impact of halogens and the alleged toxicity of antimony. Although suitable replacements have not been found for these materials in all cases, the environmental concern has served to depress their growth levels from what it would otherwise have been and/or channel the growth into alternative chemical products.

Among alternative possibilities intumescent materials have gained considerable attention because they provide fire protection with minimum of overall health hazard.[3] Since the first intumescent coating material was patented in 1938[4] the mechanism of an intumescent flame retardant refers to the forming of a foam which acts as an insulating barrier between the fire and substrate.[5] In particular, such intumescence depends significantly on the ratio of carbon, nitrogen, and phosphorus atoms in a compound.[3,5] Although intumescent coatings are capable of exhibiting good fire protection for the substrate, they have several disadvantages such as water solubility, brushing problems, and relatively high cost.[6] The fire retardation of plastic materials is generally achieved by incorporating fire-retardant additives into the plastic during processing.[7-9] Since processing requires that the additives withstand temperatures up to above 200 °C, intumescent systems without sufficient thermal stability cannot be incorporated into a number of plastics. Various phosphate-pentaerythritol systems have been investigated and developed as intumescent materials.[3] For example, a systematic study on a mixture of ammonium polyphosphate and pentaerythritol has undertaken[2,10-12] and new intumescent materials with appropriate thermal stability have been synthesised for better fire-retardance.[13,14]

The intumescent behaviour resulting from a combination of charring and foaming of the surface of the burning polymers is being widely developed for fire retardance because it is characterised by a low environmental impact. However, the fire retardant effectiveness of intumescent systems is difficult to predict because the relationship between the occurrence of the intumescence process and the fire protecting properties of the resulting foamed char is not yet understood. The physical and chemical models proposed for intumescence will now be discussed.

10.2   Physical modelling of intumescent polymer behaviour in fires

Intumescent materials provide a thermal and physical barrier to the underlying substrate or bulk polymer and thus block the high temperatures and rapid flame spread of fires. During exposure to a fire, the temperature within these materials rises, causing melting of the thermoplastic matrix. When the temperature corresponds to an appropriate value for the viscosity of the melt, an endothermic gas-producing chemical reaction is triggered. The gas collects in small bubbles, causing the material to foam. Solidification into a thick multicellular char provides an insulating layer that slows down the transport of heat and reduces the amount of material that becomes involved in the fire.[15] Published information on these materials is primarily available through the patent literature and there is limited understanding as yet of the physical and chemical mechanisms of intumescence.

Different models have dealt with intumescent behaviour as a problem in one-dimensional heat transfer, with the physical properties of the intumescent layer changing as a function of time to reflect foaming and outgassing.[16] Although these models have assisted in understanding the mechanisms providing thermal protection, they are unable to provide insight into: (1) the complicated sequence of physical, chemical, and thermal events that characterise intumescent behaviour, or (2) the effect of material properties on the performance of an intumescent system.

One of the latest three-dimensional models described by Butler et al[17] incorporated bubble and melt hydrodynamics, heat transfer, and chemical reactions to improve our understanding of intumescent mechanisms. The system is modelled as a highly viscous fluid containing a large number of expanding bubbles. The bubbles obey equations of mass, momentum and energy on an individual basis according to the values of local parameters, and their collective behaviour is responsible for the swelling and fire retardant properties of the material on a global scale.

The intumescent sample is described as an incompressible fluid whose viscosity is a function of temperature.[17] Initially, the sample is a rectangular solid containing a large number (up to 10000) of infinitesimally small bubble nucleation sites randomly distributed throughout the volume. At time $t = 0$, a specified heat flux is applied to the upper surface of the sample. The energy equation is solved to determine the temperature field in the sample. When the temperature at a given nucleation site exceeds the degradation temperature of the blowing agent, gas is produced, and the bubble begins to grow. The geometry of a bubble expanding in a local temperature gradient is illustrated in Fig. 10.1. Each expanding bubble experiences forces due to gravity, to gradients of viscosity and surface tension over its surface due to the temperature gradient, and to the motions of other bubbles. The Reynolds number for this translating motion, $Re = U\rho(2R)/\mu$, is very small due to the small bubble radius R, the low speed U, and the high kinematic viscosity of the melt μ/ρ, where μ – viscosity and ρ – density of melt.

It has been assumed that the bubble remains spherical, which is consistent with low Reynolds number flow, and that the expansion velocity is much greater than the translation, and that the flow field around a solitary

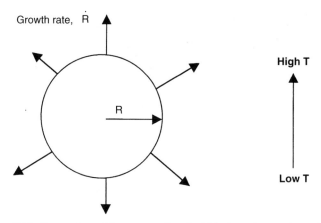

Growth rate, Ṙ

R

High T

Low T

10.1 Geometry of the expanding bubble in a thermal gradient

bubble is described by a simple Stokes equation driven by the force due to the viscosity gradient.[17] The velocity of the bubble through the melt was determined by calculating the terminal velocity resulting from a balance of forces on the bubble. A simple summation of individual flow fields provides a reasonable approximation for the total flow field if the assumption can be made that the spacing between bubbles is large compared with their size. In order to maintain a boundary condition of no normal flux across the lower surface of the sample, the field from each bubble is balanced by an identical image bubble located beneath the surface. As an example, the flow fields from four expanding bubbles are shown in Fig. 10.2.[17]

The outer surface of the intumescent sample is forced upward by the sum of forces from the bubbles expanding within the melt. As a first approximation for the surface properties of the intumescent material, the bubbles were assumed to be retained by the sample. The upper surface therefore stretches to prevent bubbles from bursting and releasing gases to the exterior. Bubble motion is influenced strongly by the local viscosity gradient. The variation of viscosity with temperature is currently estimated by the equation for polymer melts.[18] This relationship can be readily modified within the model to include other important factors such as molecular weight.

Upon exposure to the heat flux from a fire, the temperature within the intumescent sample rises, triggering gasification reactions at locations progressively farther from the outer surface. As the sample heats up, nucleation takes place at deeper and deeper sites. Modelling the protective qualities of intumescent fire retardants requires consideration of the effects of gas bubbles on heat transfer. Two separate mechanisms are responsible for

10.2 Flow field of four coplanar expanding bubbles and their images[17]

thermal protection. The degradation of the blowing agent occurs through an endothermic chemical reaction, absorbing heat during the intumescent process, and the thermal conductivity of the bubbles is much lower than that of the surrounding material, resulting in a final char that acts as an insulating layer.[17]

To observe the effects of a large number of bubbles on heat transfer, a simple analytical solution for a single bubble was proposed.[17] Under similar assumptions as those made for the hydrodynamics model, it is expected that the summation of these solutions over all bubbles will provide a reasonable approximation to the total temperature field.[17] The problem to be solved is an expanding and migrating sphere in a temperature gradient. The thermal conductivity of the sphere differs from that of the surrounding fluid, and endothermic chemical reactions take place on its surface.

In the intumescent melt, where the Reynolds number is very small, the timescales for expansion and translation are much longer than the timescale for thermal diffusion. It was therefore possible to neglect the convective terms and treat the problem as quasi-steady.[17] This reduces the energy equation to a simple Laplace equation with boundary conditions that account for the background temperature gradient and chemical heat sink. The solution to this equation is analytical, and is equivalent to a combination of sink

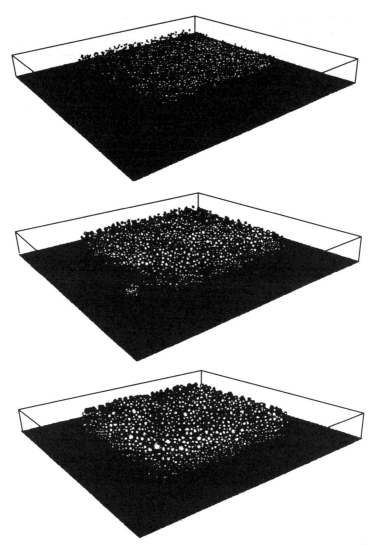

10.3 Development of 10 000 bubbles with time upon exposure to the heat flux from fire

and dipole singularities in the fluid exterior to the bubble.[19,20] The analytical solution[17] for the temperature field around a single bubble in a constant temperature gradient field is shown in Fig. 10.3.

For multiple bubbles whose separation is much larger than their radius, the total temperature field was obtained by summing the fields from individual bubbles responding to local conditions.[17] (Fig. 10.4 shows temperature contours for a single bubble). One approach to this complicated

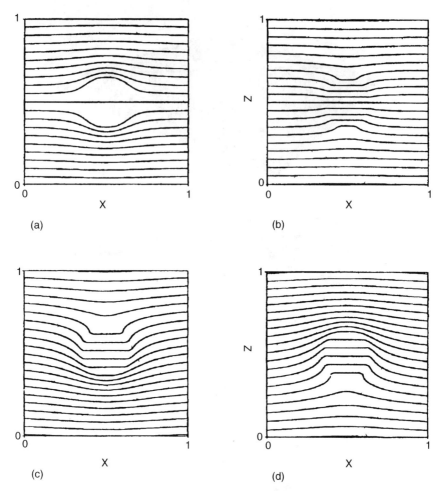

10.4 Temperature contours for single bubble. The bubble is (a) a thermal conductor, (b) a thermal insulator, (c) a heat source, (d) a heat sink

geometry is to introduce a Lagrangian coordinate system, which allows solution of the heat transfer problem in the original rectangular geometry using finite differences. The growth rate of bubbles in the intumescent material depends on the chemistry of the decomposition of the blowing agent and on the physical properties of the gas and surrounding melt. As the local temperature reaches the point at which decomposition starts, the concentration of gas in the polymeric melt begins to rise. At nearby nucleation points, bubbles begin to expand through diffusion of gas from the melt to the bubble. Initially, bubble growth is opposed by the surface tension of the melt. In a typical viscous liquid, the growth rate during later stages is controlled by liquid inertia and viscosity, by a combination of inertial and thermal effects, and, finally, by the transport of heat and mass alone. In the melted intumescent material, however, viscosity is expected to remain a dominant factor until char solidification occurs through cross-linking reactions. In the intumescent model, the locations of bubble nucleation sites are provided as inputs to the model. When the temperature at a site exceeds the temperature at which the blowing agent begins to degrade, the bubble begins to grow. Because surface tension is a relevant factor only for bubble sizes much smaller than those attained during inertia-controlled growth, the earliest stage of bubble growth is neglected. A simple analytic expression for the bubble growth rate can be obtained for a radially symmetric geometry in which diffusion dominates and the radial growth and translation of the bubble are neglected.[21]

A more realistic bubble growth submodel that includes viscous resistance to bubble expansion, temperature variations, and reaction chemistry is now being developed. In the improved model, the gasification reaction is assumed to be first order, with the rate constant determined by an Arrhenius equation using the local temperature.

10.3   Chemical aspects of intumescence

10.3.1 Flame retardants containing phosphorus and nitrogen

Phosphorus and nitrogen compounds and systems are a small but rapidly growing group of intumescent flame retardants which are in the focus of public interest concerning environmentally friendly flame retardants. Today their main applications are: melamine for polyurethane flexible foams, melamine cyanurate in nylons, melamine phosphates, ammonium polyphosphate-pentaerythryol or ethylene-urea formaldehyde polymers in polyolefins, melamine and melamine phosphates or dicyandiamide in intumescent paints, guanidine phosphates for textiles and guanidine sulphamate for wallpapers.

Their main advantages are their low toxicity in case of fire, the absence of dioxin and halogen acids as well as their low evolution of smoke. Their efficiency lies between that of halogen compounds and that of aluminum trihydrate and magnesium hydroxide. The metallic hydroxides split off water and are environmentally friendly, but their low activity requires high concentrations which change the mechanical properties of the matrix to which they are applied. In contrast to many halogen compounds, flame retardants based on nitrogen do not interfere with the types of stabilisers added to all plastic materials.

Recyclability has become important as many plastics are recycled. Flame retarded materials based on nitrogen compounds are suitable for recycling as the nitrogen flame retardants have high decomposition temperatures. In addition, flame retardants based on nitrogen do not add any new elements in addition to those already present in the polymers such as polyurethanes, nylons and ABS.

The most important inorganic nitrogen – phosphorus compound used as a flame retardant is ammonium polyphosphate which is applied in intumescent coatings and in rigid polyurethane foams. The world demand for ammonium polyphosphate is 10 000 000 kg per annum. The most important organic nitrogen compounds used as flame retardants are melamine and its derivatives which are added to intumescent varnishes or paints. Melamine is incorporated into flexible polyurethane cellular plastics and melamine cyanurate is applied to unreinforced nylons. Guanidine sulphamate is used as a flame retardant for PVC wall coverings in Japan. Guanidine phosphate is added as a flame retardant to textile fibres and mixtures based on melamine phosphate are used as flame retardants to polyolefins or glass-reinforced nylons.

Halogen–free solutions also exist for unsaturated polyesters, epoxies, saturated polyesters, polycarbonates and polystyrenes: practical applications are yet to be developed.

All the above mentioned compounds: ammonium polyphosphate, melamine, guanidine and their salts are characterised by an apparently acceptable environmental impact.

Mechanistic studies in nylon 6 with added ammonium polyphosphate (APP), ammonium pentaborate (APB), melamine and its salts were carried out using combustion and thermal decomposition approaches.[22,23] It was shown that APP interacts with nylon 6 producing alkylpolyphosphoric ester which is a precursor of the intumescent char. On the surface of burning polymer, APB forms an inorganic glassy layer protecting the char from oxidation and hindering the diffusion of combustible gases. Melamine and its salts induce scission of H–C–C(O) bonds in nylon 6 which leads to increased cross-linking and charring of the polymer.[23]

APP added at 10–30% wt to nylon 6 is ineffective in the low molecular weight polymer since the oxygen index (LOI) remains on the level of 23–24[24] corresponding to non-fire-retarded nylon 6. However, APP becomes very effective at loadings of 40 and 50% where the LOI increases to 41 and 50, respectively.

A condensed-phase fire retardant mechanism is proposed for APP in nylon 6.[24] In fact, an intumescent layer is formed on the surface of burning nylon 6/APP formulations that has an effectiveness that increases with increasing content of APP.

Thermal analysis has shown that APP destabilises nylon 6, since the thermal decomposition is observed at a temperature 70 °C lower than that of the pure nylon 6.[24] However, the intumescent layer effectively protects the underlying polymer from the heat flux and therefore in the configuration of the linear pyrolysis experiments the formulation nylon 6/APP (40%) decomposes more slowly than does the pure polymer.[24] These experiments prove the fire retardant action of the intumescent char. The mechanistic studies of the thermal decomposition in the system nylon 6 APP show that APP catalyses the degradation of the polymer and interacts with it forming essentially 5-amidopenthyl polyphosphate as is shown in Scheme 10.1.

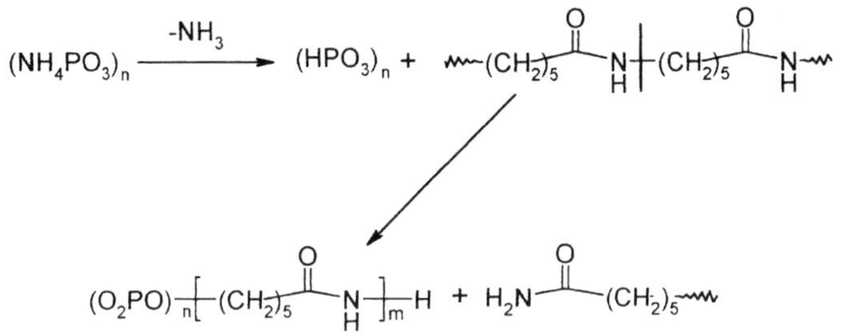

Scheme 10.1 Reaction of APP with nylon 6

On further heating, 5-amidopenthyl polyphosphate again liberates polyphosphoric acid and produces the char. The intumescent–shielding layer on the surface of the polymer is composed of the foamed polyphosphoric acid which is reinforced with the char.[24]

The effectiveness of ammonium pentaborate $NH_4B_5O_8$ (APB) in the high molecular weight nylon 6 ($M_n = 35000$) is similar to that of APP as measured by OI.[25] In contrast to APP, APB does not give an intumescent layer. Instead, a brown-black glassy-like compact layer is formed.

As thermal analysis has shown, APB destabilises nylon 6 since the latter decomposes at 50 °C lower. It is likely that freed boric acid catalyses the thermolysis of the nylon. In contrast to APP no other chemical interaction of nylon 6 and APB was found. In fact, the residue obtained for the formulations of nylon 6/APB in nitrogen during thermogravimetry experiments corresponds to that calculated on the basis of individual contributions of nylon 6 and APB to the residue.[25]

It is likely that accumulated on the surface of burning polymer is a molten glassy layer of boric acid/boric anhydride which protects the char from oxidation. This layer, reinforced by the char, creates a barrier against diffusion of the volatile fuel from the polymer to the flame which decreases combustibility of nylon 6.[25]

A systematic mechanistic study of halogen-free fire retardant nylon 6, *via* the combustion performance and thermal decomposition behaviour of non-reinforced nylon 6 with added melamine, melamine cyanurate,

Table 10.1 Limiting oxygen indices (%) for high molecular weight nylon 6 with added melamine or its salts[26]

CONTENT (% WT.)	3	5	10	15	20	30
Additive[a]						
Melamine	—	29	31	33	38	39
Dimelamine phosphate	—	23	24	25	26	30
Melamine pyrophosphate	—	24	25	25	30	32
Melamine oxalate	—	28	29	—	33	—
Melamine cyanurate	—	35	37	39	40	40
Melamine phthalate	34	48	53	—	—	—

[a] LOI for pure nylon 6 = 24%.[26]

melamine oxalate, melamine phthalate, melamine pyrophosphate or dimelamine phosphate has been reported.[26]

Melamine, melamine cyanurate, melamine oxalate and melamine phthalate promote melt dripping of nylon 6, which increases as the additive concentration increases. These formulations self-extinguish very quickly in air and their LOI increases with increasing concentration (Table 10.1).[26] The melt dripping effect is very strong in the case of melamine phthalate, where a small amount of the additive (3–10%) leads to large increases in LOI (34–53).

The combustion behaviour of melamine pyrophosphate and dimelamine phosphate is different from that of melamine and the other melamine salts (Table 10.1). The former are ineffective at low concentrations 15% and become effective at a loading of 20–30% because the intumescent char is formed on the surface of burning specimens. The mechanism of the fire-retardant action of both melamine pyrophosphate and dimelamine phosphate is similar to that of APP, since melamine, by analogy with ammonia, volatilises, whereas the remaining phosphoric acids produce esters with nylon 6, which are precursors of the char.[27] Some part of the freed melamine condenses probably forming melem and melon.[28]

Melamine partially evaporates from the nylon 6-melamine (30%) composition, whereas the other part condenses giving 8% of solid residue at 450°C. However, similar behaviour with a more thermostable residue is shown by melamine cyanurate. Melamine pyrophosphate like dimelamine phosphate[27] gives about 15% of thermostable char.

It is likely that accumulated on the surface of the burning polymer is a glassy layer of molten boric–acid boric anhydride that protects the char from oxidation. This glassy layer, reinforced by the char, creates a barrier against diffusion of the volatile fuel from the polymer to the flame which decreases combustibility of nylon 6.[25]

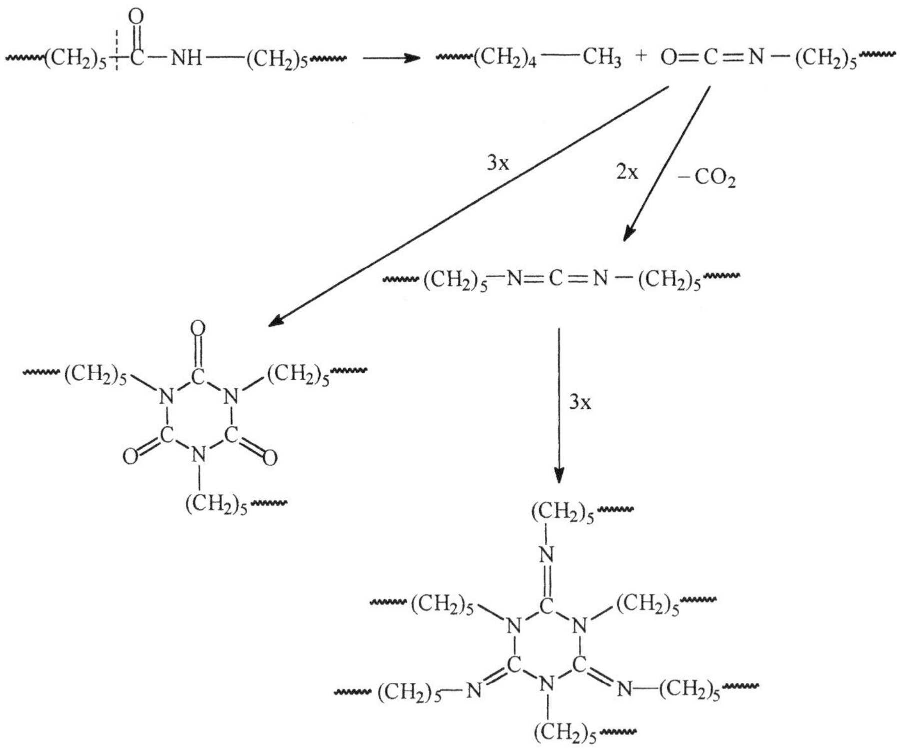

Scheme 10.2 Mechanism of thermal decomposition of nylon–6 in the presence of melamine[29]

As infrared characterisation of solid residue and high boiling products has shown[27] carbodiimide functionalities are formed during the thermal decomposition of nylon 6 with melamine and its salts. An unusual mechanism of chain scission of nylon 6 through CH_2–$C(O)$ bonds[29] is likely to become operative in the presence of melamines (Scheme 10.2). The resultant isocyanurate chain ends undergo dimerisation to carbodiimide or trimerisation to N-alkylisocyanurate. Carbodiimide can also trimerise to N-alkylisotriazine. These secondary reactions increase the thermal stability of the solid residue and increase the yield of the char.

10.3.2 Model study of char formation in ammonium polyphosphate – pentaerythritol system

In order to understand better the chemical reactions that are responsible for the intumescent behaviour of APP–PER mixtures, a thorough study of the thermal degradation of pentaerythritol diphosphate (PEDP) was undertaken.[30] PEDP is a model compound for structures identified in ammonium polyphosphate–pentaerythritol mixtures heated below 250 °C.

Using TGA five major degradation steps between room temperature and 950 °C have been identified and volatile products are evolved in each step. The formation of the foam reaches a maximum at 325 °C, corresponding to the second step of degradation; foam formation decreases at higher temperatures. There are no differences in the TGA or DSC curves in nitrogen or air up to 500 °C. Above this temperature, thermal oxidation leads to almost complete volatilisation in a single step, which is essentially completed at 750 °C. The elucidation of the chemical reactions which occur upon degradation is easier if each step is studied separately. The separation of the steps is accomplished by heating to a temperature at which one step goes to completion, and the following reaction occurs at a negligible rate.[30] The chemical reactions which occur in the first two steps lead to the initial formation of a char-like structure which will undergo subsequent graphitisation.

The first reaction is the elimination of water with the condensation of OH groups. This overlaps with the elimination of organic species when as little as 28% of the possible water has been evolved. This involves essentially complete scission of the phosphate ester bonds and results in a mixture of polyphosphates and a carbonaceous char. Three mechanisms have been proposed in the literature for this reaction:[31–36] a free-radical mechanism, a carbonium ion mechanism, and a cyclic cis-elimination mechanism. The free-radical mechanism was eliminated due to the lack of an effect of free-radical inhibitors on the rate of pyrolysis.[35] The carbonium ion mechanism is supported by acid catalysis and kinetic behaviour and may compete with the elimination mechanism.[34,35] The carbonium ion mechanism should occur exclusively if there is no hydrogen atom on the β-carbon atom, as in PEDP, which is necessary for the cyclic transition state of the elimination mechanism. The olefin is generated from the thermodynamically most stable carbonium ion. Hydride migration or skeletal rearrangement may take place to give a more stable ion of a carbonium ion of high reactivity is produced. After ring opening in the ionic ester pyrolysis mechanism, a second ester pyrolysis reaction occurs, which could also take place by the cis-elimination mechanism, as shown in Scheme 10.3.

The formation of char can occur either by free-radical- or acid-catalysed polymerisation reactions from the compounds produced in the pyrolysis. For example, the Diels–Alder reaction followed by ester pyrolysis and sigmatropic (1,5) shifts leads to an aromatised structure; this is shown in Scheme 10.4. Repetition of these steps can eventually build up the carbonaceous char which is observed. The reaction pattern shown in Scheme 10.4 should help to provide the structures observed by spectroscopy in the foamed char. These reactions probably occur in an irregular sequence and

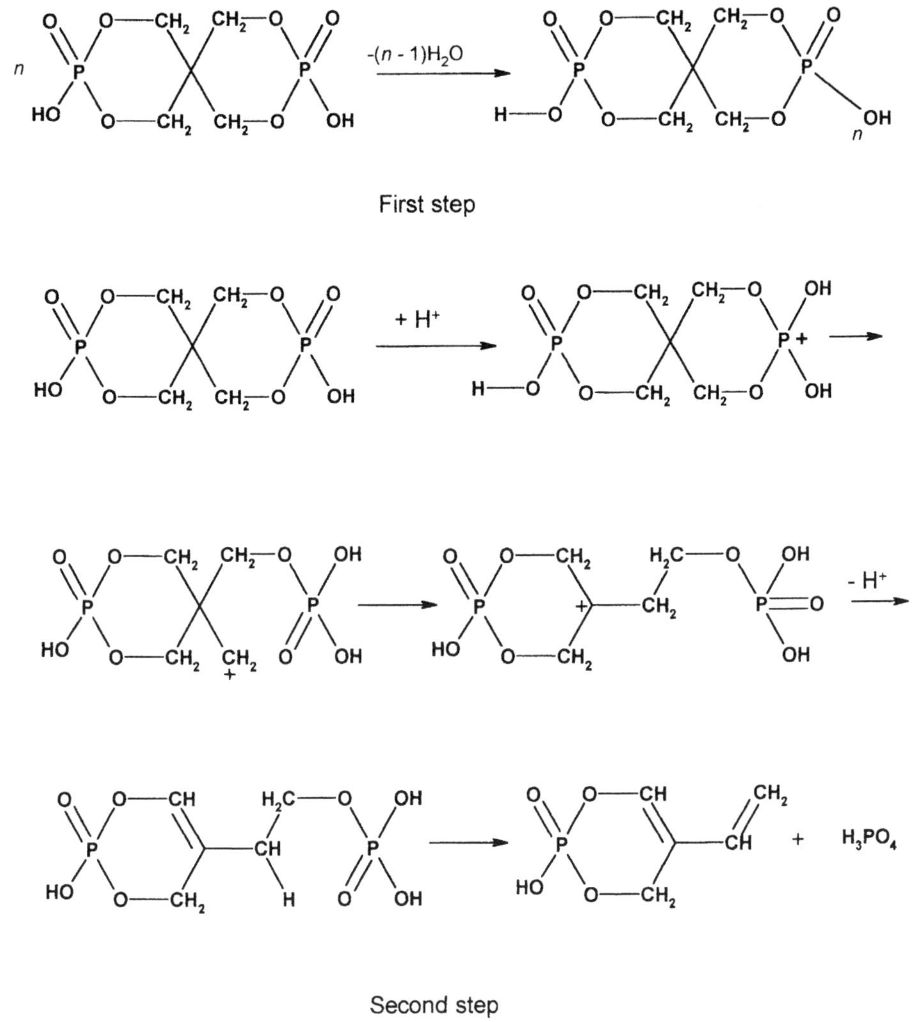

First step

Second step

Scheme 10.3 Ester pyrolysis mechanism

in competition with other processes; the final products are obtained by some random combination of polymerisation, Diels–Alder condensation, aromatisation, etc. Ester pyrolysis supplies the chemical structures which build up the charred material through relatively simple reactions (Scheme 10.3). These schemes give a better account of the charring reactions than that previously proposed in the literature[37] for similar compounds, based on the formation of intermediate carbenes. It is unlikely that carbenes would survive the strongly acidic reaction medium.

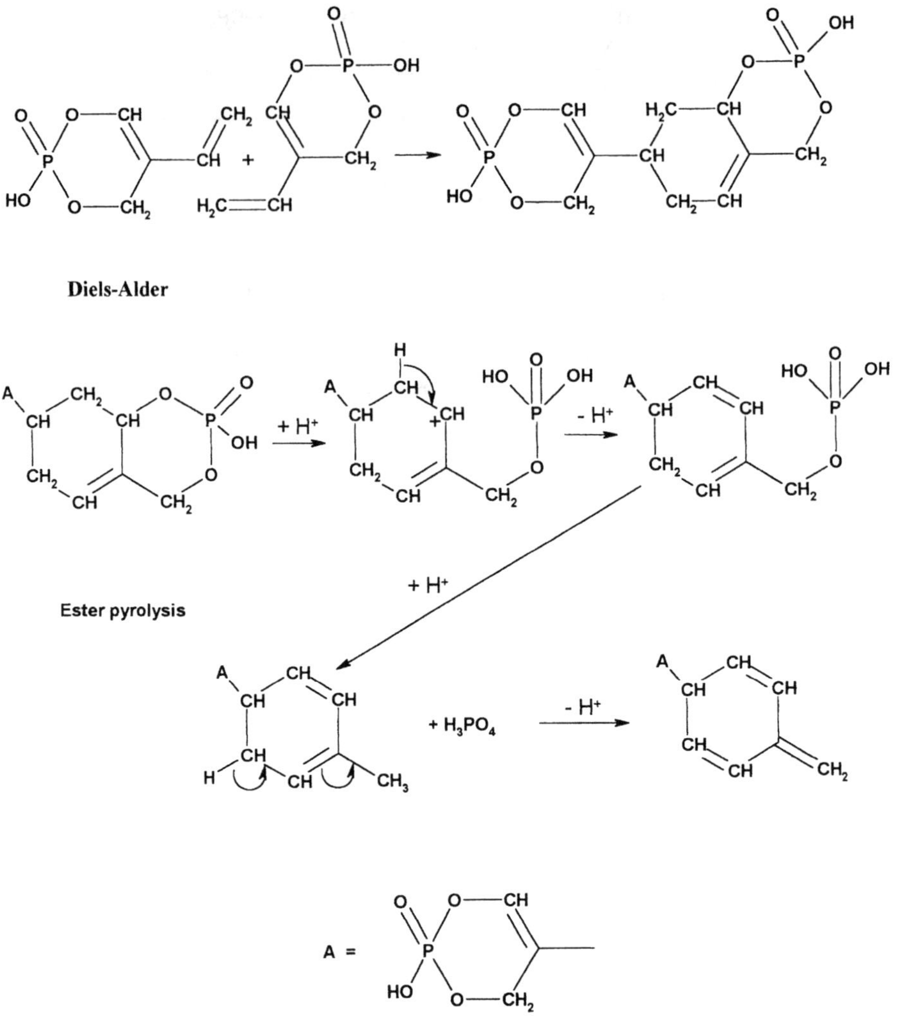

Diels-Alder

Ester pyrolysis

+ H$_3$PO$_4$ - H$^+$

A =

Scheme 10.4 Diels – Alder followed ester pyrolysis

10.4   Intumescent systems

Examples of commercially available intumescent systems are presented in Table 10.2.[38]

10.5   Conclusion

The intumescent behaviour resulting from a combination of charring and foaming of the surface of the burning polymers is being widely developed

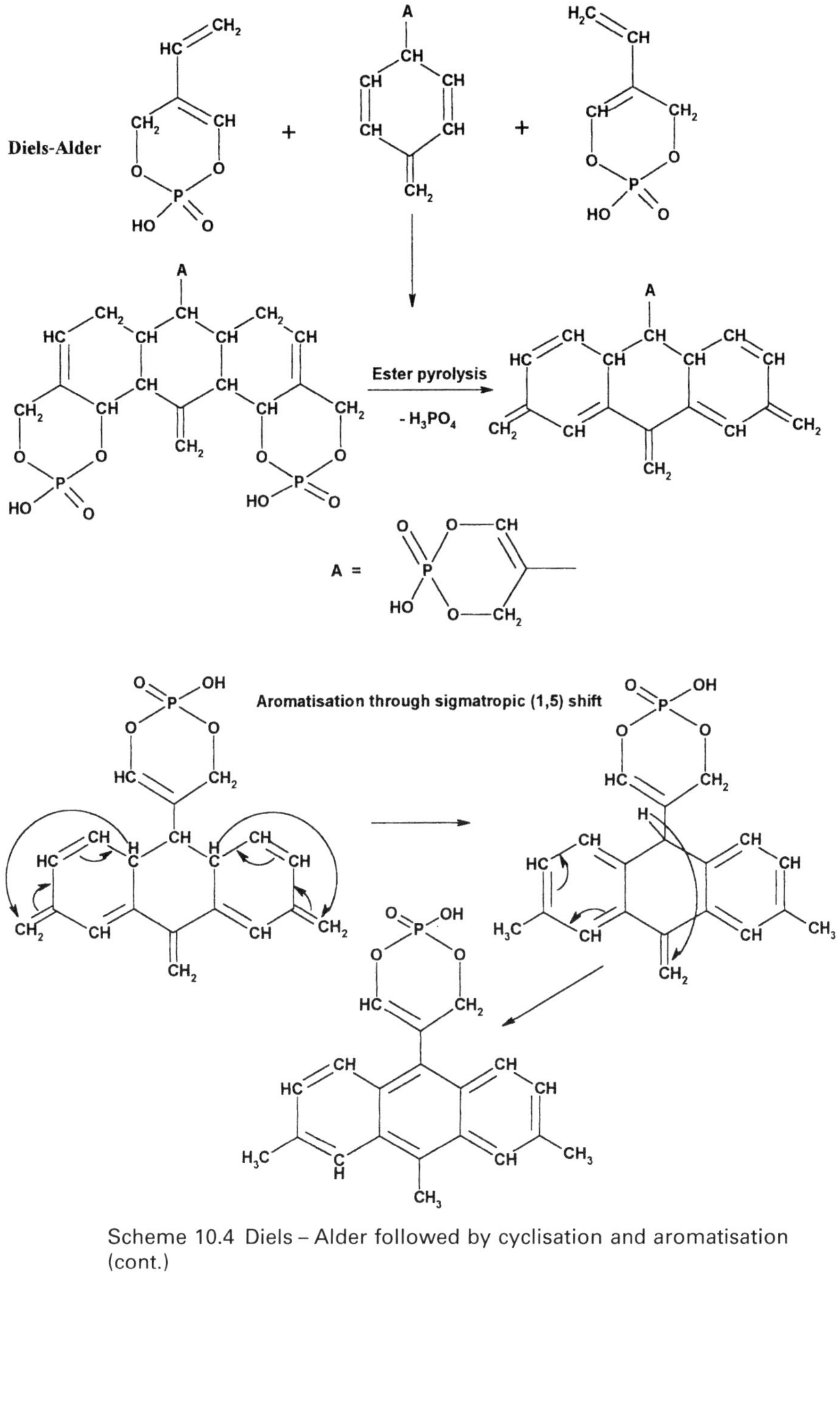

Scheme 10.4 Diels – Alder followed by cyclisation and aromatisation (cont.)

Table 10.2 Flame retardant intumescent systems[38]

Chemical name	Manufacturer	Trade name	Applications
Ammonium polyphosphates	Albright & Wilson*	Amgard MC series, TR, CL, ALBRITE	ABS, Acrylic, Epoxy, Polyester, PVA, Polystyrene, Polyethylene PVC, Uff, Phenolic Polypropylene, Ufr, Polycarbonate
	Great Lakes	FRCROS 480, 481, 484, 485	ABS, Acrylic, Epoxy, Polyester, PVA, Polystyrene, Polyethylene PVC, Uff, Phenolic Polypropylene, Ufr Polycarbonate, EVA
	Hoechst Celanese	Hostaflam AP 422, 462	Acrylic, Epoxy, Polyester, PVA, Polyethylene, PVC, Uff, Polypropylene, Ufr
	Solutia, Inc.	Phos-Chek P/30, P/40	Acrylic, Polyester, PVA, Polyethylene, PVC, Uff, Ufr
Monoammonium phosphate	Rhone-Poulenc		ABS, Epoxy, Polyester, PVA, Polystyrene, Polyethylene, PVC, Phenolic, Polypropylene
	Total Speciality	Total Phosphate Series	ABS, Epoxy, Polyester, PVA, Polystyrene, Polyethylene, PVC, Phenolic, Polypropylene
Diammonium phosphate	Rhone-Poulenc		ABS, Epoxy, Polyester, PVA, Polystyrene, Polyethylene, PVC, Uff, Phenolic Polypropylene, Ufr Polycarbonate
	Great Lakes	Ultra Carb	ABS, Epoxy, Polyester, PVA, Polystyrene, Polyethylene, PVC, Uff, Phenolic Polypropylene, Ufr
Melamine phosphate	Akzo Nobel	Fyrol MP	ABS, Polyethylene, Polypropylene, Uff, Ufr
	Albright & Wilson*	Amgard NH/ND, ALBRITE	ABS, Polyethylene, Polypropylene, Uff, Ufr
	Great Lakes	FRCROS-490	ABS, Polyethylene, Polypropylene, Uff, Ufr
	Miljac		ABS

Table 10.2 (cont.)

Chemical name	Manufacturer	Trade name	Applications
Melamine pyrophosphate	Akzo Nobel	Fyrol MPP	ABS, Polyethylene, Polypropylene, Uff, Ufr
	Great Lakes	FRCROS-491	ABS, Polyethylene, Polypropylene, Uff, Ufr
	Miljac		ABS
	StanChem	MPP	ABS, Polyethylene, Polypropylene, Uff, Ufr
Phosphorus-containing polyol	Albright & Wilson*	Vircol 82	Epoxy, Ufr, Uff
	Hoechst Celanese	Hostaflam OP 514, 515, 550	Ufr, Uff
Phosphonate esters	Akzo Nobel	Fyrol DMMP, HMP, 6, 76	Acrylic, Epoxy, Phenolic
O,O-diethyl-1-N, bis (2-hydroxyethyl) aminomethyl-phosphonate	Akzo Nobel	Fyrol 6	Ufr
Di-(polyoxyethylene) hydromethyl phosphonate	Akzo Nobel	Victastab HMP	Polyester, Phenolic, Ufr, Uff

Where: Uff – Urethane foam flexible, Ufr – Urethane foam rigid.
*Albright and Wilson is now Rhodia.

for fire retardance because it is characterised by a low environmental impact. However, the fire retardant effectiveness of intumescent systems is difficult to predict because the relationship between the occurrence of the intumescence process and the fire protecting properties of the resulting foamed char is not yet understood. The characterisation of the char is quite complex and requires special techniques for solid state characterisation.

Both of the physical and chemical models proposed for intumescence further enable one to understand the overall reaction scheme.

References

1 Lomakin S M and Zaikov G E, *Ecological Aspects of Flame Retardancy*, V S P International Science Publishers, Zeist, Netherlands, 1999, 170.
2 Halpern Y, Mott D M and Niswander R H, *Ind. Eng. Chem. Prod. Res. Dev.* 1984, **23**, 233.
3 Anderson Jr C E, Dziuk Jr J, Mallow W A and Buckmaster J, *J. Fire Sci.* 1985, **3**, 151.
4 Tramm H, Clar C, Kuhnel P and Schuff W, *US. Pat.* 1938; 2 106 938.

5 Kay M, Price A F and Lavery I, *J. Fire Retard. Chem.* 1979, **6**, 69.
6 Cagliostro D E, Riccitiello S R, Clark K J and Shimizu A B, *J. Fire Flamm.* 1975, **6**, 205.
7 Delobel R, Le Bras M, Quassou N and Alistigsa F, *J. Fire Sci.* 1990, **8**, 85.
8 Delobel R, Le Bras M, Quassou N and Descressain R, *Polym. Degrad. Stab.* 1990, **30**, 41.
9 Camino G, Costa L and Trossarelli L, *Polym. Degrad. Stab.* 1984, **7**, 25.
10 Camino G, Martinasso G, Costa L and Cobetto R, *Polym. Degrad. Stab.* 1990, **28**, 17.
11 Levchik S V, Costa L and Camino G, *Polym. Degrad. Stab.* 1992, **36**, 31.
12 Heinrich H, Deutsch. Pat. 1991; DE 4 015 490Al.
13 Rychly J, Matisova-Rychla L and Vavrekova M, *J. Fire Retard. Chem.* 1981, **8**, 82.
14 Cullis C F, Hirschler M M and Khattab M A A M, *Eur. Polym. J.* 1992, 145.
15 Vandersall H L and *J. Fire Flamm.* 1971, **2**, 97–140.
16 Cagliostro D E, Riccitiello S R, Clark K J and Shimizu A B, *J. Fire Flamm.* 1975, **6**, 205–20.
17 Butler K M, Baum H R and Kashiwagi T, International Association for Fire Safety Science. *Fire Safety Science.* Proceedings, Fifth International Symposium. March 3–7, 1997, Melbourne, Australia, Intl. Assoc. for Fire Safety Science, Hasemi Y, ed., Boston, MA, 523–34 p (1997).
18 Anderson C, Ketchum D E and Mountain W P, *J. Fire Sci.* 1988, **6**, 390–410.
19 Anderson C E, Dziuk J, Mallow W A and Buckmaster J, *J. Fire Sci.* 1985, **3**, 161–94.
20 Epstein P S and Plesset M S, *J. Chem. Phys.* 1.
21 Levchik SV, Camino G, Costa L and Levchik G F, *Fire and Mater.* 1995, **19**, 1–8.
22 Levchik S V, Levchik G F, Balabanovich A I, Camino G and Costa L, *Polymer Degrad. Stab.* 1996, **54**, 205–15.
23 Levchik S V, Costa L and Camino G, *Polymer Degrad. Stab.* 1992, **36**, 229.
24 Levchik S V, Levchik G F, Selevich A F and Leshnikovich A I, *Vesti AN Belarusi, Ser. Khim.* 1995, **3**, 34–9.
25 Levchik S V, Levchik G F, Camino G and Costa L, *J. Fire Sci.* 1995, **13**, 43.
26 Levchik S V, Balabanovich A I, Levchik G F, Camino G and Costa L, *Polymer Degrad. Stab.* 1998.
27 Costa L, Camino G and Luda di Cortemiglia M P, *In Fire and Polymers.* ed. G L Nelson, ACS Symposium, Series 425, ACS, Washington DC, 1990, 211.
28 Levchik S V, Costa L and Camino G, *Polymer Degrad. Stab.* 1992, **43**, 43–9.
29 Camino G, Martinasso G and Costa L, *Polym. Degrad. Stab.* 1990, **27**, 285–96.
30 Baumgarten H E, Setterquist R A, *J. Am. Chem. Soc.* 1957, **79**, 2605–8.
31 Canavan A E, Dowden B F and Eaborn C, *J. Chem. Soc.* 1962, 331–4.
32 Higgins C E and Baldwin V M, *J. Org. Chem.* 1965, **30**, 3173–6.
33 Berlin K D, Morgan J G, Peterson M E and Pivonka W C, *J. Org. Chem.* 1969, **34**, 1266–71.
34 Haake P and Diebert C E, *J. Am. Chem. Soc.* 1971, **93**, 6931–7.
35 Lhomme V, Bruncau C, Soyer N and Brault A, *Ind. Eng. Chem.* 1994, **23**, 98–102.
36 Rychly J, Rychla L M and Vavrekova M, *J. Fire Ret. Chem.* 1981, **8**, 82–92.
37 *Modern Plastics, Encyclopedia-99*, 62–71.
38 Butler K M, Baum H R and Kashiwagi T, *Proceedings of the International Conference on Fire Research and Engineering*, 261–6, Orlando, 10–15 September 1995.

11

Graft copolymerisation as a tool for flame retardancy

CHARLES A WILKIE

Department of Chemistry
Marquette University
Wisconsin
USA

11.1 Introduction: the production of char from polymers

For several years the work of my research group had been concentrated on understanding the mechanistic details of the interaction of a wide variety of additives with poly(methyl methacrylate) (PMMA). The expectation of this work was that it would permit one to design a suitable flame retardant for this polymer since one would know the locus of the reaction on the polymer and also know the important structural features of an additive which could allow the reaction. Additives which were studied included red phosphorus,[1,2] $(PPh_3)_3RhCl$,[3,4] Nafion–H,[5] Ph_xSnCl_{4-x} (x = 0–4),[6,7] Ph_2S_2,[8] various transition metal halides,[9,10,11,12] and copolymers of 2-sulphoethyl-methacrylate with methyl methacrylate.[13]

In general, additives interact with the carbonyl group of PMMA and some amount of non-volatile residue is usually produced. This non-volatile residue is usually referred to as char. The usual process for the interaction between the additive and the polymer involves the formation of a radical from the degradation of PMMA and its subsequent interaction with the additive. Two pathways have been used to describe this interaction: either the polymeric radical is coordinated by some additive species and then the electron is transferred to the additives or coupling of additive radicals with polymeric radicals occurs. Both of these possible pathways are illustrated in Fig. 11.1.

While the work described above enables one to understand how individual additives interact with the polymer, there is not enough information available to permit the true design of a flame retardant. An additional problem is that even if it were possible to design a useful fire retardant for PMMA, this design information would be unlikely to extend to other polymers. With these limitations in mind, we set out to develop an approach which may be generally applied to a variety of polymers. Since all of the work which we had carried out previously had involved cross-linking

11.1 Stabilisation of the tertiary PMMA radical by transition metal halides and by coupling with stabilising radicals; both remove the radical from the polymer chain (Reprinted from Polymer Degradation & Stability, 56, T J Xue and C A Wilkie, 'Thermal degradation of poly(styrene-g-acrylonitrile), 109–113, (1997), with permission from Elsevier Science)

chemistry, it was logical to think in terms of cross-linking and char-forming processes.

11.2   The importance of char

The role of a char on the surface of a polymer is to protect the underlying polymer from the heat of the fire. Ideally a char layer should be an excellent insulator; in addition it should be difficult to combust, must remain in place so that it can be effective (i.e. be adherent), must have structural integrity so that it cannot be easily ruptured, and so forth. The typical char is a carbonaceous material which is formed during pyrolysis. The best example of this type of char may be that which is produced in the degradation of polyacrylonitrile and which yields elemental carbon if appropriately treated.[14]

The process which we envisaged was to devise a way by which a polymeric precursor of a char could be attached to a polymer in such a way that the precursor would thermally degrade and offer protection to the polymer. The challenges which are faced in this endeavour are: (1) to identify suitable char-formers and (2) to develop processes by which these could be delivered to the surface of the relevant polymer. It must be remembered that this should be of utility with a wide variety of polymers so the process of delivery of the char former must be general.

11.2.1 Identification of the char formers

It was thought that the ideal char former would be inorganic, so that it could not be combusted, but with some amount of an organic component which may serve as a binder to hold the material together. This char former must thermally degrade at a temperature lower than the temperature at which the polymer degrades so that the char is available to protect when the degradation of the polymer commences. A perusal of the literature led to the discovery that McNeill and others had studied the thermal degradation of various salts of methacrylic and acrylic acids and that these salts produced relatively large amounts of char at somewhat modest temperatures.[15-25] The results of McNeill's investigations are presented in Table 11.1.

The thermal degradation of poly(methacrylic acid) begins near 200°C with the loss of water and the formation of anhydrides.[14-17] These anhydrides suffer further degradation near 400°C with the eventual formation of a small amount of carbonaceous char.

The degradation of the salts produces substantial amounts of non-volatile residue which has been identified as the metal carbonate plus some amount of carbonaceous material. For the alkali metal salts[17-18] the amount of residue is diminished when the degradation is conducted in a vacuum rather than in nitrogen. The major volatile product is carbon dioxide with minor amounts of olefins and ketones. The yield of ketones decreases and the amount of non-volatile residue increases with increasing size of the cation, suggesting that the size of the cation plays some significant role in the degradation.

The degradation of the alkaline earth salts follows a similar pattern.[17,19] Since $MgCO_3$ is thermally unstable, the residue from the thermolysis of the magnesium salt is the oxide. The onset of degradation and the yield of non-volatile residue increases with increasing size of the cation, again suggesting the important role that size plays in the degradation process. It is likely that the size effect is actually due to the increased mass of the metal which must lead to a greater residue and a lower yield of volatiles. The major volatile products are carbon dioxide, also seen for the alkali salts, and

Table 11.1 Thermal degradation of salts of poly(methacrylic acid) and poly(acrylic) acid

Cation	Onset temperature of degradation, °C	% residue at 500°C	Identity of residue
Methacrylic acid			
H^+	200	3	'C'
Li^+	400	54	Li_2CO_3 + C
Na^+	400	64	Na_2CO_3 + C
K^+	400	66	K_2CO_3 + C
Cs^+	400	82	Cs_2CO_3 + C
Mg^{2+}	200	31	MgO + C
Ca^{2+}	280	57	$CaCO_3$ + C
Sr^{2+}	320	61	$SrCO_3$ + C
Ba^{2+}	400	70	$BaCO_3$ + C
Acrylic acid			
H^+	175	12	Char
Na^+	400	64	Na_2CO_3 + C
K^+	400	60	K_2CO_3 + C
Mg^{2+}	450	49	MgO + C
Ca^{2+}	470	57	$CaCO_3$ + C
Zn^{2+}	—	48	ZnO + C
Co^{2+}	—	39	Co + C
Ni^{2+}	—	34	Ni + C
Cu^{2+}	—	55	Cu + C

dimethylketene, which is not observed for the alkali metal salts. The minor products are similar for both alkali and alkaline earth salts. For the alkali salts there are two competing pathways for degradation: degradation to monomer and formation of carbonate and ketone; for the alkaline earth salts monomer formation is unimportant and all degradation occurs by the formation of carbonates and ketones.

The thermal degradation of the ammonium salt of poly(methacrylic acid) has also been studied.[20] There is an early loss of both water and ammonia; these come from competing reactions: intramolecular reactions of cyclic anhydrides and imidisation reactions, with imidisation being the dominant reaction. Since there is no metal oxide or carbonate that may be formed, it is surprising that about 10% non-volatile residue is obtained at 500°C.

The thermal degradation of poly(acrylic acid) proceeds in a three-step process and about 12% residue is obtained at 500°C.[21] The initial reaction is the formation of anhydrides which are subsequently degraded. The major volatile products are carbon dioxide and water with minor amounts of olefins and ketones.

The thermal degradation of sodium and potassium polyacrylates produces a residue containing the metal carbonate and some carbonaceous material.[22] The major volatile products are carbon dioxide and acetone with minor amounts of carbon monoxide, methane, ketene, olefins, and ketones. There is some difference in these degradations depending upon whether the polymer was formed by the direct polymerisation of the alkali acrylate or from the neutralisation of poly(acrylic acid). The degradation of magnesium and calcium polyacrylates[23] proceeds in a three-step process and also produces a large amount of residue and a substantial amount of carbon dioxide. The other volatiles are the same as observed for the alkali metal salts. The degradation of some transition metal acrylates and polyacrylates has been studied by Gronowski and Wojtczak.[25]

The foregoing description indicates that the methacrylate and acrylate salts are suitable char-forming monomers to graft onto the target polymers. Not only do they produce a significant amount of char; they also produce reasonable amounts of carbon dioxide which can serve to quench a fire. The char consists of some inorganic salt or metal together with carbonaceous material. This combination offers an excellent opportunity to produce a thermally insulating and adherent char on the surface of the polymer. If the char-forming reactions actually involve the underlying polymer, so that it participates in char formation, this will offer a significant opportunity to render the polymer flame retardant.

11.2.2 Delivery of the char former to the polymer

A truly general process is required for applicability to a wide variety of polymers. It was felt that graft copolymerisation offers the opportunity to attach a methacrylate salt or some other char former to the surface of a polymer. Graft copolymerisation, as with any polymerisation process, may be accomplished by radical or ionic mechanisms so there should be ample opportunity to develop processes to attach the char former. It is usually agreed that the monomer may be attached onto a polymeric radical which has been produced by the interaction between the base polymer and an initiator, 'grafting from', or if the monomer interacts with the initiator to form a polymeric radical which then adds to the base polymer, 'grafting onto'. Both 'grafting from' and 'grafting onto' may occur by either an ionic or a radical process; these are illustrated for a radical process in Fig. 11.2.

Our investigations of graft copolymerisation in order to enhance the thermal stability of polymers commenced using acrylonitrile–butadiene–styrene as the base polymer with methacrylic acid as the monomer and the anthracene-photosensitised formation of singlet oxygen as the means to achieve graft copolymerisation. This process has been developed by Geuskens[26,27] and involves an initial migration of anthracene into the

$$\text{wwCH}_2-\text{CH}=\text{CH}-\text{CH}_2\text{www} \xrightarrow{\ \text{R}^{\cdot}\ } \text{wwCH}_2-\text{CH}=\overset{\cdot}{\text{C}}\text{H}-\text{CH}\text{www} \ + \text{RH}$$

$$\Big\downarrow \text{R}-\text{M}_n-\text{M} \qquad\qquad\qquad\qquad \Big\downarrow \text{n M}$$

$$\text{wwCH}_2-\text{CH}=\text{CH}-\text{CH}\text{www}$$
$$\overset{\displaystyle |}{\underset{\displaystyle \overset{\cdot}{\text{M}_n}}{}}$$

$$\text{ww CH}_2-\overset{\cdot}{\text{CH}}-\text{CH}-\text{CH}_2\text{www}$$
$$|$$
$$\text{M}$$
$$|$$
$$\text{M}_n$$
$$|$$
$$\text{R}$$

11.2 Grafting from and grafting onto schemes for graft copolymerisation

$$\text{An} \xrightarrow{\ v\ } \text{An}^*$$

$$\text{An}^* + {}^3\text{O}_2 \longrightarrow \text{An} + {}^1\text{O}_2$$

$${}^1\text{O}_2 \ + \ \text{wwCH}_2-\text{CH}=\text{CH}-\text{CH}_2\text{ww} \longrightarrow \text{wwCH}_2-\text{CH}=\text{CH}-\text{CH}\text{ww}$$
$$\underset{\displaystyle \text{OOH}}{|}$$

$$\text{wwCH}_2-\text{CH}=\text{CH}-\text{CH}\text{ww} \longrightarrow \text{wwCH}_2-\text{CH}=\text{CH}-\text{CH}\text{ww} + \overset{\cdot}{\text{OH}}$$
$$\underset{\displaystyle \text{OOH}}{|} \qquad\qquad\qquad\qquad\qquad\qquad \underset{\displaystyle \overset{\cdot}{\text{O}}}{|}$$

11.3 Anthracene photosensitised formation of singlet oxygen for graft copolymerisation

polymer. Upon irradiation of the anthracene-containing polymer, singlet oxygen is produced and this adds at an allylic position in the butadiene portion of the terpolymer to give a hydroperoxide which may be easily cleaved, by heating to about 60°C, to form a radical from which the monomer may attach. This is illustrated in Fig. 11.3.

11.2.2.1 Assessment of the efficacy of graft copolymerisation for thermal protection of polymers

There are several ways in which one may assess the efficacy of the graft copolymerisation process for the thermal stabilisation of polymers. On the one hand, the onset temperature of the degradation, as measured by thermogravimetric analysis, is the surest indicator of the primary degradation step in the pathway. In some cases this is easy to measure while in other cases, due to competing reactions, it is not possible to evaluate it. An alternative approach is to measure the fraction of non-volatile residue at high temperature. The disadvantage with using char formation as an assessment of stabilisation is that char forms as a result of secondary reactions that commence from the species which are formed in the initial step. Thus the fraction of char which is formed is some measure of the efficacy of the process but it is not truly an assessment of enhanced thermal stability.

One may imagine two ways in which graft copolymerisation may effect thermal stability, a chemical interaction between the graft layer and the substrate and simply the physical effect of the char layer insulating the substrate from the heat. In the first case one may expect to see an effect on the onset temperature of the degradation since there is a chemical interaction and the substrate is necessarily changed. If there is only the physical effect of the char insulation, no change in onset would be expected and the fraction of char is the important parameter.

11.2.2.2 Graft copolymerisation onto ABS and related polymers

The initial work in this area was to graft methacrylic acid onto acrylonitrile–butadiene–styrene terpolymer and then to convert the acid into its sodium salt using sodium hydroxide.[28] The efficacy of the process was evaluated by thermogravimetric analysis and cone calorimetry.[29] Figure 11.4

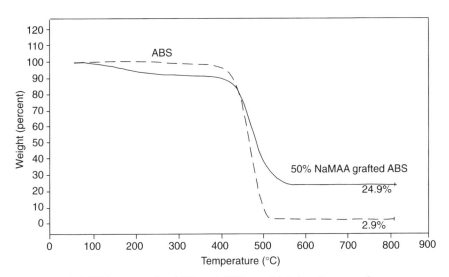

11.4 TGA curves for ABS and ABS to which has been graft copolymerised sodium methacrylate[29]

shows the TGA curve for ABS and for ABS to which has been graft copoly-merised 50% methacrylic acid (the acid was subsequently converted to its sodium salt). One can see that the degradation of the graft copolymer com-mences earlier than that of the virgin ABS; this early degradation is due to the presence of residual acid units which degrade much more easily than the salt. The main degradation step for both the graft copolymer and the unreacted ABS occurs at essentially the same temperature. One must con-clude from this that the onset temperature of the degradation is little changed and therefore the graft layer is not likely to interact chemically with the ABS substrate.

If one examines the fraction of non-volatile residue, one sees that the char formed from the graft layer does retain ABS. For a given amount of sodium methacrylate, one can calculate the fraction of non-volatile residue which may be expected. Any amount beyond this is polymer which has been trapped by the char formed from the poly(sodium methacrylate). Table 11.2 shows the calculated residue and the observed residue for ABS samples to which 0 to 20% of sodium methacrylate has been grafted. As a measure of merit, we have computed an enhancement figure which is the difference between the actual and non-volatile residue divided by the theoretical amount. The difference between the actual residue and the calculated residue at low graft levels is quite small and likely to be within experimental error. The enhancement factors exaggerate these numbers so they appear to be more significant than they really are. The only significant dif-ferences are at 20% graft level and above. This enhancement is some measure of the extent to which the char layer is able to entrap the under-lying polymer.

The results from cone calorimetry are even more exciting. If one com-pares virgin ABS with a sample to which 21% sodium methacrylate has been graft copolymerised, one finds that the time to ignition is approxi-mately doubled and the peak heat release rate is reduced by a factor of about 3. Similar improvements are evident in the other cone calorimetry results; these are presented in Table 11.3.

Table 11.2 Thermogravimetric analysis of ABS grafted with sodium methacrylate[29]

% NaMAA	Actual residue	Calculated residue	Enhancement
0	4	4	0
1	5	4	25
10	6	5	20
20	15	6	150

Table 11.3 Cone calorimetry results (heat flux is 25 kW/m^2) for virgin and grafted ABS[29]

	Virgin ABS	Grafted ABS
Time to ignition, s	285	460
Peak heat release rate, kW/m^2	901	259
Time to peak HRR, s	530	1130
Time to burnout, s	670	1400+
Energy released through 670 s, kJ	1700	150
Mass loss rate, mg/s	170	40
Mass loss, % at 20 min	92	37
Specific extinction area, m^2/kg, average over run	798	826
Extinction area, (total smoke) m^2		
Produced through 670 s	49	3.1
Produced through 20 min	58	29

Of particular note in this table is the great change in mass loss rate and in absolute mass loss; the change between the virgin ABS and the graft copolymer is remarkable. The data on extinction area at first glance does not appear that significant. It must be remembered that this information is on a per mass basis; samples which show lower mass loss must release less smoke. After 670 seconds of irradiation the virgin ABS is almost completely consumed while the graft copolymer is only 5% consumed and the smoke is one-sixteenth of that released by the virgin ABS.

These results are outstanding and have led to additional experiments in this area. These include the graft copolymerisation of acrylic acid by the benzophenone-photosensitised process[30] and to chemical initiation of graft copolymerisation.[31] As might be expected, since there is less organic mass to lose in acrylic acid than in methacrylic acid, the enhancements in the amount of char are even greater for sodium acrylate. These results are shown in Table 11.4.

When large samples appropriate for cone calorimetry were prepared, it was observed that the sodium acrylate coating did not adhere well to the ABS. In the small samples that are used for TGA, this is not observed. It is suspected that the methyl group plays an important role in the adherence of the graft layer onto the underlying polymer.

The encouraging results obtained for ABS led to attempts to extend this approach to other polymers in order to prove the generality of graft copolymerisation. Other polymers that have been examined include styrene–butadiene block copolymers,[32-35] poly(ethylene terephthalate),[36,37] polyamide-6,[38] and polystyrene.[39]

We have extensively explored graft copolymerisation onto styrene–butadiene block copolymers both by the anthracene approach[32] and by

Table 11.4 Thermogravimetric analysis of ABS grafted with sodium acrylate[30]

Sodium acrylate, %	Actual residue, %	Expected residue, %	Enhancement, %
0	4	4	0
3	7	5	40
10	15	8	88

Table 11.5 Thermogravimetric analysis of SBS and K-resin grafted with sodium methacrylate[33]

Sodium methacrylate, %	Actual residue, %	Calculated residue, %	Enhancement, %
SBS-based copolymers			
0	0	0	0
20	14	3	366
K-resin			
0	0	0	0
10	4	1	300
20	14	3	366

chemical initiation.[34,35] Two different styrene-butadiene copolymers have been studied; one was supplied by Shell Chemical and contained 75% butadiene and is called SBS while the other was supplied by Phillips Petroleum, contained 25% butadiene and is called K-resin. The TGA results for graft copolymerisation of sodium methacrylate onto each of these are shown in Table 11.5.

These enhancements are much larger than those seen for ABS and appear to indicate that this system is quite effective for these polymers as well. Unfortunately the cone calorimetry results are not as encouraging; these are shown in Table 11.6.

It is quite disappointing to observe that the change in various parameters is only around 20 to 30% whereas it is typically 60% or higher in the ABS system. This is a surprising result since the TGA gave such promising information. At this time there is no explanation for these differences.

11.2.2.3 Graft copolymerisation onto poly(ethylene terephthalate)

There are a great number of reports that describe the graft copolymerisation of acrylic monomers onto poly(ethylene terephthalate) (PET); initia-

Table 11.6 Cone calorimetry results (heat flux is 25 kW/m²) for SBS and K-resin and graft copolymers of these with sodium methacrylate[33]

	% sodium methacrylate on SBS		% sodium methacrylate on K-resin	
	0	10	0	10
Time to ignition, s	180	221	245	335
Peak heat release rate, kW/m²	991	741	590	544
Time to peak HRR, s	405	550	504	735
Time to burnout, s	465 = t	571	630 = t	829
Energy output through t, kJ	1620	1170	1555	990
Mass loss rate, mg/s	183	160	154	125
Total smoke produced through t, m²	67	43	69	38

tors such as benzoyl peroxide, transition metals ions and hydrogen peroxide have been used. We have found that it is not possible to perform such graft copolymerisation and that the actual product obtained in such a reaction is really a physical mixture of the two homopolymers.[36,37] Since a graft copolymer is not formed, the thermal stability of the polymer is not changed. The most significant observation from this work is that PET which contains poly(methacrylic acid) is degraded much more easily by aqueous sodium hydroxide than is virgin PET. Apparently the strands of poly(methacrylic acid) that flow throughout the PET permit the easy migration of sodium hydroxide to all portions of the polymers so that the entire polymer is exposed to the base and reaction occurs much more easily.

11.2.2.4 Graft copolymerisation onto polyamide-6

Graft copolymerisation of methacrylic acid onto polyamide-6 has been carried out using hydrogen peroxide, benzoyl peroxide, and azobisisobutyronitrile as the initiator.[38] The thermal stability of the polyamide-6 is much lower when the acid is present. This is presumable due to the known acid instability of polyamides. When the acid is converted to its sodium salt, the degradation begins at a lower temperature but there is an increasing amount of non-volatile fraction produced at high temperatures. It appears to be very unlikely that this process will enhance the thermal stability of polyamides.

11.2.2.5 Graft copolymerisation onto polystyrene

It is not possible radically to graft copolymerise onto polystyrene for reasons which are not obvious. One would imagine that it would be quite easy to remove the benzylic hydrogen atom and create a radical at that site but this reaction has never been shown to occur. In this laboratory we have spent a significant amount of time attempting to accomplish the reaction and have concluded that graft copolymerisation does not occur at this site. This may be due to an inability to remove the hydrogen atom or enhanced stability for the radical that does not allow graft copolymerisation to occur.[39]

It is possible to prepare graft copolymers through an anionic route in which the reaction of *n*-butyllithium on polystyrene removes the *para*-hydrogen on the benzene ring.[40] Anionic initiation is incompatible with graft copolymerisation of methacrylic acid so a new char-forming monomer was selected, acrylonitrile. Polyacrylonitrile is well known to form carbon fibres under certain degradation conditions. We have recently published some new information which sheds new light on the pathway by which polyacrylonitrile thermally degrades.[41] Using thermogravimetric analysis the graft copolymer shows enhanced thermal stability relative to both polystyrene and polyacrylonitrile.[42] It can clearly be seen in the TGA curves for polyacrylonitrile, poly(styrene-g-acrylonitrile) and polystyrene, shown in Fig. 11.5, that the onset temperature of the degradation increases in the presence of the graft layer. This implies that there must be some chemical interaction between the graft acrylonitrile and the substrate polystyrene during the thermal degradation process. The formation of char from both blends and graft copolymers of acrylonitrile and styrene have been exam-

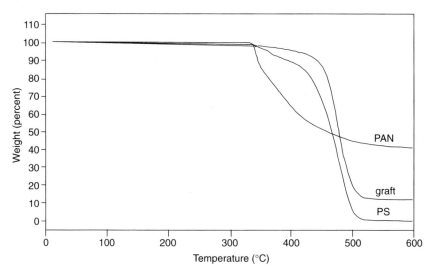

11.5 TGA curves for polyacrylonitrile, polystyrene and poly(styrene-g-acrylonitrile)

Table 11.7 Thermogravimetric analysis of blends and graft copolymers of acrylonitrile and styrene[42]

% acrylonitrile	Actual residue at 600 °C	Theoretical residue	% enhancement
Graft copolymers			
11	5	4	25
21	11	9	22
25	14	10	40
30	17	12	42
Blends			
9	5	4	25
17	7	7	0
23	10	10	0

ined and the TGA results are shown in Table 11.7. The enhanced formation of char indicates that the physical insulation mechanism is also operative for this system.

There is no enhancement for blends while some of the polystyrene is retained in the graft copolymers. The enhancements are much smaller than those observed in the ABS–sodium methacrylate system and this suggests less efficacy for polystyrene. The preparation of the graft copolymer invariably leads to the formation of the homopolymer of acrylonitrile along with the graft copolymer. The amount of actual non-volatile residue is not significantly different between graft copolymers and blends. We suggest that the difference between grafts and blends is apparently the degree of intimacy between the components. In a blend, mixing of the components will almost certainly be on the macro level and not on the molecular level. In the graft copolymers, on the other hand, mixing is likely to be molecular and the formation of char from polyacrylonitrile has the opportunity to protect thermally the underlying polystyrene layer. It is of interest to compare the graft copolymers and blends with the random styrene–acrylonitrile copolymer, SAN.[43] SAN degrades in a single step to give no non-volatile residue. The non-volatile residue which arises in the degradation of polyacrylonitrile occurs because one acrylonitrile unit attacks the next and leads to cyclisation. In the case of SAN the acrylonitrile units are not adjacent so cyclisation cannot occur.

Cone calorimetric results confirm that neither the graft copolymerisation of acrylonitrile or blends of polyacrylonitrile with polystyrene show any significant enhancement of the combustion performance compared to virgin polystyrene, as shown in Table 11.8.

Table 11.8 Cone calorimeter data for graft copolymers and blends of acrylonitrile and styrene[42]

	Heat flux = 25 kW/m²			Heat flux = 35 kW/m²		
	PS	PS-g-AN, 25%	PS-b-AN, 25%	PS	PS-g-AN, 25%	PS-b-AN, 25%
Time to ignition, s	193	148	115	85	40	60
Peak heat release rate, kW/m²	560	468	394	747	529	497
Time to peak HRR, s	548	590	453	390	425	450
Time to burnout, s	623	685	645	443	552	560
Energy released through burnout time, kJ	1480	1790	1675	1950	1860	1970
Mass loss rate, mg/s	141	123	120	178	140	152
Specific extinction area, m²/kg – average from ignition to burnout	1167	1166	1296	1198	1096	1125
Total smoke, m² – produced through burnout	68	76	79	72	77	72

The TGA work showed that the thermal stability of polystyrene was slightly improved by the graft copolymerisation process. From the cone calorimeter data one must conclude that the graft copolymer is less thermally stable than is virgin polystyrene. For instance, the time to ignition is lower for both blends and grafts than is pure polystyrene. There is a very marginal improvement in some properties but, in general, this is not a process which will enhance the thermal stability of polystyrene.

It appears that even though the onset temperature of the degradation is increased, the polymer will still degrade very easily and release large quantities of heat at high temperatures. On the other hand, since the onset temperature is raised, if the sample is subjected to only a small heat flux, it should degrade less rapidly.

The difference between the ABS system and the polystyrene system is that for ABS the char-former will produce a char with a significant inorganic component while polystyrene can only produce a carbonaceous char. We believe that the formation of an inorganic char is very desirable since it cannot burn (although in some cases it may be susceptible to oxidation) and will always be available to insulate the underlying polymer. In future work we shall be looking for new materials which can produce inorganic chars.

11.2.2.6 Other inorganic char-forming monomers

We have recently examined both the acid form and the sodium salt of vinyl-sulphonic acid, styrenesulphonic acid, and vinylphosphonic acid.[44] All of these materials show a large increase in volume as they char: the volume increase is typically about eight times. The sodium salts of all three acids produce at least 60% non-volatile residue and, surprisingly, there is little difference between reaction in inert atmosphere (N_2) and reaction in air. These materials present unique opportunities for the formation of new graft copolymers and blends which may have enhanced thermal stability.

11.3 Conclusion

Graft copolymerisation may provide an opportunity to develop systems for flame and fire retardancy which are more-or-less polymer independent and this will be a very significant advance if the process is ever fully developed. In one case, that of ABS-g-sodium methacrylate, the results from cone calorimetry are very encouraging but in other cases similar results have not been obtained. There is still much to do in this area in order to develop further the systems for graft copolymerisation and to identify new monomers which may be attached. The future appears bright but there is a considerable amount of work in front of us.

References

1 Wilkie C A, Pettegrew J W and Brown C E, 'Pyrolysis reactions of poly(methyl methacrylate) and red phosphorus: an investigation using cross polarization, magic angle NMR spectroscopy', *J Polym Sci: Poly Lett Ed*, 1981 **19** 409–14.

2 Brown C E, Wilkie C A, Smulkalla J, Cody R B and Kinsinger J A, 'Inhibition by red phosphorus of unimolecular thermal chain-scission in poly(methyl methacrylate): investigation by NMR, FT-IR and laser desorption/Fourier transform mass spectroscopy', *J Polym Sci Polym Chem Ed*, 1986 **24** 1297–311.

3 Sirdesai S J and Wilkie C A, 'Wilkinson's salt: a flame retardant for poly(methyl methacrylate)', *J Appl Polym Sci*, 1989 **37** 863–6.

4 Sirdesai S J and Wilkie C A, 'Mechanism of poly(methyl methacrylate) fire retardation by Wilkinson's salt', *J Appl Polym Sci*, 1989 **37** 1595–603.

5 Wilkie C A, Thomsen J R and Mittleman M L, 'The interaction of poly(methyl methacrylate) and nafions', *J Appl Polym Sci*, 1991 **42** 901–9.

6 Chandrasiri J A and Wilkie C A, 'The thermolysis of poly(methyl methacrylate) in the presence of phenyltin chlorides, Ph_xSnCl_{4-x}', *Polym Degrad Stab*, 1994 **45** 83–9.

7 Chandrasiri J A and Wilkie C A, 'Thermal degradation of poly(methyl methacrylate) in the presence of tin(IV) chloride and tetraphenyl', *Polym Degrad Stab*, 1994 **45** 91–6.

8 Chandrasiri J A and Wilkie C A 'Thermal degradation of diphenyl disulfide and a blend of diphenyl disulfide with poly(methyl methacrylate)', *Polym Degrad Stab*, 1994 **46** 275–84.

9 Wilkie, C A, Leone J T and Mittleman M L, 'The interaction of poly(methyl methacrylate) and manganese chloride', *J Appl Polym Sci*, 1991 **42** 1133–41.

10 Beer R S, Wilkie C A and Mittleman M L, 'The interaction of poly(methyl methacrylate) and chromium chloride: transfer of methyl groups from the ester to the main chain', *J Appl Polym Sci*, 1992 **46** 1095–102.

11 Wilkie C A, Beer R S, Leone J T and Mittleman M L, 'Comparative effects of transition metal halide additives on the thermal degradation of poly(methyl methacrylate)', *J Fire Sci*, 1993 **11** 184–92.

12 Chandrasiri J A, Roberts D E and Wilkie C A, 'The effect of some transition metal chlorides on the thermal degradation of poly(methyl methacrylate): a study using TGA/FT–IR', *Polym Degrad Stab*, 1994 **45** 97–101.

13 Hurley S M, Wilkie C A and Mittleman M L, 'Degradation of polymers and copolymers: copolymer of methyl methacrylate and 2-sulfoethyl methacrylate', *Polym Degrad Stab*, 1993 **39** 345–54.

14 Mathur R B, Bahl O P and Sivram P, Thermal degradation of polyacrylonitrile fibres, *Curr Sci*, 1992 **62** 662–9.

15 McNeill I C, 'Polymer degradation and characterization by thermal volatilization analysis with differential condensation of products', *Eur Polym J*, 1970 **6** 373–95.

16 Jamieson A and McNeill I C, 'The thermal degradation of copolymers of methyl methacrylate with methacrylic acid', *Eur Polym J*, 1974 **10** 217–25.

17 Grant D H and Grassie N, 'The thermal decomposition of poly(t-butyl methacrylate)', *Polymer*, 1960 **1** 445–55.

18 McNeill I C, 'Degradation of synthetic polymers: application of some pyrolysis methods in an investigation of the degradation behavior of ionomers containing methacrylate salt units', *Develop Polym Degrad*, 1987 **7** 1–33.

19 McNeill I C and Zulfiqar M, 'Preparation and degradation of salts of poly(methacrylic acid) Part I Lithium, sodium, potassium, and caesium salts', *J Polym Sci: Polym Chem Ed*, 1978 **16** 3201–12.

20 McNeill I C and Zulfiqar M, 'Preparation and degradation of salts of poly(methacrylic acid) – Part II: Magnesium, calcium, strontium and barium salts', *Polym Degrad Stab*, 1979 **1** 89–104.

21 McNeill I C and Zulfiqar M, 'Thermal degradation of ammonium polymethacrylate and polymethacrylamide', *J Polym Sci: Polym Chem Ed*, 1978 **16** 2465–74.

22 McNeill I C and Sadeghi S M T, 'Thermal stability and degradation mechanisms of poly(acrylic acid) and its salts: Part 1 Poly(acrylic acid)', *Polym Degrad Stab*, 1990 **29** 233–46.

23 McNeill I C and Sadeghi S M T, 'Thermal stability and degradation mechanisms of poly(acrylic acid) and its salts: Part 2 Sodium and potassium salts', *Polym Degrad Stab*, 1990 **30** 213–30.

24 McNeill I C and Sadeghi S M T, 'Thermal stability and degradation mechanisms of poly(acrylic acid) and its salts: Part 3 Magnesium and calcium salts', *Polym Degrad Stab*, 1990 **30** 267–82.

25 Gronowski A and Wojtczak Z, 'The thermal decompositions of some transition metal acrylates and polyacrylates', *J Thermal Anal*, 1983 **26** 233–44.

26 Geuskens G and Kanda M N, Surface modification of polymers – I. Grafting initiated by photo-generated hydroperoxides, *Eur Polym J*, 1991 **9** 877–9.

27 Geuskens G and Thiriaux Ph, Surface modification of polymers – II. Photo-oxidation of SBS containing anthracene and grafting initiated by photo-generated Hydroperoxides, *Eur Polym J*, 1993 **29** 351–5.

28 Suzuki M and Wilkie C A, Graft copolymerization of methacrylic acid and acrylamide onto acrylonitrile–butadiene–styrene terpolymer by photoinduced hydroperoxide, *J Polym Sci: Part A: Polym Chem*, 1995 **33** 1025–9.

29 Suzuki M and Wilkie C A, The thermal degradation of acrylonitrile–butadiene–styrene terpolymer grafted with methacrylic acid, *Polym Degrad Stab*, 1995 **47** 223–8.

30 Deacon C and Wilkie C A, Graft copolymerization of acrylic acid onto acrylonitrile–butadiene–styrene terpolymer and thermal analysis of the copolymers, *Eur Polym J*, 1996 **32** 451–5.

31 Chandrasiri J A and Wilkie C A, Chemically initiated graft copolymerization of acrylic acid onto acrylonitrile–butadiene–styrene(ABS) terpolymer and its constituent polymers, *J Polym Sci; Part A: Polym Chem*, 1996 **34** 1113–20.

32 Dong X, Geuskens G and Wilkie C A, Graft copolymers of methacrylic acid and SBS and K-Resin by the anthracene photosensitized formation of singlet oxygen, *Eur Polym J*, 1995 **31** 1165–8.

33 Wilkie C A, Suzuki M, Dong X, Deacon C, Chandrasiri J A and Xue T J, Grafting to achieve flame retardancy, *Polym Degrad Stab*, 1996 **54** 117–24.

34 Jiang D D and Wilkie C A, Chemical initiation of graft copolymerization of methyl methacrylate onto styrene–butadiene block copolymer, *J Polym Sci: Part A: Polym Chem*, 1997, **25**, 965–73.

35 Jiang D D and Wilkie C A, Graft copolymerization of methacrylic acid, acrylic acid and methyl acrylate onto styrene–butadiene block copolymers, *Eur Polym J*, 1998 **34** 997–1006.

36 Xue J and Wilkie C A, Swelling-assisted modification of poly(ethylene terephthalate) by methacrylic acid, *J Polym Sci: Part A: Polym Chem*, 1995 **33** 1019–24.

37 Xue T J and Wilkie C A, The interaction of vinyl monomers and poly(ethylene terephthalate) in the presence of various initiators produces a physical mixture, not a graft copolymer, *J Polym Sci: Part A: Polym Chem*, 1995, **33** 2753–8.

38 Langer N M and Wilkie C A, Surface modification of polyamide-6: graft copolymerization of vinyl monomers onto polyamide-6 and thermal analysis of the graft copolymers, *Polym Adv Tech*, 1998 **9** 290–6.

39 Xue T J and Wilkie C A, Graft copolymerization of acrylonitrile onto polystyrene, *J Polym Sci: Part A: Polym Chem*, 1997 **35** 1275–82.

40 Xue T J, Jones M S, Ebdon J R and Wilkie C A, Lithiation–Alkylation of polystyrene occurs only on the ring, *J Polym Sci: Part A: Polym Chem*, 1997 **35** 509–13.

41 Xuz T J, McKinney M A and Wilkie C A, The thermal degradation of polyacrylonitrile, *Polym Degrad Stab*, 1997 **58** 193–202.

42 Xue T J and Wilkie C A, Thermal degradation of poly(styrene-g-acrylonitrile), *Polym Degrad Stab*, 1997 **56** 109–113.

43 Suzuki M and Wilkie C A, The thermal degradation of acrylonitrile–butadiene–styrene terpolymer as studied by TGA/FTIR, *Polym Degrad Stab*, 1995 **47** 217–21.

44 Jiang D D, Yao Q, McKinney M A and Wilkie C A, TGA/FTIR studies on the thermal degradation of some polymeric sulfonic and phosphonic acids and their sodium salts, *Polym Degrad Stab*, 1999 **63** 423–34.

12

Performance-based test methods for material flammability

BJÖRN KARLSSON

Department of Fire Safety Engineering, Lund University, Sweden

12.1 Introduction

Fire safety regulations have a major impact on the overall design of buildings with regard to layout, aesthetics, function and cost. Due to advances in building technology and increased understanding of fire phenomena, there has been a world-wide move towards replacing prescriptive regulations by performance criteria and performance design procedures with respect to fire risks in buildings.

In order to be able to apply such methods with regards to material flammability, a comprehensive and coherent philosophy on material reaction-to-fire must be developed. This includes such steps as defining end-use scenarios, defining limit states (or critical conditions) identifying the dominant physical processes involved and developing performance-based test methods that provide the essential material properties.

This chapter starts with a description of the current situation, giving an example of a purely prescriptive test method and describing three test methods that can be said to be performance-based. A general discussion on performance-based design and performance-based test methods follows. A number of end-use scenarios and critical conditions will be discussed. The dominant physical mechanisms leading to these critical conditions will be described, indicating which material properties must be measured.

Following these discussions, the material properties that are of greatest importance for the end-use scenarios can be identified. These properties can be measured by bench-scale instruments that have already been developed and standardised by the International Standards Organisation. Examples of how the properties can be used in mathematical modelling to predict critical conditions in full-scale tests are given, showing that these properties are indeed the desired product of the test methods. Finally, recommendations for future work will be given.

12.2 Testing and ranking systems

12.2.1 The current situation

The reaction-to-fire of products used in buildings has been a concern for legislators and authorities since the advent of building fire safety regulations. In recent years there has been intense ongoing activity to develop reaction-to-fire test methods and ranking systems. This chapter discusses a number of existing test methods in order to exemplify the state of the art and to build a foundation for later discussions on performance-based test methods.

Some proposed testing and ranking systems for the reaction-to-fire behaviour of products have been based on test methods which give as output certain rating terms or arbitrary numbers. As an example, such a test method may report a 'burnt length' at a given time, or the time until the flame has reached a given length. Such numbers have a very weak or very uncertain link to material properties and the dominant physical processes involved. They cannot be used for rational classification nor for design calculations. Such test methods have been used, in conjunction with empirical approaches, to rank materials. However, the ranking has a questionable basis.

Figure 12.1 shows the results from a survey, made in the 1970s, of several European test methods which were used to rank the flammability of lining

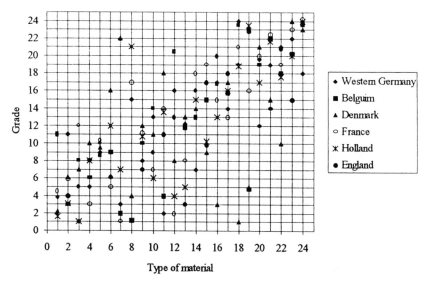

12.1 The ranking of 24 lining materials determined at 6 national testing laboratories in Europe, according to each nation's regulations in the 1970s[1]

materials used in buildings. The tests were carried out in six national laboratories according to each country's test and classification method. A low
grade indicates high risk. The figure shows that there is an alarming spread
in the results. Material 18, for example, was considered the best material
tested in Germany, while it was classed as the worst in Denmark.

All of the test methods used to obtain the data presented in Fig. 12.1 were
prescriptive test methods. The very large spread in the results is due to the
fact that in the 1970s Europe had no harmonised philosophy with regards
to reaction-to-fire testing and most of the methods used were prescriptive
in nature, i.e. measured arbitrary numbers or ranking terms that had little
to do with material flammability properties.

Vandevelde[2] and Blaciere et al[3] discussed these anomalies and this undesirable state of affairs led to a very comprehensive study, funded by the
Commission of the European Communities, on how scientific tools could
be used to classify construction products with regard to product behaviour
in fire situations. The study recommended that extensive, long term research
and testing be carried out.[4]

12.2.2 Four common tests

12.2.2.1 *The single burning item test (SBI)*

The pressing need for a harmonised European approach for the reaction-
to-fire classification of products was such that the recommendations for
long term research[4] could not be followed. In 1994, an important agreement
was made between the member countries in the European Community[5]
that stated all such countries should have the same test procedures and the
same classification system for surface lining materials used in buildings. The
classification system is mainly based on the FIre Growth RAte (FIGRA)
index, which is calculated using the parameters from the main test method,
the single burning item (SBI) test.[6]

In addition to the SBI test the non-combustibility test (ISO 1182),[7]
the gross calorific value test (ISO 1716)[8] and the ignitability test (ISO
11925-2)[9] are also used for the classification. A total of ten different
parameters are used to specify the seven Euroclasses (A1, A2, B, C, D, E
and F).

While the ambition of the SBI test developers was to measure such variables as time to ignition, flame spread and heat release rate, the method
cannot provide a meaningful measure of these. To try to observe the occurrence of ignition or the position of a flame front behind a large gas burner
flame is not very meaningful. Also, the heat release rate of a material must
be measured per unit area if the measurement is to be used in an engineering fashion. The SBI method measures heat release rate as the

pyrolysing area increases; the data is therefore only applicable to the SBI scenario and cannot be used in engineering. Neither can the data be used as a reference scenario for a real room fire since the apparatus has little relation to a typical room.

Convincing regulators in European countries to change their national testing and classification system, in order to agree on a single harmonised system, was an onerous and time-consuming task. In this sense the development of the SBI test method and the European classification system must be seen as a great political success. However, in terms of science, engineering and performance-based design the SBI method has been severely criticised.[10] The method cannot be said to be a performance-based test method.

12.2.2.2 The room corner test (ISO 9705)

The room corner test[11] is a large-scale test method for measurement of the burning behaviour of surface lining materials used in buildings. The test apparatus consists of a small compartment with one open door and a gas collection system which is supplied with necessary instruments to measure the fire gas properties, as shown in Fig. 12.2.

The lining material, which is mounted on three walls and the ceiling, is exposed to a fire placed in one of the rear corners of the compartment. The compartment measures 2.4 m × 2.4 m × 3.6 m (length × height × width) and

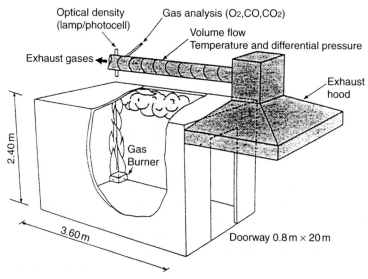

12.2 The room corner test (ISO 9705)

the opening measures 0.8 m wide and 2.0 m high. The ceiling, the floor and the walls are constructed of non-combustible material.

A propane burner is used as ignition source and has a heat output of 100 kW for the first 10 minutes, thereafter the output level is increased to 300 kW for another 10 minutes. The experiment will continue until flash-over occurs or until 20 minutes have passed. The criterion of 1000 kW for the heat release rate is said to be equal to flashover, defined as flames coming out through the doorway, if that has not occurred earlier.

The output data available from the Room Corner test are mainly the time to flashover and the following parameters as a function of time:

- Heat release rate (HRR).
- Smoke production rate (SPR).
- CO production rate.
- CO_2 production rate.
- Oxygen depletion rate.

12.2.2.3 The cone calorimeter test (ISO 5660)

The cone calorimeter test[12] is a bench-scale test used to determine the reaction to fire for surface lining materials used in buildings. The test apparatus consist basically of an electric heater, an ignition source and a gas collection system, as shown in Fig. 12.3.

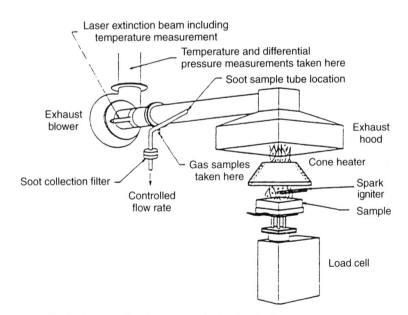

12.3 Typical set up for the cone calorimeter test

The test specimen measures 100 mm × 100 mm and has a thickness between 6 mm and 50 mm. During the test the specimen is mounted horizontally on a low heat loss insulating ceramic material. The orientation of the specimen can also be vertical, but this position is mainly used for exploratory studies.

After the test specimen has been mounted and placed in the right position, it is exposed to a heat flux from the electric heater. The output from the heater can be chosen in the range of 0–100 kW/m², but usually the heat output is in the range of 25–75 kW/m². When the mixture of gases above the test specimen is higher than the lower flammability limit, it is ignited by an electric spark source. The duration of the test is normally 10 minutes but is not fixed and can vary, depending on the material.

Many variables are measured but the main results from each test are:

- Time to ignition (TTI).
- Mass loss rate (MLR).
- Heat release rate (HRR).

If a gas analyser is added to the test equipment it is also possible to take the production of smoke and toxic gases into account.

12.2.2.4 The LIFT apparatus test (ISO 5658)

Both the International Standards Organisation (ISO) and the American Society for Testing and Materials (ASTM) have developed test methods to characterise lateral flame spread properties of materials.[13,14] The test consists of two procedures; one to measure ignition and one to measure lateral flame spread, sometimes referred to as opposed flow or creeping flame spread. This mode of flame spread is relatively slow in comparison with upward flame spread, also referred to as concurrent flow flame spread.

The test apparatus has been named LIFT, for lateral ignition and flame spread test, and is shown in Fig. 12.4 Vertically mounted specimens are exposed to the heat from a vertical air-gas fuelled radiant-heat energy source inclined at 15° to the specimen.

For the ignition test, a series of 155 mm by 155 mm specimens are exposed to a nearly uniform heat flux and the time to ignition, using a pilot flame as igniter, is recorded. For the flame spread test, a specimen measuring 155 mm by 800 mm is exposed to a heat flux that gradually decreases along the horizontal length of the sample. After ignition, the flame spread velocity along this length of the specimen is recorded. The data from the two procedures is then correlated with a theory of ignition and flame spread for the derivation of the following material flammability properties:

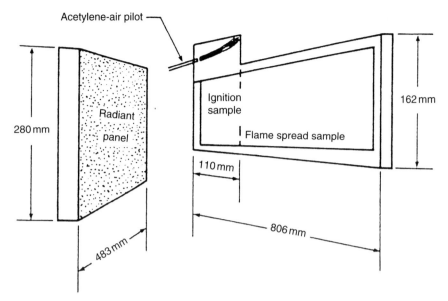

12.4 A schematic view of the LIFT apparatus

- Critical heat flux for ignition, per unit area of specimen.
- Ignition temperature.
- Apparent thermal inertia.
- Flame heating parameter.

12.2.3 Concluding remarks on test methods

This section has given an example of a recently developed, purely pre-scriptive test method (SBI). Very little data relevant to engineering and design can be retrieved from the test results.

Three examples of test methods that have been used as a basis for the performance-based approach have also been described. One of these is fre-quently used as a reference scenario to real-world fires for the reaction-to-fire of lining materials (room corner test). Tests for other such full-scale scenarios have also been developed such as those for for testing cables and floorings.

Two of the methods, the cone calorimeter and the LIFT apparatus, are very frequently used to approximate material properties with regards to flammability, especially the cone calorimeter. Section 12.5 will discuss how the results from these bench-scale methods have been used by scientists and engineers to predict the reaction-to-fire behaviour of materials in full-scale scenarios. Sections 12.3 and 12.4 describe the philosophy behind the performance-based concept and how the test methods are a part of the performance-based system.

12.3   Performance-based test methods

During the last decades rapid development in modern building technology has resulted in unconventional structures and design solutions; the physical size of buildings is continually increasing; there is a tendency to build large underground car parks, warehouses and shopping complexes. The interior design of many buildings with large light shafts, patios and covered atriums inside buildings, connected to horizontal corridors or malls, introduces new risk factors concerning spread of smoke and fire. Past experiences or historical precedents, which form the basis of current prescriptive building codes and regulations, rarely provide the guidance necessary to deal with fire hazards in new or unusual buildings.

At the same time there has been rapid progress in the understanding of fire processes and their interaction with humans and buildings. Advancement has been particularly rapid where analytical fire modelling is concerned. Several different types of such models, with varying degrees of sophistication, have been developed in recent years and are used by engineers in the design process.

As a result, there is a world-wide movement to replace prescriptive building codes with those based on performance. Instead of prescribing exactly which protective measures are required (such as prescribing a number of exits for evacuation purposes), the performance of the overall system is presented against a specified set of design objectives (such as stating that satisfactory escape should be effected in the event of fire). Application of the performance concept in buildings is driven by the need for a more flexible global building market, and the elimination of barriers to innovation.

In this context, there is considerable international interest in developing performance-based test methods and classification systems for building products. The interest has specifically been directed towards lining materials for interior surfaces, but studies of other fixed interior products in buildings, such as floorings and cables, have also been conducted. We shall in this chapter describe methods developed for interior lining materials as a good example of how performance-based test methods for reaction-to-fire of products can be developed.

12.3.1 Criteria for performance-based test methods

The CIB (acronym for the former name 'Conseil International du Bâtiment', now renamed as 'International Council for Research and Innovation in Building and Construction') has for over two decades been working with the performance concept in building design. Two groups within CIB mainly work with these concepts; Task Group TG37 (Performance Based Building Regulation Systems) and Working Commission W060 (Performance Concept in Building). These are assisted in their task by the many other CIB task groups and working commissions. In a publication from 1982, CIB Working Commission W060 discusses a number of criteria for performance test methods.[15] These criteria can be summarised as follows:

- Conditions of test under which the behaviour of the article is being assessed must be realistic in relation to the expected conditions of use, or related to them in some known way.
- There needs to be a clear scientific basis for relating the results of performance testing under simplified conditions to conditions in practice.
- It is important to consider – and to reconsider – whether the method will be suitable for predicting the behaviour of the product under real conditions of use.

The report[15] also states that although it may theoretically be desirable that a performance-based test method should be independent of the material or construction tested, it is difficult to respect this principle in all cases. Furthermore, it mentions that the method should ideally be simple but that simplification should not go so far that the method fails to provide a reasonable simulation of conditions of use.

12.3.2 Philosophy for reaction-to-fire test methods

In 1995 CIB organised a workshop on 'New Developments in Performance Test Methods',[16] where Karlsson and Kokkala[17] discussed developments in the Nordic countries (Denmark, Sweden, Norway and Finland) with respect to performance-based test methods for assessing the fire safety of products. The Nordic development is briefly described since the process is typical of the efforts being made internationally.

The Nordic philosophy on material reaction-to-fire is based on a long tradition of harmonising test methods and regulations. In addition to testing, the use of calculation methods has been promoted. In many cases, the use of calculation methods is already recognised in the national building regulations. To minimise the problem of getting input data for calculations, the policy of the Nordic fire researchers has been to support methods in which the output data can be used both to classify products for prescriptive codes and as input data in calculations. As an example, all the Nordic national fire laboratories have been actively supporting the development and use of methods applying oxygen consumption techniques to quantitative measurement of energy release rates.

In the late 1980s Wickström[18] compared the testing of mechanical behaviour of building elements with the reaction-to-fire of interior surface lining materials. His comparison is shown schematically in Fig. 12.5. The left-hand column (mechanical behaviour) shows how the real end-use condition is simplified and represented by a beam with supports, tested in full scale. A part of the material can be tested in small scale to give material properties. These properties can then be used as input to a mathematical model to calculate critical conditions in large scale and the results compared to design criteria.

Similarly, for surface lining materials, the real end-use condition is represented by a full-scale test method with lining material attached to the

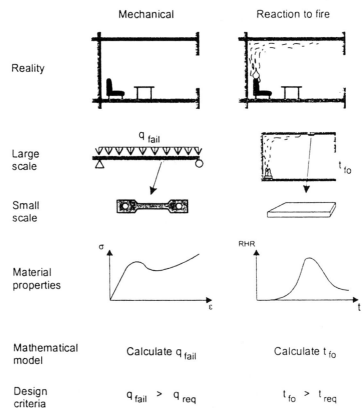

12.5 A schematic comparing the philosophy for testing mechanical behaviour of building elements with that of the reaction-to-fire of interior surface lining materials. σ is stress, ε is strain, q_{fail} is the failure load, q_{req} is the load required by design, RHR is the rate of heat release, t is time, t_{fo} is time to flashover and t_{req} is time required for safe rescue. Concept from Wickström[18]

enclosure surfaces. An ignition source is provided and time to flashover, t_{fo}, assumed to be the critical condition, is recorded. A piece of the material is tested on the small scale to give material properties that are used in a mathematical model to calculate full-scale behaviour.

A Nordic fire research programme, EUREFIC (European Reaction to Fire Classification) was carried out between 1989 and 1991 inclusive in order to speed up the development of performance-based test methods for reaction-to-fire of interior building products.[19] The programme focused on two test methods; the full-scale ISO 9705 – room corner test[11] and the bench-scale ISO 5660 – cone calorimeter.[12] The main motivation for this programme was to avoid a change to methods of no better technical quality, and instead facilitate a change towards more technically advanced methods.

It is evident that there are products and processes that cannot be reliably tested on a small scale. These may include some composite materials and such phenomena as melting and dripping and mechanical failure of the product. The system must then offer a possibility to test the product in the large-scale reference scenario. The principles behind using the cone calorimeter (ISO 5660) and the room corner test (ISO 9705) for testing and classifying flammability of products are shown in Fig. 12.6.

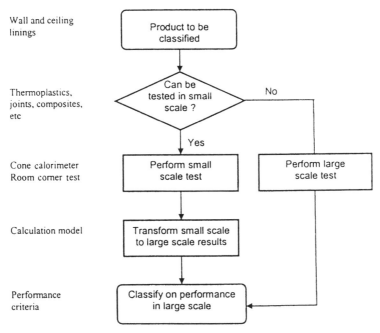

12.6 Principles of testing and classification scheme employing the cone calorimeter and the room corner test, according to Wickström[20]

12.4   End-use scenarios, critical conditions and controlling processes

Before any performance-based test method can be devised, the details behind the testing philosophy must be developed. Specifically, in order to work out a rational ranking system for the flammability of products, which to some extent reflects the hazards encountered in an end-use scenario, one must:

- Define one or more end-use scenarios (large room, small room, ignition source, openings, etc.) and develop a standardised test for that purpose.
- Define one or more limit states or critical conditions (i.e. time until a certain temperature is attained, time to flashover, time until a certain concentration of gases is attained, etc.).
- Use knowledge of the end-use conditions and limit states to define the controlling physical processes involved.
- Use engineering methods and simplifications in order to allow practical use of bench-scale test results to estimate hazards in the end-use condition.
- Design a performance-based bench-scale test method that gives as results, actual or estimated, the flammability properties which are needed for input in the engineering methods.

Some of these aspects are discussed further below, namely end-use scenarios, critical conditions for humans and controlling physical processes.

12.4.1 End-use scenarios

Ranking materials with respect to fire safety should reflect a certain end-use condition, or a full-scale test. There are many conceivable scenarios:

- A small room, a large room or a corridor.
- The ignition source can be in the corner or by the wall and can be large or small.
- The material may be mounted on walls, ceiling or both.
- The scenario may be well ventilated (open door) or poorly ventilated.

The end-use scenario must be well defined if one wants to identify the physical processes controlling the environmental results of a fire. This is of considerable importance since, having identified them, one can single out the most important material parameters involved in the processes, which lead to a certain critical condition. Additionally, the importance of each of the material parameters can be assessed.

One or more well-defined end-use scenarios must therefore be chosen. A material can then be ranked according to each of these.

12.4.2 Critical conditions

There are certain critical conditions induced by fire beyond which humans are at great risk of losing their life and they must be specified and quantified in order to allow identification of the processes leading towards them. The conditions for human safety may, for example, be defined as

- A certain critical gas temperature.
- A certain critical incident heat flux.
- A certain concentration of CO or other gases.
- A certain optical density.

Other criteria have also been used, such as time to flashover in a small room. This is quite a reasonable criterion (even though human beings would not survive it) since it gives a measure of how quickly the fire grows and thus how quickly conditions in adjoining rooms will become hazardous to humans.

All of the above conditions depend very strongly on the end-use scenario and on the heat release rate (HRR), which in turn is a result of how fast the flame spreads over the material. Babrauskas and Peacock[21] argue convincingly that heat release rate is the single most important variable in fire hazard. Flame spread is in this case one of the fundamental processes since it, to a large extent, controls how fast the associated phenomena (such as heat generation, gas production, change in optical density) happen.

12.4.3 Controlling physical processes

The physical processes which are of greatest interest with regards to material hazards are those of ignition and of flame spread, especially upward (or concurrent flow) flame spread.

The physical processes dominating upward and downward flame spread are considerably different. In order to develop as simple a ranking scheme as possible, it is advantageous to consider the two processes separately, since in some end-use scenarios it is clear that one of these processes will dominate.

In the first case the density differences drive the flame upwards, preheating the unburned material mainly through flame radiation, often resulting in rapid development of heat and gases. This type of flame spread can occur not only on walls but also under ceilings.

In the second case, preheating is either dominated from an external source (e.g. a hot gas layer) or from the small tip of the flame front. This downward or lateral flame spread is often termed slow, creeping flame spread, so this process appears to be less hazardous than upward flame spread. However, in a room fire the walls will slowly be heated up and after a considerable time (often getting close to flashover) lateral and downward flame spread can occur very rapidly.

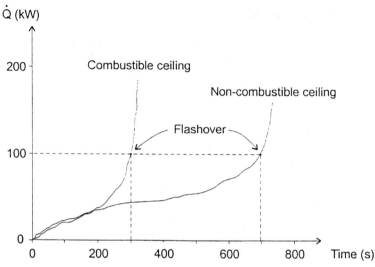

12.7 HRR in a room corner test with particleboard mounted on walls only, and with material mounted on both walls and ceiling (from Karlsson and Quintiere[22]). Flashover in this room occurs at a heat release rate of 100 kW

An example of how the two processes behave in a one-third scale room corner test can be seen in Fig. 12.7. Two tests were performed where particleboard was mounted on walls and ceiling in one test and the walls only in the other.

The two tests show how the HRR initially increases in both cases due to upward flame spread in the corner of the room. In the case where combustible material is only mounted on the walls, the flame spread (and HRR) is more or less halted until the hot gas layer has heated the walls sufficiently for lateral flame spread to occur, resulting in flashover after roughly 12 minutes. In the case where the material is mounted on both walls and ceiling, the concurrent flow flame spread continues directly under the ceiling, causing flashover in roughly 4 minutes.

The example shows that in many end-use scenarios one would mainly be concerned with upward flame spread. This mode of flame spread is in most cases more hazardous than downward flame spread and we shall therefore look more closely at upward flame spread.

12.5 Models for calculating upward flame spread and fire growth

In the last decade several groups of scientists, working separately in various countries, have developed flame-spread theories that can be used in an engineering fashion to calculate upward flame spread and the resulting fire growth. These methods are of various degrees of sophistication and complexity. Some give approximate answers for specified end-use scenarios, can be used by non-experts and require simple data input. Others are more general, but may require expert knowledge and a large amount of input data.

There have been mainly two types of method for such predictions, in practical end-use scenarios, proposed in the literature. Firstly, purely thermal models for upward flame spread have been used, with input data from the cone calorimeter, to predict flame spread on a large scale and the resulting heat release rate. Secondly, more fundamental work has been carried out using CFD (computational fluid dynamics) models and pyrolysis models to predict fire growth.

12.5.1 Upward flame spread models based on thermal theories

Ignition of a solid material can be defined as the attainment of a given surface temperature, called the ignition temperature. Purely thermal theories can then be used to calculate surface temperatures on a solid and as soon as an element reaches the ignition temperature, that element is assumed to be pyrolysing. Often, the element is then assumed to release a certain amount of energy, usually linked to heat release rate measurements from the cone calorimeter. Such an approximation eliminates the need for calculating the mass flow rate of combustion gases from the solid element and there is no need to take chemical kinetics into account.

Very many different approaches to such modelling have been made where the results have been compared to experiments involving practical building materials. All of these require that the flame morphology, specifically the flame length, be estimated as well as the heat flux from the flame to the solid materials. It is generally difficult to estimate these variables and many workers have therefore opted for making relatively simple assumptions with respect to flame lengths and flame heat fluxes.

Hasemi[23] used a variable flame heat flux to analyse the temperature rise of the unburned fuel ahead of the pyrolysis front. Delichatsios et al[24] and Beyler et al[25] also used expressions for a variable heat flux over the flame height to calculate the upward flame spread velocity and fire growth.

One of the most straightforward approaches is characterised by assuming a simple relationship between flame length and heat release rate and assuming a constant flame heat flux over this length, Saito et al.[26] This led to an analytical mode for upward flame spread velocity involving a Volterra-type integral. Thomas and Karlsson[27] solved the Volterra equation and Karlsson[28] used this approach to develop a model for predicting flame spread and fire growth in several geometries, including the room corner test. This model requires that the material be tested in the cone calorimeter at a number of different heat flux levels in order to derive an apparent thermal inertia, kρc, which is used to calculate time to ignition. The heat release rate data from the cone calorimeter and the kρc value are then used to calculate flame spread velocity and heat release rate in the large-scale test (for example the room corner test).

Several models of this type have been described in the literature and only a few are mentioned here as examples. Cleary and Quintiere[29] developed a method that allowed both upward and lateral flame spread to be calculated, using data from the cone calorimeter and the LIFT apparatus. Baroudi and Kokkala[30] developed a computer program to solve the Volterra-type integral equation and Kokkala et al[31] have tested it against experiments, using cone calorimeter data as input.

Many of the applications have only compared the calculated results with a very limited number of full-scale experiments, but Karlsson[32] used 22 different room corner test experiments and compared calculated and experimentally measured heat release rates. The experimental data originates from two series of experiments, the S-series[33] and the E-series.[19] Figure 12.8 shows the heat release rate history of four of these materials, showing good agreement with experimental measurements.

Figure 12.9 shows calculated and experimentally measured time to flashover for all 22 materials tested. Only 2 out of 22 materials deviate significantly. Some of the materials did not go to flashover in the room corner test: this is indicated by the longest bars in Fig. 12.9.

Many other models have also been used for predicting full-scale fire growth using input data from the cone calorimeter. This section has concentrated on upward flame spread on practical surface lining materials, and results have only been shown displaying their behaviour in the room corner test. Very many other applications have been developed; Grant and Drysdale[34] used these methods to model flame spread in warehouse fires; van Hees and Thureson[35] have used this technique for predicting flame spread on cables; and Kokkala and Baroudi[31] to study flame spread on vertical wooden materials, to name only a few studies.

12.5.2 Flame spread models in CFD codes

The most sophisticated deterministic models for simulating enclosure fires are termed CFD (computational fluid dynamics) models. The volume under consideration is divided into a very large number of sub-volumes and the basic laws of mass, momentum and energy conservation are applied to each of these. A handful of such models have been specially developed to simulate fires in compartments.

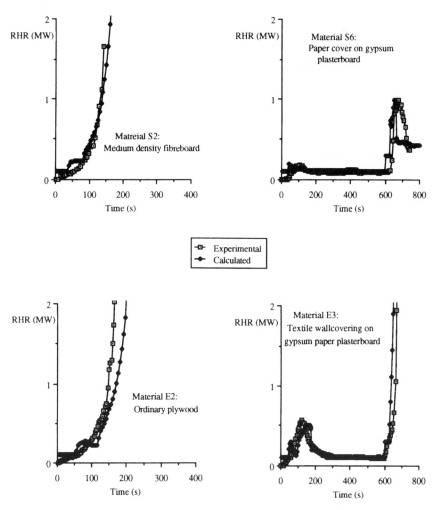

12.8 Comparison of calculated and measured heat release rates from the room corner test for 4 different surface lining materials. (After Karlsson[32])

Since CFD models allow variables to be calculated locally in a very fine mesh, there is no need to make the very rough assumptions on flame height and heat flux made by the thermal models. This opens up possibilities for more sophisticated models, both for calculating solid material temperatures and mass flow rate of pyrolysis products from the solid material and the subsequent combustion. However, for some applications, data from the cone calorimeter can be used instead of calculating the pyrolysis and the combustion process.

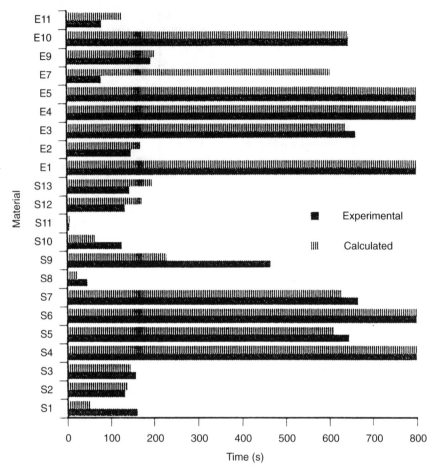

12.9 Calculated and experimental time to flashover in the room corner test for 22 different surface lining materials (After Karlsson[32])

As an example of this approach, Yan and Holmstedt[36] presented a pyrolysis model embedded in a CFD code for predicting flame spread on a vertical PMMA slab. Both the turbulent combustion of the gas phase and the pyrolysis of the solid fuel were numerically simulated.

Tuovinen et al[37] have implemented this model into the well known CFD code SOFIE[38] and tested it for other types of materials, showing good results.

The model is based on a one-dimensional numerical heat transfer model that uses a standard numerical solver for the heat conduction equation. Each numerical heat conduction strip is then divided into a number of sub-strips to which a simple pyrolysis model is applied. The input parameters with respect to the thermal and pyrolysis model are:

- Ignition temperature (K), only of interest for non-charring materials.
- Pyrolysis temperature (K).
- Heat of pyrolysis (J/kg).
- Heat of combustion (J/kg).
- Virgin density (kg/m³).
- Char density (kg/m³).
- Specific heat (J/kg K).
- Thermal conductivity (W/m K).

Alternatively, the flame spread model can directly use heat release rate data from the cone calorimeter for each cell at the surface of the material. The pyrolysis model must therefore not be applied. This simplifies the input data requirements substantially.

Figure 12.10 shows the results obtained when the model calculations were compared to two identical experiments carried out by Yan[39] and Andersson[40] in a one-third scale of the room corner test, where particleboard was attached to both walls and ceiling. The figure shows the calculated heat release rate inside the room (marked 'inside') and the total heat release (marked 'total'), since some of the heat is released outside the room. The figure also shows the difference between using the full pyrolysis and combustion model (marked 'p') and using cone calorimeter results to estimate heat release rate from each solid element (marked 'c').

The flame spread and pyrolysis models used in CFD codes are still being developed and they have not been put to much practical use as yet.

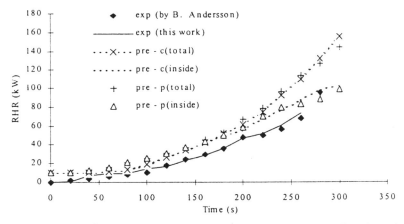

12.10 Calculated and measured heat release rates in the 1/3 scale scenario. The symbol (c) denotes using cone calorimeter data as input and (p) using the pyrolysis model. From Yan[39]

However, this type of flame spread model will in the near future be an invaluable tool for researchers and engineers, since it can be used for a very wide variety of end-use scenarios and a wide variety of products.

12.6 Conclusions and recommendations

One of the major differences between a prescriptive rule and a performance requirement stems from the method of assessing its fulfilment. For the prescriptive rule one must observe and verify that every detail in the prescriptive legislation is observed: this can be done during the design process, the construction process or after construction is completed. For the performance criterion this must be done using evaluation tools that measure or predict the relevant properties and performance level. Such tools are test methods and simulation models. Such models will need material parameters as input. These parameters must be measured by carefully developed test methods and the results from the test methods must be relevant to the end-use condition. Testing and modelling nearly always involve considerable approximation or simplification of real conditions of use.

This chapter has discussed the need for developing a sound engineering philosophy for testing and ranking products with respect to fire hazards and has given certain recommendations on how to achieve this goal. Examples have been given of bench-scale and full-scale test methods that can be said to be a part of a performance-based testing system. Further, a number of mathematical models of various degrees of sophistication and complexity have been described. Some require simple input data and give approximate answers for certain end-use conditions; others are more general but require non-standard input data and expert knowledge.

During the last decades much work has been concentrated on developing bench-scale and full-scale performance-based test methods for predicting reaction-to-fire behaviour of products. Much less work has been concentrated on developing models that allow data from bench-scale methods to be used for full-scale calculations. Many of the available methods have only been validated for a very narrow range of materials and often only a single large-scale scenario.

One of the most urgent issues for the future is the need to make the existing fire growth models available to engineers and scientists in the manufacturing and building industries. The existing models must not only be developed further, but their validity must be checked against a wider range of products and their user-friendliness and availability to engineers must be increased.

Even though performance-based bench-scale and full-scale test methods have been successfully developed during the last decade, there is a need to consider a wider variety of full-scale scenarios for a wider range of products.

Once experiments have been carried out in such scenarios, the engineering models can be evaluated against these in order to pave the way toward a more rationally based classification and performance-based engineering design with regard to reaction-to-fire behaviour of products.

References

1 Emmons H W, 'Fire and fire protection', *Scientific American*, 1974, **231** (1) 21–7.
2 Vandevelde P, 'Comparison between different reaction to fire test methods on wall lining and flooring materials', *Fire Mat*, 1981 **5** (2) 77–84.
3 Blaciere G, Tephany H and Trottein Y, *Fire Reaction Tests in the EEC*, Aux. IRBat Report 744/III/85-EN, 1985.
4 Cox G, Curtat M, Di Blasi C and Kokkala M, *Reaction to Fire of Construction Products, Area B: Fire Modelling*, Construction Research Communications Ltd, London, 1996.
5 European Commission, 'Commission decision of 9 September 1994, implementing Article 20 of Directive 89/106/EEC on construction products', *Official Journal of the European Communities*, 16 September, 1994, Brussels.
6 Proposed European Standard, prEN SBI 1999, *Reaction to Fire Tests for Building Products – Building Products Excluding Floorings Exposed to the Thermal Attack by a Single Burning Item*, Technical Committee CEN/TC 127, Fire Safety in Buildings, BSI, London, June 1999.
7 ISO 1182:1990, 'Fire tests – Building materials – Non-combustibility test', International Standards Organisation, Geneva, Switzerland, 1990.
8 ISO 1716:1973, 'Building materials – Determination of calorific potential', International Standards Organisation, Geneva, Switzerland, 1973.
9 ISO 11925-2:1997, 'Reaction to fire tests – Ignitability of building products subjected to direct impingement of flame – Part 2: Single flame source test', International Standards Organisation, Geneva, Switzerland, 1997.
10 Babrauskas V, 'Redefining the value of π in the European Union', Editorial, *Fire Safety J*, 1997, **29**, 4, ix.
11 ISO 9705:1993 Fire tests – *Full-scale Room Test for Surface Products*, International Standards Organisation, Geneva, Switzerland, 1993.
12 ISO 5660-1:1993 Fire tests – *Reaction to Fire – Part 1: Rate of Heat Release from Building Products – (Cone Calorimeter Method)*, International Standards Organisation, Geneva, Switzerland, 1993.
13 ASTM E 1321-93, 'Standard test method for determining material ignition and flame spread properties', *Annual Book of ASTM Standards*, ASTM, Philadelphia, 1993.
14 ISO 5658-2:1996 *Reaction to fire tests – Spread of Flame – Part 2: Lateral Spread on Building Products in Vertical Configuration*, International Standards Organisation, Geneva, Switzerland, 1996.
15 *CIB Working with the Performance Approach in Building*, CIB Report 64, CIB, Rotterdam, The Netherlands, 1982.
16 *CIB International Workshop on New Developments in Performance Based Test Methods*, Proceedings publication 179, CIB, Rotterdam, The Netherlands, 1996.

17 Karlsson B and Kokkala M, 'New developments in performance based test methods for fire safety assessment of products', *International Workshop on New Developments in Performance Based Test Methods*, Proceedings publication 179, CIB, Rotterdam, The Netherlands, 1996.

18 Wickström U, Personal communication, SP, Borås, 1992

19 *Proceedings of the International EUREFIC Seminar*, Copenhagen 11–12 September 1991, Interscience Communications Limited, London, 1991.

20 Wickström U, 'The Future of European Fire Testing', *Proc. Of Flame Retardants '90*, 17–18 January, London, Elsevier, 1990.

21 Babrauskas V and Peacock R D, 'Heat release rate: the single most important variable in fire hazard', *Fire Safety J.* 1992, **18** 255–72.

22 Karlsson B and Quintiere J Q, *Enclosure Fire Dynamics*, CRC Press LLC, New York, 1999.

23 Hasemi Y, 'Thermal modeling of upward wall flame spread', *Proceedings of the First International Symposium, Fire Safety Science*, Hemishpere, 1986.

24 Delichatsios M M, Mathews, M K and Delichatsios M A, 'An upward flame spread and growth simulation', *Proceedings of the Third International Symposium, Fire Safety Science*, Elsevier Science, London, 1991.

25 Beyler C L, Hunt S P, Ibqal N and Williams F W, 'A computer model of upward flame spread on vertical surfaces', *Proceedings of the Fifth International Symposium, Fire Safety Science*, IAFSS, Melbourne, 1997.

26 Saito K, Quintiere J G and Williams F A, 'Upward turbulent flame spread', *Proceedings of the First International Symposium, Fire Safety Science*, Hemisphere, Washington, 1986.

27 Thomas P H and Karlsson B, *On Upward Flame Spread on Thick Fuels*, SE-LUTVDG/TVBB-3058, Department of Fire Safety Engineering, Lund University, 1990.

28 Karlsson B, *Modelling Fire Growth on Combustible Lining Materials in Enclosures*, Report TVBB 1009, Department of Fire Safety Engineering, Lund University, 1992.

29 Cleary T and Quintiere J G, 'A framework for utilizing fire property tests', *Proceedings of the Third International Symposium, Fire Safety Science*, Elsevier Science, London 1991.

30 Baroudi D and Kokkala M, *Analysis of Upward Flame Spread, VTT Publications 89*, VTT, Espoo, Finland 1992.

31 Kokkala M and Baroudi D, 'Upward flame spread and heat release rate of wood products: Experiments and Numerical Modelling', *Proceedings of the Fifth International Symposium, Fire Safety Science*, IAFSS, Melbourne, 1997.

32 Karlsson B, 'Models for calculating flame spread on wall lining materials and the resulting heat release rate in a room', *Fire Safety J*, 1994, **23** (4) 365–86.

33 Sundström B, 'Full scale fire testing of surface materials', *SP-Report* 1986:45, SP, Borås, Sweden, 1986.

34 Grant G and Drysdale D, 'Numerical modelling of early flame spread in warehouse fires,' *Fire Safety J*, 1995, **24** 247–78.

35 Van Hees P and Thureson P, 'Burning behaviour of cables – modelling of flame spread', SP-Report 1996:30, SP, Borås, Sweden, 1996.

36 Yan Z and Holmstedt G, 'CFD and experimental studies of room fire growth on wall lining materials', *Fire Safety J*, 1996, **27** 201–38.

37 Tuovinen H, Van Hees P, Axelsson J and Karlsson B, 'Implementation of a physical flame spread model in the SOFIE CFD model', *SP-Report* 1999:32, SP, Borås, Sweden, 1999.

38 Lewis M J, Moss J B and Rubini P A, 'CFD modelling of combustion and heat transfer in compartment fires', *Proceedings of the Fifth International Symposium, Fire Safety Science*, IAFSS, Melbourne, 1997.

39 Yan Z, *Numerical Modeling of Turbulent Combustion and Flame Spread*, Department of Fire Safety Engineering, TVBB-1018, Lund University, Lund, 1999.

40 Andersson B, *Model Scale Compartment Fire Tests with Wall Lining Materials*, Report LUTVDG/TVBB-3041, Dept. of Fire Safety Eng., Lund University, Lund, 1988

13

Fire safety design requirements of flame-retarded materials

DOUGAL DRYSDALE

Fire Research Group
Department of Civil and Environmental Engineering
University of Edinburgh

13.1 Introduction: drawbacks of tests and measurements

Numerous methods have been used since historical times to render combustible materials less hazardous in the event of fire. These range from the application of coatings of non-combustible material, such as mortar over thatch, to various forms of chemical treatment.[1,2] More recently, we have seen the appearance of new materials which are inherently flame resistant as a consequence of their chemical structures.[3]

The effectiveness of a coating is dependent on its retaining its integrity as a physical barrier during exposure to fire and in principle this may be quantified relatively easily. However, while the effect of a chemical treatment may be easy to explain in terms of the chemical mechanism, its effectiveness with respect to fire performance is much more difficult to describe, particularly if a quantitative measure is required. There is little doubt that flame retardant treatments can greatly reduce the hazard to life: there have been several full-scale demonstrations of how flame-retarded materials can reduce the rate of fire growth and prevent or delay the onset of flashover and the fully developed fire.[4,5] This is equally true whether considering wall-lining materials[5] or the contents of buildings, such as furnishings.[4,6]

Such demonstrations show that for one particular fire scenario (distribution and configuration of combustible material and size and location of the ignition source), the use of flame-retarded materials can delay flashover significantly, providing precious time for the occupants of the building to move to a place of safety.[4] Unfortunately, these demonstrations do not provide any guidance to fire behaviour of other scenarios, for example if the ignition source was larger, located somewhere else, or the distribution of combustible items was different. These are extremely important issues that have to be considered in fire safety design, whether selecting a wall lining material or specifying the building contents. They pose questions that are difficult to answer, but not impossible. The advances that have been

378

made in fire safety science over the past two decades[7–10] can now be harnessed to help resolve these issues and provide the architect and the designer with a more rigorous but flexible approach to the selection of materials. Closer involvement of the fire safety engineer in the design process will lead to greater safety in the built environment.

At the fundamental level, major problems are encountered in expressing the fire hazard of any combustible material in a quantitative manner. When it was first developed, the Limiting Oxygen Index (LOI) Text was seen as a breakthrough because it provided a 'number' which appeared to have some relevance to 'fire'.[11] Unfortunately, this was a delusion; not only is the LOI apparatus dependent (in particular, the result depends on the sectional area of the sample) but the sample is burned in the least challenging configuration – candle-like, under ambient conditions. For example, a woollen fabric can have an LOI > 22% oxygen, but it will burn in normal air if suspended vertically and ignited at the bottom edge. The results can also be misleading: an additive which reduces the viscosity of a polymer melt will increase the LOI when compared to the parent thermoplastic, simply because molten polymer will flow away from the tip of the sample, effectively removing heat from the point of application of the ignition flame. The treated polymer may in fact be more hazardous under fire conditions than the untreated material.

The LOI test is an extreme example, but in fact all the so-called 'tried and tested' standard methods of test are of very limited value: like the full-scale demonstrations to which reference has already been made, they address only one small-scale fire scenario and the results cannot be generalised or (in most cases) quantified.[12] Each of these tests is only capable of ranking materials according to their relative performance in that test. The ranking order is test-specific, as was shown clearly by Emmons in the Round-Robin comparison of six European standard tests, each purportedly designed to assess the hazard of combustible wall-lining materials. There was no correlation between the six ranking orders,[13] a problem that has come to haunt the European single market.

This type of test still forms the basis of the traditional Building Regulations and Building Codes which exist in every country. Design requirements are specified strictly according to guidelines based on a comparison ('correlation' is too strong a term) between behaviour of materials in the appropriate standard test and 'experience' of the performance of the same materials under fire conditions (Fig. 13.1).[14] Because the materials have to be ranked according to their performance, arbitrary divisions have to be set to distinguish different classes of performance. For example, the British Standard Surface Spread of Flame Test (BS 476 Part 7[15]) divides materials into Classes 1–4, from best to worst. The German Standard DIN 4102 defines two classes, A and B (non-combustible and combustible, respec-

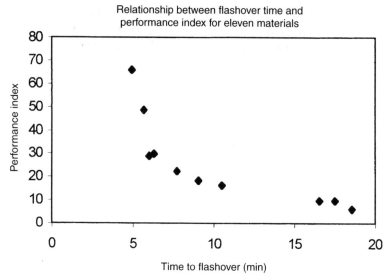

13.1 Relationship between time to flashover and the performance index for eleven wall lining materials tested according to BS 476 Part 6 (the Fire Propagation Test)[13]

tively), defining materials of low flammability (Class B1) as those which pass the stringent requirements of the Brandschacht test.[16] In general, this system appears to have worked satisfactorily but it is clear from Emmons' study[13] that the ranking order in a standard test is unlikely to bear any resemblance to the ranking order in a real fire (if such can ever be defined).

There are other problems. The tests on which the regulations are based were developed more than 30 years ago, before the fundamental principles of fire science had been developed and widely disseminated, and before synthetic materials began to gain a major foothold in the market. The tests may work reasonably well with traditional materials, but many modern materials melt and flow when they are heated, and cannot be classified simply because they do not remain in the apparatus for the duration of the test. There is at least one case in which a material was initially excluded from use because it was unclassifiable in BS 476 Part 7, yet in a series of full-scale tests it proved to be inherently safe in comparison with a composite formulation which achieved the required rating in the same test.[17]

The above observations illustrate some of the problems and hidden difficulties that lie behind strict adherence to the prescriptive approach to fire safety in buildings. Under the prescriptive regulations, design requirements can be ascertained simply by reference to the relevant tables in the appropriate section of the regulations. While the statistical record suggests that this approach has served us well in the past, it would be unreasonable to assume that an alternative approach cannot be used in which materials are selected more rigorously on the basis of their end-use configuration and the fire scenario to which they may be exposed. It is possible that in some cases current requirements are too severe, but the fire safety engineer is more concerned about the identification of situations in which the potential fire exposure is significantly greater than expected and in which materials deemed to be acceptable are found to perform badly because exposure in the test is not severe enough. Such an example, in which a PVC flock wall covering was ignited by a fire involving an item of upholstered furniture and led to multiple life loss, is described by Reveneugh, Mowrer and Williamson.[18]

The Fire Safety Engineer must take a holistic approach to the problem of fire safety and attempt to meet the prescribed objectives which may be stated very simply, e.g. 'In the event of fire, all occupants of a building must be able to leave the building safely'. The Engineer can achieve this by the application of standard fire safety design systems which may include all or some of the following: sprinklers, other suppression systems, automatic detection and alarm, smoke control, building layout to facilitate escape, strict control over linings and contents, but the Engineer must be able to demonstrate quantitatively that the required level of safety will be achieved. He or she must apply the growing body of scientific knowledge which forms the basis of fire safety engineering.[19] Much of this is contained in the Draft British Standard which is currently in preparation.[20]

It is argued that the most important parameter that must be considered in a fire is the rate of heat release (or power, in kW or MW).[21] This has been shown to correlate with flame height, temperatures under a ceiling, the rate of formation of smoke and toxic species and, for a fire in an enclosure, the onset of flashover.[9,20] The fire safety engineer must be able to estimate how quickly a fire is likely to achieve certain critical rates of heat release, such as those required to activate a sprinkler or lead to flashover. In current practice, the fire growth rate is frequently assumed to follow a simple power law, the 't-squared fire':

$$\dot{Q} = \alpha_f t^2 \tag{13.1}$$

where $\dot{Q}$ is the rate of heat release (kW), t is the time from the end of the incubation (or growth) period (s), and α_f is a growth rate coefficient (kW/s^2) whose value depends on the type and distribution of fuel present (Table 13.1).[9,23] This represents a very crude and conservative approximation to reality, but it does emphasise the importance of the growth rate. This must be taken into account when designing a fire safety system as it will determine the response time required of a detector or sprinkler system to

Table 13.1 Parameters used for t-squared fires[23]

Description	Typical Scenario	α_f (kW/s)
Slow	Densely packed paper products Solid wooden cabinets	0.0029
Medium	Traditional mattress/boxspring Traditional armchair	0.0117
Fast	PU mattress (horizontal) Wooden pallets 1.5 m high PE Pallets, stacked 1 m high	0.0469
Ultra-fast	High rack storage PE rigid foam stacked 5 m high	0.1876

prevent untenable conditions being created at points remote from the seat of the fire while people are still moving towards a place of safety. Widespread use of non-combustible materials would be a safe option but this is seldom acceptable for reasons of comfort and general appearance. Control over the combustible fabric and contents is essential, particularly for buildings in which there is the potential for high life loss, although issues relating to the safety of individuals in their own home cannot be ignored.

13.2   Some fundamentals: flame spread and flashover

Fire Safety Engineers have tended to focus their attention on the development of quantitative approaches to the traditional 'active and passive' methods of protection. They have been increasingly successful in designing systems which suit specific hazards more closely, thereby providing the means of reducing the risk to acceptable levels more economically. This is certainly true of property protection where very large losses can occur, particularly in warehouses with high rack storage. It is also true for life safety, although in this case the driving force is public concern. The life safety objective is to maximise the chances that everyone within a building will escape safely, or in some circumstances be able to survive the fire in a safe refuge. The rate at which the fire develops becomes of critical importance: this is best illustrated by reference to the simple equation:

$$t_p + t_a + t_{rs} < t_u \qquad [13.2]$$

where t_p is the time delay between ignition and detection; t_a is the delay between detection and the start of the 'escape activity'; t_{rs} is the time taken to reach a place of relative safety; and t_u is the time taken by the fire to generate conditions that are untenable.[24] The objective will be achieved if the above inequality holds for everyone within the building. The terms t_p, t_a and t_{rs} can be reduced respectively by (a) installing a reliable detection and alarm system to warn the occupants of a building of the existence of a fire as soon after ignition as possible; (b) ensuring that everyone knows what to do in the event of a fire; and (c) providing well-marked, clear escape routes of appropriate length. However, it can be argued that measures by which t_u may be increased are of greater importance.

This term can conveniently be related to the time to flashover in the room or compartment of origin. Flashover will occur if heat output of the fire increases beyond a critical power ($\dot{Q}_{FO}$) which is a function of the size of the compartment and its ventilation opening, and the thermal properties of the compartment boundaries. A number of approximate equations have been derived for estimating ($\dot{Q}_{FO}$), but the one derived by McCaffrey et al[22]:

$$\dot{Q}_{FO} = 610(h_k A_T A_W \sqrt{H})^{1/2} \qquad [13.3]$$

is an empirical correlation based on a large set of experimental data. Here, h_k is an effective heat loss coefficient which depends of the thermal properties of the boundaries (kW/m²), A_T is the internal surface area of the compartment (m²) and A_W and H are the area (m²) and height (m) of the ventilation opening respectively. The key issue is then how quickly the fire will develop to achieve this rate of heat release. The risk to life will be greatly reduced if flashover is delayed for as long as possible: this risk may be reduced to zero if flashover can be prevented, either by ensuring that the materials present cannot produce a high rate of heat release, or by the activation of an automatic sprinkler system. Indeed it could be argued that the activation of a sprinkler represents a failure in the original selection of materials and/or the fire safety management of the premises.

Materials can be selected by using existing tests methods, but this approach carries with it many hidden problems, as discussed above. Most tests were developed for wall-lining materials as these present a particular hazard in a building. Not only are they in an optimal configuration for rapid (upward) flame spread and fire growth, but they also present large surface areas at which active burning can take place, leading to very high rates of heat release. It is probable that flashover will occur shortly after a combustible lining material becomes involved in fire. Consequently, materials which can be ignited easily and contribute to a fire early in its development stage are deemed unsuitable for this purpose, particularly in escape routes and places of public assembly. The British Standard 'Fire Propagation Test'[25] was designed to allow such materials to be identified by determining at what stage during an increasingly severe fire exposure (albeit on a small scale) a sample would contribute to the rate of heat release. The results of this test correlate well with carefully designed flashover experiments on a large scale (Fig. 13.1),[14] but these are both 'single scenario' tests.

What is required is an approach which allows the response of materials under different fire conditions to be analysed quantitatively. This can only be achieved through modelling the material's response to various scenarios, an aspect of fire science that has been under active development for over a decade. The research effort has been concentrated on wall linings and has shown considerable progress (e.g. see Karlsson[26]). Vertical spread is a *relatively* simple process compared to the problem of flame spread over surfaces of complex geometry, such as those found on items of upholstered furniture.[6,27] This issue will be discussed further below.

Flame spread can be modelled as a propagating ignition front, but first the material must be exposed to an ignition source. This must be of sufficient strength to cause ignition, although self-extinction may occur if the ignition source is removed.[10] This will depend not only on the strength of the ignition source but also how long it is applied to the material. If the ignition source remains in place, it is more likely than not that burning will be sustained if ignition occurs. Possible exceptions to this are materials which intumesce or otherwise build up a layer of char which insulates the unaffected material below.

The ignition source is – in essence – a source of heat, but in this argument it is assumed that once the surface temperature of the material has reached a critical value (which may be referred to as the firepoint, by analogy with combustible liquids) the flammable vapours that are evolved from the surface can be ignited and will sustain a flame. The source of heat may also provide the source of ignition for the vapours but, if not, it is necessary to assume that there is an independent source of ignition for the vapours (see Fig. 13.2). This should be distinguished from spontaneous ignition in which the evolved vapours are sufficiently hot to flame spontaneously when they mix with air (see Fig. 13.3). Inevitably, this requires a higher surface temperature and a more powerful source of heat.

From this, it is clear that ignitability is not a simple material property. Not only is it a function of the strength of the ignition source, but also of the geometry and thickness of the material. In an ideal world, if we could define the maximum ignition source strength to which a material could be subjected, it would be possible to select a combustible material which was intrinsically safe[a] in that environment. This could only be possible in strictly controlled environments, but in reality the best we can do at present is to identify materials which have an inherent resistance to ignition. This is the

---

[a] This concept, in reverse, is encountered in the Process Industries in which low power electrical equipment can be designed to be intrinsically safe with respect to the ignition of flammable vapour/air mixtures.

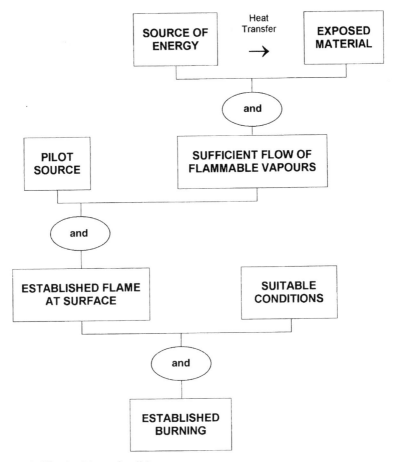

13.2 Pilot ignition of solids

approach that has been taken with upholstered furniture on sale to the general public in certain countries. In the UK, items must pass simple ignition tests (the application of a smouldering cigarette and a small flame, equivalent to a cigarette lighter).[28] However, the requirement for domestic furniture does not extend to larger ignition sources which, *in principle*, could give rise to a sustained, developing fire. Much greater ignition resistance is required in specific occupancies, such as hospitals and prisons.

Ignition leading to sustained burning will be followed by further development of the fire involving flame spread over the surface (or item) first ignited. In general, materials which are ignition resistant will also be resistant to flame spread but, once sustained burning has been established, the situation changes. Heat transfer from the established flame will drive the propagation of the spreading flame front. This is critically dependent on the orientation

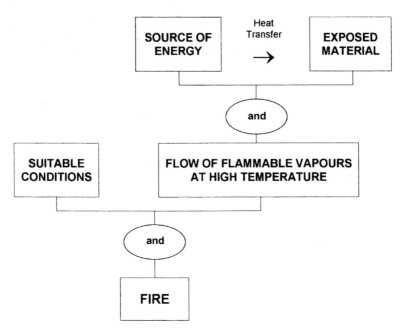

13.3 Spontaneous ignition of solids

of the surface: the worst scenario is upward flame spread on a vertical surface as the material above the burning area is exposed to flame and the rising, hot combustion products fill the buoyancy-induced boundary layer at the surface which is thereby heated rapidly. High rates of spread can occur. The flow characteristics associated with the spreading flame are extremely important. Any change of configuration of the surface, or surfaces, that enhances the rate of heat transfer from the flame to the material will increase the rate of spread. Experience has alerted us to most hazardous configurations, such as the vertical duct lined with combustible material, but any configuration which provides confinement and enhances boundary layer flow of flame and hot gases represents a hazard. The wooden escalator that became involved in the King's Cross fire of 18 November 1987[29] is an example of a type of hazard that has only recently been recognised.

13.3 Selection of materials as part of the design process

13.3.1 Relevant tests and models

Clearly, the way in which a material responds in a fire is determined by many factors, some of which are specific to the end-use of the material and the configuration in which it is used. These will affect the severity of the fire to which a material is exposed and it is necessary to bear this in mind when a selection is being made.

In the case of wall linings, prescriptive Building Regulations state unambiguously what classification of combustible material is required for specific locations. The classification is arbitrary and based on performance in a standard test, the result being apparatus dependent. As discussed above, this has led to major problems regarding free trade between the member nations of the European Community. There are two ways forward. The first remains in the past by adopting a set of new tests acceptable to each member state, with a new classification system. Within Europe, the development of the SBI (Single Burning Item) test is following this route.[30] Like those that have preceded it, it defines a single fire scenario, produces a ranking order for a range of materials, with arbitrary divisions, and provides no quantifiable data which can be used to predict behaviour under other conditions. The second approach takes advantage of the new generation reaction-to-fire tests which *can* provide quantitative data. The most important of these is the cone calorimeter.[31] The results are still apparatus dependent, but the apparatus has been designed to reduce this problem as much as possible. It allows time to (piloted) ignition and the rate of heat release to be measured at different radiant heat fluxes, thus generating data over a wide range of conditions. Used properly, it has the potential to give a much more complete picture of how a material is likely to behave in a fire. Tests of this nature do not lend themselves to producing simple ranking orders, although several sophisticated proposals have been put forward in which data from the cone can be combined to give performance indices which correlate well with data on time to flashover in certain large scale tests (e.g. the corner-wall test and the room-corner test).[32] This is a correlation with a single fire scenario (the test configuration) but it may provide an alternative to performing large-scale tests which have now been standardised.[33] However, correlations with other scenarios should be sought to enable this approach to be made much more general.

The alternative is to develop mathematical models of fire spread which use data from the cone calorimeter to check behaviour under different scenarios. Early attempts at this have been encouraging.[34] The rate of vertical spread on combustible wall linings has been modelled successfully using numerical techniques in which data from the cone calorimeter are used directly.[26,35] The spreading flame preheats the fuel ahead of the burning region which ignites under the heat flux imposed by the flame (approximately 25kW/m^2). The time to ignition is measured in the cone at a heat flux of 25kW/m^2 and used in the model. The rate of heat release from the burning area is assumed to be equivalent to that measured at 50kW/m^2 in the cone, which determines the size of the flame filling the boundary layer.

Upward spread is thereby modelled as a process by which successive elements of surface are raised to the temperature at which they contribute actively to the burning process, i.e. they have ignited.

In principle, such models would be able to define the rate of growth of the fire in the early stages: the objective would then be to limit the rate of development of the wall fire to a level which is considered to be of low hazard in relation to the egress time of the occupants of the building. Again, in principle, different scenarios could be considered if the relevant heat transfer characteristics could be quantified and appropriate cone data selected for use in the model.

13.3.2 The properties of the materials

It will be some time before we have a sufficient understanding of fire processes to be able to quantify the behaviour of materials under a wide range of fire scenarios from data obtained from tests such as the cone calorimeter. While many of these complexities will have to be addressed by future research, it is clear that the current approach to the selection of materials has the potential to fail unless the end-use scenario is taken into account. An excellent illustration of this is presented in the paper by Abbott.[17]

When selecting materials for life safety, the fire safety objective is to increase t_u to an acceptable value, consistent with the escape parameters defined above.[24] Whether it is to be used on its own or as part of a composite, the requirement should be that the fire performance is consistent with this goal, either through (a) resistance to ignition, or (b) once ignited, a low rate of heat release. A collateral objective is to avoid materials which – following ignition – will achieve their maximum rates of heat release rapidly, but it is likely that this will be met if (a) and/or (b) are achieved. 'Resistance to ignition' and 'rate of heat release' may be considered as components of what is loosely called 'material flammability' (for which there is no simple definition).

It is necessary to consider fundamental material properties which are relevant to fire behaviour and how these can be modified beneficially by the use of flame retardants. They include both physical properties, such as thickness, density, thermal capacity and thermal conductivity, and chemical properties, although the latter are less easily defined and quantified. They relate specifically to pyrolysis and combustion processes. Heat of combustion (kJ/kg) is relevant, but only when combined with a rate of burning (kg/s) to determine the rate of heat release. However, textbook values based on combustion bomb calorimetry are not necessarily relevant as they refer to complete combustion of the material to fully oxidised products (of which CO_2 and H_2O are the dominant species). Even under well-ventilated conditions, combustion is never complete, either because some fuel remains behind as char or because the gas-phase reactions are not sufficiently fast for oxidation to be completed within the diffusion flame. Indeed, flame retardants act either by promoting char formation (at the expense of releasing flammable volatiles) or by inhibiting the gas phase oxidation reactions, both reducing the amount of heat released during flaming.

It is appropriate at this point to consider 'ease of ignition' and 'rate of heat release' to identify the strategies that may be adopted when selecting flame-retarded materials as part of fire safety design.

13.3.3 Ease of ignition vs rate of heat release

The limiting condition at which a flame can be established at the surface of a combustible solid is analogous to the firepoint of a combustible liquid. It can be defined approximately as a critical surface temperature above which stable flaming will occur after the vapours released from the surface have been ignited (Fig. 13.2).[36] The ease with which ignition will occur will therefore be strongly influenced by physical factors (principally, thickness and thermal inertia ($k\rho c$)) which determine how rapidly the surface temperature will respond to an imposed heat flux.

At or just above the firepoint the flame is weak but it will gradually strengthen and grow in size as it contributes more and more heat to the surface, thus providing the energy to maintain and increase the flow of volatile pyrolysis products. The stability of the flame at the surface will depend on the flammability, or reactivity, of the fuel vapours. Rasbash[37] proposed a means by which this could be estimated, in terms of a critical mass flux of volatiles necessary to support a flame at the surface ($\dot{m}_{cr}$). There is a limited amount of data which supports this hypothesis: certainly, critical mass fluxes can be measured[38,39,43] (Table 13.2), but different experimental techniques give different results. The reason for this has still to be investigated thoroughly, but there is a consistency within each data set. For example, the critical mass fluxes are related inversely to the heat of combustion of the fuel – or more specifically, of the vapour released by the fuel. The values for hydrocarbon fuels (typically, $\Delta H_c = 45\,kJ/g$) are about half of those for a partially oxygenated fuel such as PMMA ($\Delta H_c = 26.2\,kJ/g$) (Table 13.2). In general, the values of the critical mass flux for flame retarded materials are higher than for the parent (non-flame retarded) materials (Table 13.2), at least for retardants that are gas-phase active and inhibit the flame reactions. An increase is also observed if the flammable vapours are diluted by water vapour (e.g. from alumina trihydrate, ATH) as this will reduce the effective heat of combustion of the volatiles (Table 13.3[40]). There is evidence that some of the retardant additive may be lost, or 'cooked-out', if the material is heated slowly to the firepoint and thus

Table 13.2 Critical mass fluxes (g/m^{2}s) at the firepoint[38]

| Radiant flux | 13 kW/m^2 | 19 kW/m^2 | 25 kW/m^2 | 33 kW/m^2 |
Material[a]				
PMMA	1.90	1.96	1.87	2.04
PMMA (FR)	NI	4.48	4.32	5.19
POM	1.83	1.71	1.64	1.73
PP	1.03	1.12	1.10	1.20
PP (FR)	NI	NI	2.34	3.58
PS	0.93	1.01	1.07	0.91

[a] PMMA = polymethylmethacrylate; POM = polyoxymethylene; PP = polypropylene; PS = polystyrene; FR = fire retarded (undefined brominated agents). NI = no ignition. Note that Tewarson and Pion[39] obtained values which are significantly higher than the values reported by Thomson and Drysdale.[36] For example, they found 3.2 g/m^{2}s for PMMA and 3.0 g/m^{2}s for PS (see text).

Table 13.3 Critical mass fluxes and times to ignition for a cast polyester filled with ATH[40]

Heat flux (kW/m^2)	Loading (ATH)	0%	10%	25%	40%	50%
15	t_{ig} (s)	480	521	603	NI[a]	NI[a]
	$\dot{m}_{cr}$ (g/m^{2}s)	1.97	2.31	2.92	NI[a]	NI[a]
26	t_{ig} (s)	240	253	292	339	350
	$\dot{m}_{cr}$ (g/m^{2}s)	1.83	2.71	3.05	3.20	3.39
37	t_{ig} (s)	61	69	80	96	112
	$\dot{m}_{cr}$ (g/m^{2}s)	2.11	2.66	3.20	3.38	3.51

[a] NI = no ignition.

maintained at subcritical temperatures for an extended period. This may adversely affect the rate of heat release post-ignition, but it is unlikely to be an issue for retardants which are active in the solid phase (e.g. phosphates in wood).

The following conclusions may be drawn from theoretical and experimental studies of ignition: combustible materials will be more resistant to ignition if:

1 They are thermally thick.
2 The density (ρ), thermal capacity (c) and thermal conductivity (k) are high (i.e. high thermal inertia, $k\rho c$).
3 The vapours released by the solid have a low heat of combustion.

Table 13.4 Data on the time to ignition and rate of heat release for flame-retarded and non-flame-retarded materials as obtained in the cone calorimeter (Hirschler)[43]

Heat Flux (kW/m²)	Material[a]	TTI[b] (s)	Peak RHR[c] (kW/m²)	Average RHR[c] (kW/m²)
20	PTFE	∞	3	0
	ABS	236	614	—
	ABS-FR	212	224	—
	PS	417	723	—
	PS-FR	244	277	—
40	PTFE	∞	13	0
	ABS	69	944	544
	ABS-FR	66	402	215
	PS	97	1101	504
	PS-FR	90	334	201
70	PTFE	252	161	53
	ABS	48	1311	628
	ABS-FR	39	419	225
	PS	50	1555	797
	PS-FR	51	445	275

[a] PTFE = polytetrafluoroethylene (Du Pont); ABS = acrylonitrile butadiene styrene terpolymer (Borg-Warner 'Cycolac CTB'); ABS-FR = ABS terpolymer flame retarded with bromine compounds (Borg-Warner 'Cycolac KJT'); PS = polystyrene (Huntsman 333); PS-FR = flame retarded PS (Huntsman 351) (flame-retardant species not given, but likely to be gas-phase active).
[b] TTI = Time to Ignition.
[c] RHR = Rate of Heat Release.

4   The vapours contain an inhibiting agent which has been released from the flame retardant (e.g. bromine-containing agents).

5   The energy required to release the fuel vapours (L_v kJ/kg) is high.

Thermal thickness of the material (1) is determined by a combination of physical thickness and duration of heating and cannot be altered by a flame-retardant treatment. In principle, all the others can. Inert fillers will increase the thermal inertia, and may also affect (3) (e.g. release of water vapour from ATH) (Table 13.3) and possibly (5). Species which act as gas-phase inhibitors raise the firepoint temperature and are likely to reduce the rate of heat release (Table 13.4). Retardants which are solid-phase active, in general, alter the pyrolysis process and increase the yield of char: this will reduce the amount of fuel available for flaming combustion and for some materials may affect the composition of the volatiles in an advantageous manner. Thus, phosphates and borates promote char formation in wood and wood products. This reduces the amount of volatiles released, but of greater importance, it reduces the heat of combustion of the pyrolysis products considerably as char formation is accompanied by the release of more carbon dioxide and water vapour.

The effect can best be explained by reference to the following equation:

$$\dot{m}'' = \frac{\dot{Q}''_{flame} - \dot{Q}''_{loss}}{L_v} > \dot{m}''_{cr} \qquad [13.4]$$

where $\dot{m}''$ is the mass flux from the surface (kg/m²), $\dot{Q}''_{flame}$ is the heat flux from the flame to the surface (kW/m²), $\dot{Q}''_{loss}$ is the rate of heat loss from the surface, expressed as a heat flux from the surface (kW/m²) and L_v is the 'heat of gasification' of the combustible solid (kJ/g) (cf. latent heat of evaporation of a liquid). The subscript cr refers to the critical condition (the firepoint). Sustained burning is possible only if $\dot{m}'' > \dot{m}''_{cr}$, i.e. if:

$$\dot{Q}''_{flame} - \dot{Q}''_{loss} > L_v \dot{m}''_{cr} \qquad [13.5]$$

The heat flux from the flame to the surface ($\dot{Q}''_{flame}$) represents a relatively small proportion of the rate of heat release in the flame. If the heat of combustion of the fuel vapours is low, then $\dot{Q}''_{flame}$ may be insufficient to support sustained combustion without an external heat flux ($\dot{Q}''_{external}$) applied to the surface. This is illustrated very well by the results of Tewarson and Pion[41] who compared values of $\dot{Q}''_{flame}$ and $\dot{Q}''_{loss}$ in what has come to be called the Factory Mutual flammability apparatus. Table 13.4 shows that under steady burning conditions, $\dot{Q}''_{flame} > \dot{Q}''_{loss}$ for most common materials including the four thermoplastics cited in the table. On the other hand, a thick sample of a typical species of wood, such as Douglas Fir, has $\dot{Q}''_{flame} = \dot{Q}''_{loss}$: according to Equation 13.5, it cannot sustain burning on its own. This is why thick sections of wood (e.g. logs) cannot burn in isolation, but will burn in a fire when cross-radiation between the burning surfaces increases the net heat flux to the surface. Sustained burning requires:

$$\dot{Q}''_{flame} + \dot{Q}''_{external} - \dot{Q}''_{loss} > L_v \dot{m}''_{cr} \qquad [13.6]$$

Even the most ignition resistant material can be ignited if there is a source of external heating (e.g. from a nearby burning object) of sufficient strength. The imbalance between $\dot{Q}''_{flame}$ and $\dot{Q}''_{loss}$ for the flame retarded plywood quoted in Table 13.5 is very substantial, but it will begin to burn if $\dot{Q}''_{external}$ is large enough, as it would be in a post-flashover fire. Such ignition resistance can be assessed in the cone calorimeter: non-flame-retarded materials will ignite and burn at heat fluxes of c. 10–15 kW/m², while materials with a resistance to ignition will require much higher levels. This may be illustrated by reference to polytetrafluoroethylene (PTFE) (Table 13.4[42]). This material has an intrinsic resistance to ignition and will not support flame under ambient conditions. It will not ignite in the cone calorimeter at

Table 13.5 Values of $\dot{Q}''_{flame}$ and $\dot{Q}''_{loss}$ measured in the Factory Mutual flammability apparatus[41]

	$\dot{Q}''_{flame}$ (kW/m²)	$\dot{Q}''_{loss}$ (kW/m²)
PMMA[a]	38.5	21.3
POM	38.5	13.8
PP (solid)	28.0	18.8
PS	61.5	50.2
Wood (DF)	23.8	23.8
FR Plywood	9.6	18.4

[a] PMMA = polymethylmethacrylate; POM = polyoxymethylene; PP = polypropylene; PS = polystyrene; DF = Douglas Fir; FR = fire retarded (undefined treatment, solid phase active).

40 kW/m², but will burn at 70 kW/m². In the real world, a large, sustained ignition source may provide such a heat flux. There is a reported case in which a PVC flock wallpaper on the walls of a hotel lobby was ignited by a fire which started on an adjacent settee.[18] The wall covering was deemed to be acceptable for this purpose, but the subsequent rapidity of spread was responsible for multiple life loss.

It is clearly desirable that $\dot{Q}''_{flame} < \dot{Q}''_{loss}$. In addition to application of flame retardants to reduce $\dot{Q}''_{flame}$, in some cases it is possible to increase $\dot{Q}''_{loss}$. This is the reason why wallpaper and similar wall coverings neither ignite nor burn easily when they have been stuck on plaster or plasterboard. Provided that the adhesion is sound, heat is lost from the rear surface to the plaster and the paper may only char slowly during a fire. A cautionary remark should be made about multiple layers of paint: although heat will be lost to the wall, multiple layers exposed to a relatively small fire may form blisters, each blister then acting as a thin fuel at which ignition will occur. A wall fire may then develop very rapidly.

In some cases, application of a flame retardant may actually reduce the time to ignition under a given heat flux, but the peak rate of heat release is in general significantly reduced (Table 13.4). There is a sound argument that this is a much more important factor[21] and that time to ignition, however measured, is of less relevance. Nevertheless, for certain applications the resistance of a material to ignition is important. For example, if it is known that a material may be exposed to a given heat flux, a material can be selected accordingly on the basis of results from a test such as the cone calorimeter (the ISO ignitability test might be an alternative). PTFE is an example of a material which has an intrinsic resistance towards ignition, but many common combustible materials require the addition of flame retardants to improve their ignition resistance. This is illustrated by some results presented in Table 13.5, and is particularly true for wood and wood products (such as plywood) which have been treated with phosphates or borates.

The rate of heat release from a burning solid ($\dot{Q}_{combustion}$) will be determined by the expression:

$$\dot{Q}_{combustion} = \dot{m}''\chi\Delta H_c A_f \qquad [13.7]$$

where χ is the combustion efficiency, ΔH_c is the heat of combustion of the solid (kJ/g) and A_f is the area of the burning surface (m^2). It is this quantity that should be controlled. The rate of burning ($\dot{m}''$) is given by:

$$\dot{m}'' = \frac{\dot{Q}''_{flame} + \dot{Q}''_{external} - \dot{Q}''_{loss}}{L_v} \qquad [13.8]$$

where the quantities refer to quasi-steady-state conditions.

These equations give an insight into the problems associated with selecting materials for use in buildings. We can select materials which will have a low value of $\dot{Q}''_{flame}$, or use them in a manner in which $\dot{Q}''_{loss}$ is enhanced, but the externally applied heat flux is determined by the scenario in which the material is used. We are some way from being able to specify materials in terms of the fundamental properties listed above and must still rely on well designed test methods to give data which are capable of identifying hazards when the results are examined in the light of our understanding of the fire process. However, there is now the opportunity to develop quantitative techniques which will enable us to specify the correct material to employ in a defined end-use configuration where the worst-case fire exposure (perhaps expressed as $\dot{Q}''_{external}$) has been predicted. By using the most appropriate flame retardant, a substantial reduction in the hazard can be achieved.

13.4   Conclusions

Flame-retarded materials are widely used in buildings to reduce the risks associated with the outbreak of fire. The suitability of these materials is currently assessed by subjecting them to standard tests which expose samples to a single fire scenario which may bear little relationship to the end-use of the material. While these tests have served us well in the past, new methods have to be developed which will put to better use the new generation of tests which are capable of providing quantitative data on fire performance. This will inevitably require the use of mathematical models of ignition and fire spread to explore how a given material will perform under different fire scenarios. There is a challenge to develop a new generation of flame-retarded materials on the basis of a deeper understanding of the fire process and thus produce materials to meet increasingly stringent requirements which will be demanded for fire safety in the future.

References

1 Rossotti, H, *Fire*, Oxford University Press, Oxford, 1993.
2 Lyons, J W, *The Chemistry and Use of Fire Retardants*, John Wiley, New York, 1970.
3 Tesoro, G C, 'Fire resistance in advanced engineering thermoplastics' *Fire and Polymers: Hazards Identification and Prevention*, ACS Symposium Series 425, G L Nelson (ed.) American Chemical Society, Washington, 1990, 241–52.
4 Babrauskas, V et al, *Fire Hazard Comparison of Fire-retarded and Non-fire-retarded Products* NBS Special Publication 749, 1988.
5 Moghaddam, A Z, Shields, T J, Silcock, G W H and Azakesan, M A, 'A comparison of fire retarded and non-fire retarded wood-based materials in room lining tests', *Fire Mat* **23** 17–25, 1999.
6 Sundstrom, B (ed.), *Fire Safety of Upholstered Furniture – the Final Report on the CBUF Research Programme*, European Commission Measurements and Testing Report EUR 16477 EN, Interscience Communications Ltd, London, 1995.
7 *International Symposium on the Fire Safety of Combustible Materials* (Edinburgh University, 1975).
8 *Proceedings of the International Fire Science and Engineering Conferences*, Interflam Series 1–8, Interscience Communications, 1979–1999.
9 *Proceedings of the International Symposia on Fire Safety Science Nos 1–6*, International Association for Fire Safety Science, 1985–1999.
10 Drysdale, D D, *Introduction to Fire Dynamics*, 2nd ed., John Wiley, Chichester, 1998.
11 *Standard Method of Test for Measuring the Minimum Oxygen Concentration to Support Candle-like Combustion of Plastics (Oxygen Index)*, ASTM D 2863–87, 1987.
12 Thomas, P H, 'Perception and reflections on Fire Science' *Interflam '99, Proceedings of the 8th International Fire Science and Engineering Conference*, Interscience Communications Ltd, 1999.
13 Emmons, H W, Scientific American **231** 21–7, 1974.
14 Rogowski, B F, *The fire propagation test – its development and application*, Fire Research Note 739, 1968.
15 British Standards Institution, *Fire Tests on Building Materials and Structures. Part 7. Method for Classification of the Surface Spread of Flame of Products* BS 476 Part 7: 1987.
16 Deutsches Institut fur Normung *DIN 4102, Part 1: Classification and Test Methods for Building Materials. Brandschacht.*
17 Abbott, J G, 'Standard laboratory tests: how meaningful are they in assessing the fire performance of insulating materials'. *Fire and Cellular Polymers*, Buist, J M, Grayson, S J and Woolley, W D, eds. Elsevier Applied Science, 1986 199–218.
18 Williamson, R B, Revenaugh, A and Mowrer, F W, 'Ignition sources in room fire tests and some implications for flame spread evaluation' *Proceedings of the Third*

International Symposium on Fire Safety Science, Elsevier Applied Science, London, 1991 657–66.

19 SFPE Handbook of Fire Protection Engineering, 2nd ed., P J Di Nenno et al (eds.), Society of Fire Protection Engineers, Boston, 1995, 4.167–4.173.

20 British Standards Institution, *Draft for Development: Fire Safety Engineering in Buildings. Part 1. Guide to the Application of Fire Safety Engineering Principles.* BSI DD 240: Part 1: 1997.

21 Babrauskas, V and Peacock, R, 'Heat release rate: the single most important variable in fire hazard', *Fire Safety J* **18** 255–72, 1992.

22 McCaffrey, B J, Quintiere, J G and Harkleroad, M F, 'Estimating room temperatures and the likelihood of flashover using fire test data correlations', *Fire Technology* **17** 98–119, 1981.

23 Heskestad, G, 'Engineering relations for fire plumes', *SFPE Technology Report*, 82–8, 1982.

24 Pauls, J, 'Movement of People' *SFPE Handbook of Fire Protection Engineering*, 2nd ed., P J di Nenno et al (eds.), Society of Fire Protection Engineers, Boston, 1995 3.263–3.285.

25 British Standards Institution *Fire tests on building materials and structures. Part 6. The fire propagation test* BS 476 Part 6: 1989.

26 Karlsson, B, 'Models for calculating flame spread on wall lining materials and the resulting heat release rate in a room', *Fire Safety J* **23** 365–86, 1994.

27 Babrauskas, V and Walton, W D, 'A simplified characterisation of upholstered furniture heat release rates', *Fire Safety J* **11** 181–92, 1986.

28 *The Upholstered Furniture (Safety) Regulations (1980) and subsequent amendments.*

29 Fennel, D, '*Investigation into the King's Cross Underground Fire.*' HMSO, London, 1988.

30 Smith, D A and Shaw, K, 'The single burning item (SBI) test, the Euroclasses and transitional arrangements' *Interflam'99, Proceedings of the 8th International Fire Science and Engineering Conference*, Interscience Communications Ltd, 1999 1–9.

31 American Society for Testing and Materials, '*Standard method of test for heat and visible smoke release rates for materials and products using an oxygen consumption calorimeter*'. ASTM E1354, ASTM, Philadelphia, 1993.

32 Östman, B A L and Nussbaum, R M, 'Correlation between small-scale rate of heat release and full scale room flashover for surface linings', *Proceedings of the 2nd International Symposium on Fire Safety Science*, Hemisphere Publishing Corporation, New York, 1988 823–32.

33 ISO 9705 *Fire tests – full-scale room test for surface products*, International Organisation for Standardisation, Geneva, 1993.

34 Magnusson, S-E and Sundstrom, B, 'Combustible linings and room fire growth – a first analysis', *Fire Safety Science and Engineering* ASTM STP 882, T Z Harmathy (ed.), American Society for Testing and Materials, Philadelphia. 1985, 45–69.

35 Grant, G B and Drysdale, D D, 'Numerical modelling of early fire spread in warehouse fires', *Fire Safety J* **24** 247–78, 1995.

36 Thomson, H E and Drysdale, D D, 'Flammability of polymers I. Ignition temperatures', *Fire Mat* **11** 163–72, 1987.

37 Rasbash, D J, 'Relevance of firepoint theory to the assessment of fire behaviour of combustible materials', *International Symposium on the Fire Safety of Combustible Materials*, Edinburgh University, 1975 169–78.

38 Drysdale, D D and Thomson, H E, 'Flammability of plastics II. Critical mass flux at the firepoint', *Fire Safety J* **14** 179–88, 1989.

39 Tewarson, A and Pion, R F, Factory Mutual Research Serial UI 1ARGR1.RC (1978).

40 Drysdale, D D and Thomson, H E, 'The ignitability of flame retarded plastics', *Proceedings of the Fourth International Symposium on Fire Safety Science* International Association for Fire Safety Science, Boston, 1994 195–204.

41 Tewarson, A and Pion, R F, 'Flammability of plastics. I Burning intensity', *Comb Flame* **26** 85–103, 1976.

42 Hirschler, M M, 'Heat release from plastic materials', in *Heat Release in Fires* V Babrauskas and S J Grayson (eds.), Elsevier Applied Science, Barking, 1992, 375–422.

43 Tewarson, A, 'Generation of heat and chemical compounds in fires', *SFPE Handbook of Fire Protection Engineering, 2nd ed.* P J Di Nenno et al (eds.), Society of Fire Protection Engineers, Boston, 1995 3.53–3.124.

Mathematical modelling

J STAGGS
Fuel and Energy Department, University of Leeds,
Leeds, UK

14.1 Introduction

Modelling physical phenomena such as flame spread and ignition is central to the basic ethos of Fire Science. These problems are complex in the extreme, involving the coupling of fluid dynamics, chemistry and conjugate heat transfer between solid, liquid and gas phases. The fluid dynamics part of the problem is relatively well understood insofar as it is possible to write down a governing set of equations with appropriate boundary conditions.[1] However, the problem in the solid phase is not quite so well defined. Consequently our ability to formulate accurate mathematical models of ignition and flame spread for polymers depends crucially on our understanding of the phenomena occurring during degradation of the polymer. The purpose of this chapter is to review the work done on modelling degradation of synthetic polymers such as polyethylene and polymethylmethacrylate. The discussion will concentrate on simple models, although reference will be made to char-forming and filled polymers. The physical situation we have in mind is that of a horizontal sample of material exposed to an external heat flux, as in the standard cone calorimeter test. A basic model for pyrolysis of a simple polymer is introduced and developed in later sections so that pyrolysis of filled polymers and simple char-forming systems can be approximated.

Models of degradation of solids abound in the literature (Di Blasi gives a comprehensive review[2]), but most involve one of two approaches. The simplest approach assumes that the solid decomposes to volatile products directly at a critical temperature, often denoted T_p. The critical temperature is a parameter of the model and it is often assumed to be a fundamental material property which is invariant to changes in heating rate or heat flux. This approach (often termed the ablation model, which is interesting since the term 'ablation' was first applied to the thermal erosion of glaciers in the 1850s.[3]) is mathematically analogous to the Stefan problem,[4] which is covered in its melting guise in the classic paper of Landau[5] and later by many other workers, of which references[6-13] are representative. The second approach involves incorporating a kinetic mechanism for the degradation process, often inferred from thermo-gravimetric analyses, where the solid decomposes over a characteristic temperature range. This involves the use of one or more rate equations to model the rate of change in mass of a small sample as a function of remaining mass (or some mass-fraction scalar) and temperature. A representative selection of papers using this approach is to be found in references.[14-20] Common features of many approaches are to neglect change of volume of the sample during degradation[2] and to assume that volatile gases escape as soon as they are formed (with the notable exception of Wichman's contribution[18]). So-called 'Shrinking Core' models have been applied in the field of chemical engineering to many processes involving gas-solid reactions,[21-24] where change of volume during conversion of the reactant is approximately modelled. In the present context, these models are similar to the ablation approach introduced above and so will not be discussed in detail.

14.2   A basic model for single-step pyrolysis

In this section, we develop a mathematical model for the single-step decomposition process *Polymer → Volatile Products*, where the only significant latent contribution to the energy equation is from the formation of volatiles. This approach may be used to model the degradation of a simple non-charring polymer such as PMMA under moderate to high heat fluxes. The role of thermogravimetric analysis (TGA) and the insights this can give to the decomposition mechanism of a small sample of polymer are discussed first. This theme is further developed, and a mathematical model for pyrolysis of thermally thick specimens is derived.

14.2.1 Characterisation of decomposition kinetics

Thermogravimetric analysis (TGA)[26] is an invaluable tool for investigating thermal decomposition of a small sample of material and it is natural that the discussion should start with a brief overview of the key parameters obtainable from this procedure. Typically, a sample of the order of 10 mg or less (small enough to ensure that temperature gradients can be neglected throughout the specimen) is exposed to a programmed heating curve in a carefully controlled atmosphere. One of the simplest techniques is the constant heating rate method, where the temperature of the furnace is increased linearly with time. Consider a TG curve for a typical single-step decomposition process as measured in a constant heating rate TG experiment. The mass fraction μ, defined as the ratio of the sample mass to initial mass, describes a curve similar to that shown in Fig. 14.1.

The characterisation shown in Fig. 14.1 involves two parameters. T_c is the *characteristic temperature* (which we shall abbreviate as CT) of the decomposition process. It is defined as the temperature at which the mass fraction has reduced to a specified value c ($c = 0.5$ in the figure). ΔT_c is the *characteristic temperature range* (the absolute value of the reciprocal of the gradient of the TG curve at T_c, and will be abbreviated by CTR). For more details of this characterisation, consult Hatakeyama and Quinn[26] and Staggs.[27] Other representations of the characteristic temperature have been used,[27,28] e.g. the temperature of the inflexion point of the TG curve.

Now, suppose that the decomposition process may be expressed in terms of the kinetic rate equation $d\mu/dt = -f(\mu,T)$. Many forms for f may be found in the literature for various materials decomposing under various conditions. The usual approach is to express f in the separable form $f(\mu,T) = \Phi(\mu)\Theta(T)$, where $\Theta(T) = A\exp(-T_A/T)$. Unlike the theory of statistical mechanics of gas particles, in which a physical meaning may be assigned to the Arrhenius parameter A (being a collision rate), A has no physical significance in the context of polymer decomposition. Table 5.1 of

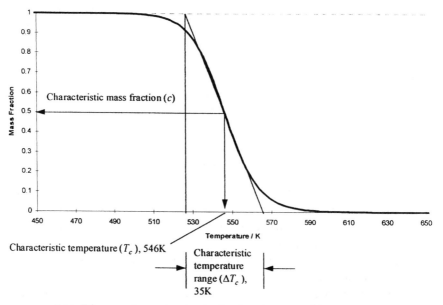

14.1 TG curve for a simple single-step decomposition process

Adrover and Giona[22] give formulae for Φ for various processes. Typically in the case of solid–phase reactions, a power-law relationship is assumed, $\Phi(\mu) = \mu^n$.

The problem of characterising a single-step decomposition process reduces essentially to a non-linear curve-fitting exercise, where the best values of the parameters A, n, and T_A must be chosen such that the predicted mass loss curve matches the experimental TG curve as closely as possible. Many practical methods exist for this procedure (Hatakeyama and Quinn[26] review some of the most popular in §5.3.1–§5.3.3), but perhaps the most directly applicable is that of Freeman and Carroll.[29]

For the case of thermal decomposition of a variety of solids, it transpires that one may assume $n = 1$.[27] This simplifies considerably the problem of finding the remaining Arrhenius parameters from a TG curve. A simple method for determining these parameters directly from the characteristic temperature and characteristic temperature range of the TG curve is detailed in Staggs.[27] In short, the activation temperature T_A can be determined from the solution of the equation $\beta = 1 - \lambda \exp(\lambda)E(\lambda)$, where $\beta = -c\Delta T_c \log(c)/T_c$, $\lambda = T_A/T_c$ and E is the exponential integral, $E(x) = \int_x^\infty \exp(-t)/t\,dt$. Once T_A has been found, the pre-exponential factor is given by $A = H\exp(\lambda)/c\Delta T_c$, where H is the heating rate. These formulae are particularly convenient, since there are many excellent approximations for the exponential integral which may be employed in engineering calculations. Abramowitz and Stegun[30] give rational polynomial approximations valid for all positive values of x.

When the activation temperature is much larger than the characteristic temperature, the asymptotic relation $x\exp(x)E(x) = 1 - 1/x + O(1/x^2)$ may be used to obtain the approximate result

$$T_A \approx -\frac{T_c^2}{c\Delta T_c \log c}.$$ [14.1]

A practical criterion for this approximation to be valid is $T_c/\Delta T_c \gg 1$. For the example shown in Fig. 14.1 above, the exact activation temperature is $23\,508\,\text{K}$ and the approximation gives $24\,577\,\text{K}$, which is within 5% of the exact result.

It is interesting to observe the behaviour of solutions of the kinetic rate equation $-Hd\mu/dt = A\mu\exp(-T_A/T)$ as $\Delta T_c \to 0$, for fixed T_c. Under these conditions approximation (14.1) becomes increasingly accurate and it can be shown that solutions observe the asymptotic relation $\log\mu \sim -\theta^2\exp\{T_c(1 - 1/\theta)/c\Delta T_c\}$, as $\Delta T_c \to 0$, where $\theta = T/T_c$, and we have chosen

$c = 1/e$ for convenience. Consequently, as $\Delta T_c \rightarrow 0$, μ will tend to the Heaviside function:

$$\mu = \begin{cases} 1, T < T_c \\ 0, T > T_c \end{cases} \qquad [14.2]$$

Hence as the characteristic temperature range reduces to 0, for fixed characteristic temperature, the mass loss curve is replaced by a step function which is invariant to changes in the heating rate. Physically, this would correspond to a material which remained inert at temperatures below T_c, but decomposed completely at T_c. This type of approximation to the mass loss curve recovers the ablation model introduced above.

14.2.2 A thermally thick model

Having now briefly described how the decomposition of a small amount of solid in the TG analyser is modelled, we go on to consider decomposition of larger samples under the action of an external heat flux. In this case, the solid is sufficiently thick for temperature gradients to be important. The discussion will refer to a standard cone calorimeter experiment,[25] where a horizontally-orientated sample is exposed to a uniform heat flux at its upper surface. Note that it is assumed that heat transfer occurs only in the direction normal to the sample's exposed surface, and so edge effects are neglected and the top surface of the sample remains horizontal throughout the test.

The net heat flux $\dot{q}''_{net}$ at the top surface of the sample (in the absence of a flame) is given by:

$$\dot{q}''_{net}(T_s) = \varepsilon\dot{q}''_0 - h(T_s - T_a) - \varepsilon\sigma(T_s^4 - T_a^4), \qquad [14.3]$$

where ε is the emissivity of the top surface, $\dot{q}''_0$ is the heat flux from the cone heater, h is the convection heat transfer coefficient, T_a is ambient temperature, T_s is the temperature of the top surface and σ is the Stefan-Boltzmann constant. For materials with small CTRs, the net heat flux at the surface will be approximately $\dot{q}''_{net}(T_c)$ and consequently the ratio of radiative to convective heat loss will be $\varepsilon\sigma T_c^3\{1 + T_a/T_c + O(T_a/T_c^2)\}/h$. For a typical polymer, $T_c \sim 600\,K$ and for conditions in the cone calorimeter $h \sim 15\,W\,m^{-2}\,K^{-1}$: consequently taking $\varepsilon = 1$, we see that this ratio is not small and radiative heat losses cannot be neglected.

The main consequence of applying a finite-rate decomposition mechanism throughout the solid is that for non-zero CTRs, material may volatilise throughout the volume of remaining material. Of course in practice volatilisation will be restricted to a surface layer of characteristic thickness δ. This in turn implies that we must be able to describe the manner in which volatile gases escape from interior regions to the surrounding atmosphere. The usual assumption here is to assume that gases escape as soon as they are formed. Kashiwagi summarises the complexities of this situation in his review paper.[1] Attempts to include volatile transport in mathematical models of pyrolysis are limited, and the models of Wichman[18] and Butler[31-33] are representative of the efforts to date.

The approximate size of δ may be estimated by matching the net heat flux at the top surface to the temperature gradient. The result is that

$$\delta \sim \frac{\kappa \Delta T_c}{\dot{q}''_{net}(T_c)},$$ [14.4]

where κ is the thermal conductivity of the polymer. Consequently, for a given polymer, we would expect δ to increase in inverse proportion to the external heat flux, i.e. halving the external heat flux should approximately double δ.

Fig. 14.2 shows results taken from the experiments of Thomson and Drysdale.[34] On the left are cross sections through acrylic sheets that have been ignited at different heat fluxes and then immediately quenched. The figure clearly shows the bubble structure and the depth of the gasifying layer. On the right is a plot of the reciprocal of the heat flux against an approximate bubble layer depth (relative to the depth at $\dot{q}''_0 = 12\,\mathrm{kWm}^{-2}$), verifying the approximate relationship suggested by Equation 14.4.

Even in the absence of bubbles, volatilisation of material from sub-surface regions with instantaneous removal of gases imposes a net *flow* of material through the degrading solid. To see this, consider the following. Imagine that we are able to label a particular element of material in the solid, and that we can track its progress as degradation occurs. Let the labelled material element be located a distance y

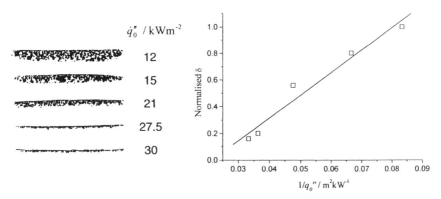

14.2 In-depth bubble formation

below the initial location of the top surface, and let the initial thickness of material be l. In a given time step, some material in the region $(y,l]$ will volatilise resulting in a net reduction of thickness. We would also observe that the material element at y would itself reduce in thickness, but would also move to a lower location, as material below y instantaneously fills the gaps left by escaping gases. In fact, if we track a material element that starts at the top surface, we would be able to follow its progress until finally some small portion of it arrives at $y = l$. This movement is characterised by defining a *displacement field* $v(y,t)$ relative to the fixed ordinate y.

The energy equation, applied throughout the interior of the remaining solid, takes the form

$$\rho c \left(\frac{\partial T}{\partial t} + v \frac{\partial T}{\partial y} \right) = \frac{\partial}{\partial y} \left(\kappa \frac{\partial T}{\partial y} \right) + \dot{Q}_v''', \qquad [14.5]$$

where ρ denotes the density, c the specific heat capacity and $\dot{Q}_v'''$ is a source term to account for the latent heat of vaporisation of the polymer. The instantaneous position of the top surface is $y = s(t)$, and the bottom surface is at $y = l$.

The kinetic equation describing the degradation of a small slice of material, when applied throughout the interior of the solid, is

$$\frac{\partial \mu}{\partial t} + v \frac{\partial \mu}{\partial y} = -f(\mu, T), \qquad [14.6]$$

and the source term in the energy equation is

$$\dot{Q}_v''' = \frac{\rho \Delta H_v}{\mu} \frac{D\mu}{Dt} = -\frac{\rho \Delta H_v}{\mu} f(\mu, T). \qquad [14.7]$$

Summing the changes of thickness of all material elements below a given 'target element' gives the overall rate of change of position with time of the target element as:[35]

$$v(y,t) = -\int_y^l \frac{1}{\mu} \frac{D\mu}{Dt} dy = \int_y^l \frac{f(\mu, T)}{\mu} dy, \qquad [14.8]$$

and the regression rate of the top surface is given implicitly by $ds/dt = v(s,t)$.

Equations [14.5] to [14.8] above constitute the mathematical model for pyrolysis of a polymer, assuming that the decomposition process occurs in a single step and the volatiles escape as soon as they are formed. The model equations, together with boundary and initial conditions appropriate for a cone calorimeter experiment are restated below for a sample of initial thickness l:

Conservation of energy: $\rho c \left(\dfrac{\partial T}{\partial t} + \dfrac{\partial T}{\partial y} \int_y^l \dfrac{f(\mu,T)}{\mu} dy \right)$

$$= \dfrac{\partial}{\partial y}\left(\kappa \dfrac{\partial T}{\partial y} \right) - \dfrac{\rho \Delta H_v}{\mu} f(\mu,T)$$

Mass fraction of polymer: $\dfrac{\partial \mu}{\partial t} + \dfrac{\partial \mu}{\partial y} \int_y^l \dfrac{f(\mu,T)}{\mu} dy = -f(\mu,T)$

Regression rate: $\dfrac{ds}{dt} = \int_s^l \dfrac{f(\mu,T)}{\mu} dy$

Boundary conditions: $-\kappa \dfrac{\partial T}{\partial y} = \dot{q}''_{net}(T)$, on $y = s(t)$,

$$\kappa \dfrac{\partial T}{\partial y} = h_B (T_a - T), \text{ on } y = l.$$

Initial conditions: $T(y,0) = T_a$, $\mu(y,0) = 1$, for $0 \le y \le l$, $s(0) = 0$.

P1: Mathematical model for simple pyrolysis of a polymer.

Note that a heat loss term has been included in the boundary conditions to model heat losses from the back of a sample, through the use of a heat transfer coefficient h_B. The mass flux of volatiles is $\dot{m}'' = \rho ds/dt$.

Of course, for first-order Arrhenius kinetics, the equations in system P1 simplify considerably. In this case, $f(\mu,T) = \mu A \exp(-T_A/T)$ and consequently μ is effectively removed from the model, leaving the simpler system P10:

Conservation of energy: $\rho c \left(\dfrac{\partial T}{\partial t} + A \dfrac{\partial T}{\partial y} \int_y^l \exp(-T_A/T) dy \right)$

$$= \dfrac{\partial}{\partial y}\left(\kappa \dfrac{\partial T}{\partial y} \right) - \rho \Delta H_v A \exp(-T_A/T)$$

Regression rate: $\dfrac{ds}{dt} = A \int_s^l \exp(-T_A/T) dy$

Boundary conditions: as in P1

Initial conditions: $T(y,0) = T_a$, for $0 \le y \le l$, $s(0) = 0$.

P10: Mathematical model for first-order pyrolysis of a polymer.

It transpires that for constant external heat fluxes, the assumption of first-order kinetics does not introduce large errors into the model.[27] Hence we shall adopt the reduced system P1a as our basic model of simple polymer pyrolysis.

Even the simplified model P1a is highly non-linear and exact solutions cannot, in general, be found. Therefore numerical solution methods must be used in order to obtain mass loss rate and temperature predictions. Fortunately, the system is amenable to standard numerical methods. Refer to Staggs[35] for a discussion of the numerical solution of system P1 for the nth order Arrhenius case $f(\mu,T) = \mu^n A \exp(-T_A/T)$.

Fortunately, the detailed numerical solution is not needed in order to obtain useful engineering estimates of the surface temperature and mass loss rate under steady conditions. Replacing temperature T by the dimensionless variable $\theta = T/T_a - 1$, assuming that the bottom face of the sample is well insulated, i.e. $h_B = 0$, integrating the energy equation from $y = s(t)$ to $y = l$ and assuming steady conditions, the following equation is obtained:

$$A \int_{s(t)}^{l} \left(\theta + \frac{\Delta H_v}{c\Delta T}\right) \exp\left(-\frac{\theta_A}{\theta+1}\right) dy = \frac{\dot{q}_{net}''(T_s)}{\rho c \Delta T}. \tag{14.9}$$

Here $\theta_A = T_A/T_a$ and T_s is the temperature on the top surface $y = s(t)$. Under thermally thick conditions, the main contribution to the integral on the left-hand side is in the vicinity of the top surface. Consequently, converting the variable of integration from distance to temperature the approximate relation

$$\int_{0}^{(T_s-T_a)/T_a} \left(\theta + \frac{\Delta H_v}{cT_a}\right) \exp\left(J - \frac{\theta_A}{\theta+1}\right) d\theta \approx \left(\frac{l\dot{q}_{net}''(T_s)}{kT_a}\right)^2, \tag{14.10}$$

is obtained, where $J = \ln(l^2 A/\alpha)$ and $\alpha = \kappa/\rho c$. This integral equation may be readily solved using nothing more sophisticated than standard spreadsheet software. Once solved for T_s, a remarkably good approximation for the surface temperature during steady pyrolysis is obtained. Once T_s has been found, the steady mass flux $\dot{m}''$ may be approximated using the formula

$$\frac{\dot{m}''l}{\alpha\rho} \approx \frac{kT_a}{l\dot{q}_{net}''(T_s)} \int_{0}^{(T_s-T_a)/T_a} \exp\left(J - \frac{\theta_A}{\theta+1}\right) d\theta. \tag{14.11}$$

Kindelan and Williams[17] applied a different method to obtain approximate formulae for the steady surface temperature and mass loss rate neglecting radiation. The formulae above do not neglect radiation heat losses and consequently are more relevant to the pyrolysis of polymers.

The interested reader should consult Staggs[36] for details of the derivations above. The quasi-steady approximations compare well with the

Table 14.1 Quasi-steady estimates

$l\dot{q}_0''/\kappa T_a$	$\dot{m}''l/\alpha\rho$	Average MLR
4	0.94	1.01
6	1.59	1.68
8	2.25	2.36
10	2.90	3.03
12	3.55	3.70

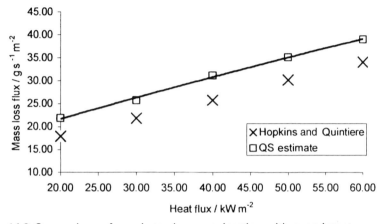

14.3 Comparison of quasi-steady approximation with experiment

average mass flux and surface temperature, as Table 14.1 shows. Here the dimensionless quasi-steady approximation in the second column is compared with the average dimensionless mass loss rate as computed numerically from the model P1a, for various values of dimensionless external heat flux (in the first column).

When compared with experimental results, the quasi-steady approximation still gives reasonable agreement as Fig. 14.3 shows. Here the quasi-steady approximations are compared with the steady-state results of Hopkins and Quintiere,[37] where they use PMMA samples of 25 mm thickness in a cone calorimeter in air. The graph in Fig. 14.3 shows the quasi-steady approximation plotted against heat flux (assuming an additional heat flux of 35 kW m⁻² for the flame, in line with measured values of Rhodes and Quintiere[38]) and also shows the experimental results.

14.3   Extensions of the simple model

Having discussed a simple model for pyrolysis of a polymer, it is now appropriate to develop some useful extensions.

14.3.1 Filled polymers

We begin the discussion by looking at the case of a homogeneous mixture of polymer and filler. Two popular fillers used with polyethylene, polypropylene and ethylene-vinyl acetate (EVA) are aluminium hydroxide (also known as alumina trihydrate, ATH) and magnesium hydroxide, two representatives of a class of hydrated mineral fillers which liberate water vapour on endothermic decomposition, leaving a porous residue of the metal oxide. When present in relatively large amounts, typically 60% by mass, these fillers act as effective flame retardants and also have the additional benefit of suppressing smoke in many different polymers during combustion.[39–45] Of course, when present in such quantities, the filler can compromise the properties which made the polymer a good choice for the end-use application and care has to be taken if this is to be avoided.[46,47] Simpler types of inert fillers, such as metal oxides, silicates and industrial carbon, which do not influence the mechanism of degradation of the host polymer, or degrade themselves, have also been used and are reviewed, together with hydrated fillers in Hausmann and Flaris.[48]

The production of volatiles from the polymer-filler mixture depends on many factors. In a highly filled material, if the polymer melts and flows before gaseous products are formed then the melt may flow through the porous filler residue and be extruded through the open pores at the exposed surface. The melt may subsequently degrade to lighter molecular weight species. As the process continues, production of gaseous species may start to occur within the filler residue, and escape of the volatiles would be determined by how they move through the porous residue. Darcy's law[49,50] has been used by some authors in related problems where a well-developed, static, porous residue is present (the pyrolysis of wood and composite materials has been described[51–55]), but it is difficult to see how to apply this to the present case where the filler may not be structurally strong enough to support pressure gradients large enough to drive the flow or when the permeability of the residue is constantly changing. Furthermore, as the residence time of volatiles in the filler increases, the possibility of further degradation/reaction/oxidation exists.

Also, there is the possibility that the internal structure of the filler residue will change as a result of the degradation process. Consider two extreme cases: a lightly filled solid where the filler is structurally weak and a highly filled solid where the filler is structurally strong. In the first case, the degradation process is likely to result in the gradual accumulation of a layer of filler dust at the surface (assuming that no filler escapes with the out-gassing of volatiles). In the latter case, the internal structure of the filler residue will remain unchanged and the volatile products will move through the structure and subsequently escape. The two cases are characterised by the change in volume of the solid as gases are produced. The first case (similar to the unfilled case where the solid consists entirely of polymer) clearly involves a large change of volume during degradation: the initial volume of material will be ultimately reduced to a small volume of filler dust at the end of the process. The second case is characterised by no volume change: the porous filler residue left at the end of the process occupies the same volume as the original solid.

Given the complexity of the processes which can occur within the solid, a complete description of the transport processes by which volatile products escape is unlikely at the present time and no satisfactory treatment exists in the literature to date. Instead, the following simplifying assumptions are made in order to make some progress:

- Volatiles formed from the polymer escape from the solid as soon as they are formed.
- The instantaneous density of the mixture depends on the mass fraction of polymer remaining in the solid and how the residue packs together during decomposition.
- The thermal conductivity of the mixture is a volume-weighted average of the relative volume fractions of polymer and filler remaining in the mixture.
- The specific heat capacity of the mixture is a mass-weighted average of the relative mass fractions of polymer and filler remaining in the mixture.
- The decomposition of the polymer occurs in a single step, according to a first-order Arrhenius process.

As above, decomposition of the polymer is modelled by a global, in-depth, single-step reaction of Arrhenius form, i.e. $-Dm_1/Dt = A_p \exp(-T_p/T)m_1$, where m_1 is the instantaneous mass of a small amount of polymer. In terms of the mass fraction $\lambda = m_1/m$ of polymer in the solid/filler mixture, where m is the total mass of polymer and filler, the rate equation becomes:

$$\frac{1}{\lambda}\frac{D\lambda}{Dt} = A_p(\lambda - 1)\exp\left(-\frac{T_p}{T}\right). \qquad [14.12]$$

This equation is augmented by an appropriate initial condition of the form $\lambda(y,0) = \lambda_0$, the initial mass fraction of polymer in the mixture.

Now consider the problem of how the mixture changes volume during pyrolysis of the polymer. Assume that the polymer and filler are closely packed in the solid, so that the total volume is the sum of the volume of the filler and the volume of the polymer. If no filler is lost, then the change in volume of a small mass of material containing mass

m_1 of polymer will be just $(1/\rho_1)dm_1/dt$, where ρ_1 is the density of the polymer. In order to model the structural effects of the filler on change of volume, we take a factor $\chi \in [0,1]$ of the maximum possible volume change. Hence χ is set to 0 for no volume change (the filler then provides a super-structure which preserves the volume of the solid) and is set to 1 when sur-rounding material completely fills in gaps left by escaping volatiles. It can be shown (consult Staggs[56] for details) that the density of the mixture is given by:

$$\frac{\rho}{\rho_0} = \frac{1-\lambda_0}{1-\lambda-\chi\rho_0(\lambda_0-\lambda)/\rho_1},$$ [14.13]

where the initial density, assuming the polymer and filler are closely packed in the solid is $\rho_0 = \{\lambda_0/\rho_1 + (1-\lambda_0)/\rho_2\}^{-1}$.

The basic problem with this approach is to choose a realistic value for the constant χ. This will clearly depend on the initial mass fraction of polymer in the mixture. We would expect that when $\lambda_0 = 1$, a reasonable value for χ would be 1; when $\lambda_0 = 0$ a reasonable value would be $\chi = 0$. From the expression for density above, it follows that the final volume V_∞, after all the polymer has vaporised, assuming that no filler has been lost, must be $V_\infty/V_0 = 1 - \chi v_0$, where $v_0 = \lambda_0\rho_0/\rho_1$ is the initial volume fraction of polymer in the mixture. Hence, if the final volume can be determined experimen-tally as a function of initial filler loading, this data may be used to obtain an empirical relationship for χ.

In the meantime, in the absence of empirical data and for the sake of illustration, we shall assume the physically reasonable form $\chi = \{1 - \tanh(2\gamma(v_{f0} - v^*))\}/2$, for $\gamma \gg 1$, where v_{f0} is the initial volume fraction of filler, v^* is the percolation threshold for v_{f0} and γ is the absolute gradi-ent of χ at the percolation threshold.

As for the simpler case of pyrolysis of a simple polymer above, removal of material imposes a displacement of material within the mixture, defined by the displacement field

$$v(y,t) = \chi A_p \int_y^l \exp\left(-\frac{T_p}{T}\right)\frac{\lambda}{\rho_1}\rho dy.$$ [14.14]

The energy equation takes the form of Equation (14.5) with $\dot{Q}_v'' = -\rho\lambda\Delta H_v^{(p)}A_p\exp(-T_p/T)$, and is augmented by boundary and initial condi-tions as in the model P1a. Here $\Delta H_v^{(p)}$ is the latent heat of vaporisation of the polymer. The thermal properties c and κ are given by mass-weighted and volume-weighted averages:

$$c = \lambda c_1 + (1-\lambda)c_2, \quad \kappa = \kappa_1\frac{\lambda\rho}{\rho_1} + \kappa_2\frac{(1-\lambda)\rho}{\rho_2}.$$ [14.15]

(The subscript 1 denotes a polymer property and the subscript 2 denotes a filler property).

The curves in Fig. 14.4 show mass loss rate predictions from the model at different filler loadings for a low density, low thermal conductivity filler. In this case the density of the filler was $200\,kg\,m^{-3}$ and the thermal conductivity was $0.05\,Wm^{-1}K^{-1}$; the corresponding values for the polymer were $900\,kg\,m^{-3}$ and $0.3\,Wm^{-1}K^{-1}$ respectively. Note that there is a characteristic double peak structure to the mass loss curves. Vaporisation of polymer at surface regions causes the initial peak to form and a layer of filler to build up shielding material below. The filler layer acts as a thermal barrier, delaying vaporisation of material in subsurface regions. Eventually the temperature in subsurface regions increases enough to switch on the degradation reaction, giving the second peak. Note also the dilution effect of adding more filler: the mass loss rate reduces in approximately direct proportion to the initial mass fraction of filler.

It is worthy of note that for the case of highly filled systems where there is little volume change during pyrolysis, and for polymers with small characteristic temperature ranges of degradation (so that the ablation approximation is valid), the problem is analogous to the classic melting problem where the solid and liquid phases have different thermal properties. There are many excellent approximations to the solution of this problem, and the analysis of Goodman[57] could be directly applied, resulting in a much simpler system of ordinary differential equations. This problem is discussed by Staggs.[58]

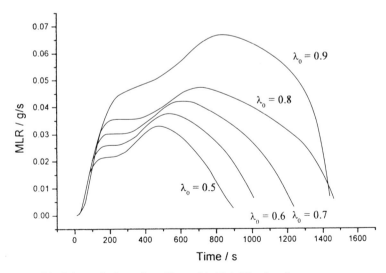

14.4 Model predictions for effect of initial filler loading

This model can be extended to include the case where the filler itself decomposes, leaving a porous residue, such as in the case of ATH. Assume that a mass m_f of filler contains mass m_w of water and mass m_s of inert residue so that $dm_f/dt = dm_w/dt = -m_w A_w \exp(-T_w/T)$. So, if λ is the mass fraction of polymer and η is the mass fraction of water, the equations describing the evolution of λ and η are symmetric and of the form

$$\frac{1}{\lambda}\frac{D\lambda}{Dt} = \eta A_w \exp\left(-\frac{T_w}{T}\right) - (1-\lambda)A_p \exp\left(-\frac{T_p}{T}\right), \qquad [14.16]$$

$$\frac{1}{\eta}\frac{D\eta}{Dt} = \lambda A_p \exp\left(-\frac{T_p}{T}\right) - (1-\eta)A_w \exp\left(-\frac{T_w}{T}\right). \qquad [14.17]$$

Applying the structural model for the inert filler discussed above, and detailed by Staggs,[56] the mixture density evolves according to the equation

$$\frac{1}{\rho}\frac{D\rho}{Dt} = -A_p \exp\left(-\frac{T_p}{T}\right)\lambda\left(1 - \chi\frac{\rho}{\rho_1}\right) - A_w \exp\left(-\frac{T_w}{T}\right)\eta, \qquad [14.18]$$

where ρ_1 is the polymer density. The displacement field will be given by Equation (14.14) and the energy equation is of the same form as Equation (14.5), with $\dot{Q}_v''' = -\rho\{\lambda\Delta H_v^{(p)}A_p \exp(-T_p/T) + \eta\Delta H_v^{(w)}A_w \exp(-T_w/T)\}$. Here $\Delta H_v^{(w)}$ is the latent heat associated with the production of water vapour from the filler. The specific heat capacity of the mixture will be $c = \lambda c_1 + \eta c_2 + (1 - \lambda - \eta)c_3$, where c_2 is the specific heat capacity of water and c_3 is the specific heat capacity of the filler residue. In order to complete the heat transfer model the thermal conductivity κ must be specified. Ideally this should be determined empirically as a function of λ and η, but in the absence of experimental data the form

$$\kappa = \kappa_1 \frac{\lambda\rho}{\rho_1} + \left(1 - \frac{\lambda\rho}{\rho_1}\right)\left\{\kappa_{FH}\frac{\eta}{\eta_0} + \left(1 - \frac{\eta}{\eta_0}\right)\kappa_{FD}\right\}, \qquad [14.19]$$

is assumed, where η_0 is the initial mass fraction of water in the mixture (equal to approximately $0.35(1 - \lambda_0)$ for ATH), κ_{FH} is the thermal conductivity of the hydrated filler and κ_{FD} is the thermal conductivity of the dehydrated filler.

14.3.2 Char-forming polymers

Di Blasi[2] gives an excellent review of some of the problems associated with modelling char-forming systems without change of volume, particularly when a well-established char layer is present. In this section, a brief description is given of a simple mathematical model for char formation including the effects of change of volume. This treatment is intended as the starting point for a description of the dynamically evolving char layer, such as might occur during the degradation of PVC for example. Consequently detailed and complex char-forming mechanisms are not considered explicitly.

One of the primary problems with char-forming systems is to represent adequately the kinetic mechanism for the various decomposition processes occurring within solid and liquid phases. The simplest starting point is the competitive reaction scheme shown in Fig. 14.5, where two reactions (a volatile-producing reaction and a char-producing reaction) compete for the remaining polymer.

Let the volatile-producing reaction have rate constant k_1 and the char-forming reaction have rate constant k_2. A step is also included for the decomposition of char (shown by the dashed arrow in Fig. 14.5), although this complication will not be included in the analysis below. This scheme (and modifications of it) has been used to model the decomposition of cellulose to char and tar,[2] although more complex schemes have been used.[59] This scheme has also been used in thermally thin models for pyrolysis and ignition of char-forming polymers.[60]

Anthony[61] has recently proposed a scheme for the pyrolysis of PVC, and this is shown schematically in Fig. 14.6. Additional kinetic analyses for PVC and various other polymers including polystyrene have also been discussed.[62–68]

Adopting the model scheme, if m_1 is the mass of virgin polymer and m_2 is the mass of char, then the kinetic equations are $dm_1/dt = -(k_1 + k_2)m_1$ and $dm_2/dt = k_2 m_1$. Hence if $\lambda(y,t)$ is the instantaneous mass fraction of polymer in the remaining solid at position y and time t, then λ will evolve according to

$$-\frac{1}{\lambda}\frac{D\lambda}{Dt} = k_1(1-\lambda) + k_2, \qquad [14.20]$$

with initial condition $\lambda(y,0) = 1$. The change of volume in the solid, as virgin polymer is converted into char, gives the displacement field as

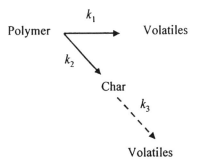

14.5 Simple competitive char-forming mechanism

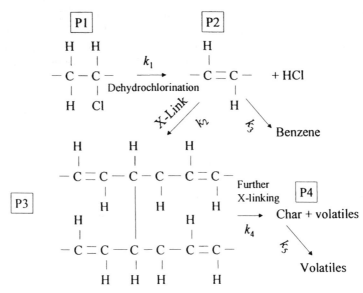

14.6 Char-forming mechanism for PVC as proposed by Anthony[61]

$$v(y,t) = \int_{y}^{l} \left\{ k_1 + \left(1 - \frac{\rho_1}{\rho_2}\right)k_2 \right\} \lambda \hat{\rho} \, dy, \qquad [14.21]$$

where $\hat{\rho} = \rho/\rho_1$. Furthermore, the dimensionless mixture density will be $\hat{\rho} = \{\lambda + \rho_1(1 - \lambda)/\rho_2\}^{-1}$. The other thermal properties will be as above for the filled polymer, i.e. Equation (14.15). Assuming that the formation of char and volatiles involves latent heat, conservation of energy takes the form of Equation (14.5) with $\dot{Q}_v''' = -\rho\lambda(\kappa_1\Delta H_1 + k_2\Delta H_2)$. Here ΔH_1 is the latent heat associated with the formation of volatiles and ΔH_2 is the latent heat associated with the formation of char. The energy equation is augmented by boundary and initial conditions as in the model P1a.

The graph in Fig. 14.7 illustrates the effect of variation in the characteristic temperature of the char-forming step on char yield and mass loss rate. Here, it has been assumed that the polymer forms a low density (100 kg m^{-3}), low thermal conductivity char (0.05 W m^{-1} K^{-1}). The characteristic temperature of the volatile-producing step was set to 600 K. The kinetic parameters for the model are shown in Table 14.2.

Note that the benefits of producing the char before the volatilisation reaction switches on are clear: the mass loss rate reduces greatly together with increased char yield. Also the production of a low density, low thermal conductivity char is clearly advantageous: most of the temperature drop occurs across the char, implying that the virgin polymer beneath is shielded from the heat to such an extent that both the volatilisation and char-forming

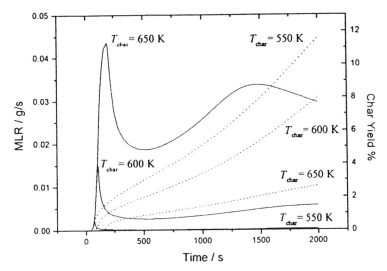

14.7 Effect of characteristic temperature of char-forming step on MLR (solid curves) and char yield (broken curves)

Table 14.2 Kinetic parameters

Characteristic Temperature/K	Pre-Exponential Factor/s^{-1}	Activation Temperature/K
550	5.846×10^{10}	16 389
600	1.038×10^{12}	19 608
650	1.854×10^{13}	23 115

reactions slow down. This is shown in Fig. 14.8 where the temperature distribution and polymer mass fraction are plotted against dimensionless distance from the sample bottom (x) at 1000 s for the case where the characteristic temperature of the char was 550 K.

Nelson and Brindley[60] have developed a model for char-forming systems in thermally thin conditions using the competitive reaction scheme considered above. Other work using the ablation approximation has been carried out by Leung et al.[14] A comprehensive review of the various char-forming models, some useful engineering approximations and a full treatment of the ablation approximation may be found in Leung.[69]

14.4   Ignition

Under certain circumstances, ignition of polymers may be modelled by considering only solid-phase processes.[70] The modelling of a flame is consider-

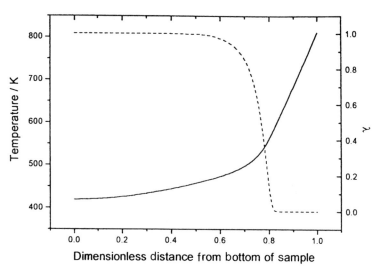

14.8. Distributions of temperature (solid curve) and polymer mass fraction (broken curve) at 1000 s for $T_{char} - 550\,K$

ably simplified in these approaches, reducing it essentially to an additional heat flux which is switched on or off. In order for such a model to be useful, a criticality condition must be applied, which determines when the flame is switched on or off. Typically this is done through the use of a critical surface temperature (CST) or a critical mass flux (CMF). Other criteria have been suggested, such as critical char depth, critical average solid temperature, gas-phase temperature gradient reversal,[71-76] but have not proved as popular.

In order to implement a CST or CMF criterion, the boundary condition on the exposed surface of the polymer is modified:

CST:

$$-\kappa \frac{\partial T}{\partial y} = \begin{cases} \dot{q}''_{net}(T), T < T_{ig} \\ \dot{q}''_{net}(T) + \dot{q}''_f, T \geq T_{ig} \end{cases},$$

CMF:

$$-\kappa \frac{\partial T}{\partial y} = \begin{cases} \dot{q}''_{net}(T), \dot{m}'' < \dot{m}''_{ig} \\ \dot{q}''_{net}(T) + \dot{q}''_f, \dot{m}'' \geq \dot{m}''_{ig} \end{cases} \text{ on } y = s(t),$$

[14.22]

where $\dot{q}''_f$ is the heat flux from the flame, T_{ig} is the CST and $\dot{m}''_{ig}$ is the CMF. Typically, $\dot{q}''_f$ is taken as a constant, although it is not immediately obvious that this should be the case. If the sample produces volatiles at a rate $\dot{m}''$ and ΔH_c is the heat of combustion of the fuel, then it seems reasonable that a fraction of the heat released in the combustion of the gaseous fuel $\dot{m}''\Delta H_c$ will be transferred back to the sample, i.e. $\dot{q}''_f = \phi \Delta H_c \dot{m}''$. The parameter ϕ

will depend on the size and shape of the flame, the combustion efficiency and the mode of heat conduction from the flame to the sample, among other things. In order to investigate this, the total heat flux at the sample surface needs to be measured simultaneously with mass loss rate both inside and outside the quasi-steady burning regime. Studies published to date such as those by Hopkins and Quintiere[37] and by Rhodes and Quintiere[38] are limited and do not attempt to correlate total heat flux at the sample surface with mass loss rate.

Of the two approaches, the CMF is probably the most physically realistic – especially in cone calorimeter experiments. Experimental values for the CMF obviously depend on the particular polymer, but Thomson and Drysdale[34] report values in the range $0.80–2.0\,gs^{-1}m^{-2}$ for thermoplastics, Atreya and Wichman[74] report a value of $1.8\,gs^{-1}m^{-2}$ for cellulosic solids and Rasbash et al[75] and Deepak and Drysdale[76] report values in the range $3–6\,gs^{-1}m^{-2}$ for PMMA. The graph in Fig. 14.9 shows values for the CMF for ethylene-vinyl acetate filled with various levels of ATH[77] in cone calorimeter tests on 10 cm by 10 cm plaques.

Although useful, the CMF has obvious drawbacks. The flame is included only as a switch and consequently modelling gas-phase active flame retardants is not possible. Furthermore, there is the irritating problem that at room temperature, the criterion predicts that any polymer will ignite if it is sufficiently massive! This follows because the mass loss rate is non-zero at temperatures above absolute zero. In practice this is not a serious problem because the mass loss rate will be negligible at room temperature for any realistic values of the Arrhenius parameters. To illustrate this point, consider a polymer with a characteristic temperature of 600 K and a characteristic temperature range of 100 K. The Arrhenius parameters in Table 14.2

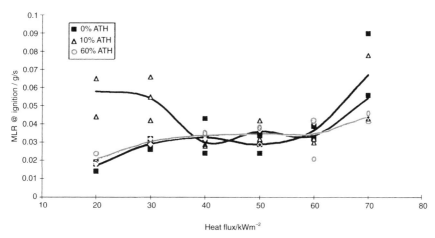

14.9 Mass loss rate at ignition for EVA filled with ATH

suggest that in order to get ignition at 293 K, assuming a CMF of $4\,\mathrm{gs^{-1}\,m^{-2}}$, the initial polymer mass must be approximately equivalent to the mass of the largest fragment of comet Shoemaker–Levy which collided with Jupiter in 1994!

Nelson et al[78–82] have approached the problem of ignition using the powerful tools of bifurcation theory. The approach is similar to that of Rychly and Costa.[83] They simplify the physics in the gas phase, assuming well-mixed conditions in a spatially-uniform, constant volume reaction zone. Thermally thin conditions are assumed in the solid. The gaseous reaction zone is coupled to the solid via radiation and convection heat transfer mechanisms and of course through the flow of volatiles from the solid to the reaction zone. The assumptions of the model lead to a system of five coupled non-linear ordinary differential equations (which the authors are able to reduce to three) describing the temperature in the solid, the temperature in the gaseous reaction zone, the rate of accumulation of mass in the reaction zone, the rate of loss of mass from the solid and the rate of outflow from the reaction zone of gaseous combustion products.

Although the physics of this approach may be limited, the mathematical formulation results in useful insights into the qualitative and quantitative behaviour of the system. For example, Nelson et al[82] found two ignition mechanisms by which a steady flame could be established: monotonic ignition, where the temperature in the reaction zone increases in time, and an oscillatory mode involving periodic excursions of the reaction zone temperature to high values. Within the framework of their model, the authors were able to suggest ways in which additives may be used to reduce flammability.

Other more physically complex models of ignition exist, and the reader is referred to the literature[84–88] for further details. These models typically remove the spatially uniform assumption in the gas phase, necessitating the need for some sort of computational fluid dynamics approach. This is a huge field of combustion research and is outside of the scope of this chapter.

14.5   Discussion

The models above for filled and char-forming polymers serve only as a starting point for the analysis of these systems. Many physical features are neglected which may be important. For example, the transport of volatiles and oxygen through the residual layer is neglected. As the layer increases in thickness, the residence time of combustible volatiles produced from the degradation of the polymer becomes increasingly long. This in turn increases the chance of oxidation of the volatiles (or some other chemical reaction) within the residual layer itself, depending on the local oxygen concentration. In order to model this situation, a transport mechanism for the flow of combustible gases and oxygen through the residue must be included. When the layer is well developed and static, one could apply the ideas of combustion in porous media,[39,51–53] using Darcy's law and diffusion, to model the flows. However, during the development of the layer the problem is far more complicated. The flow of gases and the characteristics of the residue will be coupled. For example, in the case of a developing intumescent char, the internal structure of the char will be greatly influenced by the flow of gases through it. Unfortunately, the problem is even worse than this. Consider the case of a simple polymer. The transport mechanism for the flow of volatiles through the melt is complex in the extreme. Any realistic model of a developing char-forming system must include a transport mechanism for volatiles which in the early stages of pyrolysis resembles the mechanism for a simple polymer, but becomes more percolation-like as a residual layer forms.

At present, there is no mathematical model available that adequately includes all of the important physics, although attempts have been made to model the transport of volatiles through a polymer melt. Wichman[18] modelled a quasi-steady situation in an infinitely thick polymer; Butler[31–33] has also attempted to model aspects of the situation in the context of intumescent char-forming systems.

The dynamical systems approach to modelling ignition, as developed by Nelson et al[78–82] and Rychly and Costa,[83] although lacking in physical sophistication, is extremely interesting. There is no doubt that significant insights into ignition phenomena will be gained from this approach, although quantitative agreement with experiment may always be difficult.

All mathematical models published to date incorporate some form of compromise. In order to construct a satisfactory model of flame spread and ignition of solids, which has engineering importance to the Fire Science community, several elements need to be in place, including the following:

- Production of volatiles from the solid.
- Adequate description of transport phenomena.
- Adequate gas-phase chemistry.
- Heat-transfer in the gas phase, including radiative heat transfer through a participating medium.
- Conjugate heat transfer between the gas and the solid, including gas-phase conduction and short-range convective heat transfer.

At present no complete model exists, although there are some excellent models which partially describe some of the important physics, such as transport phenomena in the gas phase. The solid-phase models, describing evolution of volatiles, are particularly troublesome for polymers for all the reasons discussed above and it is in this area that much remains to be done.

References

1 Kashiwagi T. 'Polymer combustion and flammability – role of the condensed phase. *25th. Symp. (int). On Combustion*, The Combustion Inst. 1994 Pittsburgh, 1423.

2 Di Blasi C. 'Modelling and simulation of combustion processes of charring and non-charring solid fuels', *Prog. Energy Combust. Sci.* 1993, **19** 71.

3 D'Alelio G F and Parker J A. *Ablative Plastics.* Marcel Dekker, New York, 1971.

4 Carlslaw S and Jaeger J C. *Conduction of Heat in Solids.* Oxford University Press, Oxford, 1984.

5 Landau G. 'Heat conduction in a melting solid', *Quart. J. App. Maths.* 1950, **8**(81).

6 Billings M J, Warren L and Wilkins R. 'Thermal erosion of electrical insulating materials', *IEEE Trans. on Elect. Ins.* 1971, **6**(2) 82–90.

7 Andrews G and Atthey D R. 'Analytical and numerical techniques for ablation problems', in *Moving Boundary Problems in Heat Flow and Diffusion* (Ockendon J. (ed.) Clarendon Press, Oxford, 1975, 38.

8 Rastegar S, Motamedi M, Welch A J and Hayes L J. 'A theoretical study of the effect of optical properties in laser ablation of tissue', *IEEE Trans. on Biomed. Eng.* 1989, **36**(12) 1180–7.

9 Ricci C. 'Travelling wave solutions of the Stefan and the ablation problems', *SIAM J. Math. Anal.* 1990, **21** 1386.

10 Whiting P, Dowden J M, Kapadia P D and Davis M P. 'A one-dimensional mathematical model of laser induced thermal ablation of biological tissue', *Lasers in Med. Sci.* 1992, **7** 357–68.

11 Delichatsios A and Chen Y. 'Asymptotic, approximate and numerical solutions for the heatup and pyrolysis of materials including reradiation losses', *Combust. Flame* 1993, **92** 292.

12 Quintiere J and Iqbal N. 'An approximate integral model for the burning rate of a thermoplastic-like material', *Fire Mat.* 1994, **18** 89.

13 Staggs J E J. 'A discussion of modelling idealised ablative materials with particular reference to fire testing', *Fire Safety J.* 1997, **28** 46.

14 Leung C H, Staggs J E J, McIntosh A C and Brindley J. 'Modelling of a one-dimensional ablation process with char formation', *Comb. Sci. Tech.* 1996, **119** 301.

15 Leung C H, Staggs J E J, McIntosh A C and Brindley J. 'Heat transfer to an infinite solid through a moving boundary', *Int. Comms. in Heat and Mass Transfer* 1997, **24**(2) 181.

16 Leung C H, Staggs J E J, McIntosh A C and Brindley J. 'Modelling of polymer ablation including a substrate base', *Comb. Sci. Tech.* 1997, **126** 53–70.

17 Kindelan M and Williams F A. 'Theory for endothermic gasification of a solid by a constant energy flux', *Comb. Sci. and Tech.* 1975, **10** 1.

18 Wichman I S. 'A model describing the steady-state gasification of bubble-forming thermoplastics in response to an incident heat flux', *Comb. Flame* 1986, **63** 217.

19 Vovelle C, Delfau J and Reuillon M. 'Experimental and numerical study of the thermal degradation of PMMA', *Comb. Sci and Tech.* 1987, **53** 187.

20 Staggs J E J and Whiteley R H. 'Modelling the combustion of solid-phase fuels in cone calorimeter experiments', *Interflam '96. 7th. Int. Fire Science & Eng.*

Conf., Interscience Communications, University of Cambridge, Cambridge, UK, 1996, 103.

21 Carabin P and Berk D. 'Analysis and modelling of the isothermal and non-isothermal shrinking core model with non-linear kinetics', *Chem. Eng. Science* 1992, **47**(9–11) 2499–504.

22 Adrover A and Giona M. 'Solution of unsteady-state shrinking-core models by means of spectral/fixed-point methods: nonuniform reactant distribution and nonlinear kinetics', *Ind. Eng. Chem. Res.* 1997, **36** 2452–65.

23 Abdoulaye Ouedraogo, Mulligan J C and Cleland J G. 'A quasi-steady shrinking core analysis of wood combustion', *Comb. Flame* 1998, **114** 1–12.

24 Liliedahl T and Sjöström K. 'Heat transfer controlled pyrolysis kinetics of a biomass slab, rod or sphere', *Biomass and Bioenergy* 1998, **15**(6) 503–9.

25 *Standard Test Method for Heat and Visible Smoke Release Rates for Materials and Products Using an Oxygen Consumption Calorimeter.* ASTM E 1354, American Society for the Testing of Materials, Philadelphia 1992.

26 Hatakeyama T and Quinn F X. *Thermal Analysis: Fundamentals and Applications to Polymer Science.* Wiley, Chichester 1994, §5.3.

27 Staggs J E J. 'Thermal degradation of solids using equivalent first-order kinetics', *Fire Safety J.* 1999, **32** 17–34.

28 Nelson M I, Brindley J and McIntosh A C. 'Ignition properties of thermally thin materials in the cone calorimeter: a critical mass flux model', *Combust. Sci. and Tech.* 1996, **113–114** 221.

29 Freeman E S and Carroll B. *Journal of Physical Chemistry* 1958, **62** 394.

30 Abramowitz M and Stegun I A (eds.), *Handbook of Mathematical Functions (9th. ed.).* Dover, New York 1972, §5.1.

31 Butler K M, Baum H R and Kashiwagi T. 'Three-dimensional kinetic model for the swelling of intumescent materials', *NIST Ann. Conf. of Fire Res. Oct. 17–20 1994*, Gaithersburg MD, 1994, 109.

32 Butler K M, Baum H R and Kashiwagi T. 'Three-dimensional modelling of intumescent behaviour in fires', *Proc. of the 5th. Int. Symp. on Fire Safety Science* 1995, 523.

33 Butler K M. 'Physical modelling of intumescent fire retardant polymers', *Proc. American Chem. Soc. ACS Symp. Series* 1997, **669** 214.

34 Thomson H E and Drysdale D D. 'Flammability of plastics. I: Ignition temperatures', *Fire Mat.* 1987, **11** 163.

35 Staggs J E J. 'A theoretical investigation into modelling thermal degradation of solids incorporating finite rate kinetics', *Combust. Sci. Tech.* 1997, **123** 261.

36 Staggs J E J. 'Quasi-steady thermal degradation of solids', *Fire Mat* 1998, **22** 109–18.

37 Hopkins Jr and Quintiere J G. 'Material fire properties and predictions for thermoplastics', *Fire Safety J.* 1996, **26** 241.

38 Rhodes B T and Quintiere J G. 'Burning rate and flame heat flux for PMMA in a cone calorimeter', *Fire Safety J.* 1996, **26** 221.

39 Gokhale S S and Krier H. 'Modelling of unsteady two-phase reactive flow in porous beds of propellant', *Prog. Energy Combust. Sci.* 1982, **8** 1–39.

40 Levendusky T L and Musselman L L. 'Alumina trihydrate as a flame retardant and smoke suppressant filler for polypropylene: development of a commercial flame retardant polypropylene', *Plastics Engineering* 1983, **39**(3) 32.

41 Hirschler M M and Thevaranjan T R. 'Effects of magnesium oxide/hydroxide on

flammability and smoke production tendency of polystyrene', *Eur. Polym. J.* 1985, **21**(4) 371–5.

42 Hornsby P R and Watson C L. 'Magnesium hydroxide – a combined flame retardant and smoke suppressant filler for thermoplastics', *Plastics and Rubber Processing and Apps.* 1986, **6**(2) 169–75.

43 Hornsby P R and Watson C L. 'Mechanism of smoke suppression and fire retardancy in polymers containing magnesium hydroxide filler', *Plastics and Rubber Processing and Apps.* 1989, **11**(1) 45–51.

44 Wan Hanafi W Z A and Hornsby P R. 'Smoke suppression and fire retardancy in unsaturated polyesters containing hydrated fillers', *Plastics, Rubber and Composites Processing and Applications* 1993, **19**(3) 175–84.

45 Molesky F, Schultz R, Midgett S and Green D. 'Smoke reduction for flexible PVC-high surface area: alumina trihydrate vs. magnesium hydroxide', *J. Vinyl & Additive Tech* 1995, **1**(3) 159–61.

46 Chiu S-H and Wang W-K. 'The dynamic flammability and toxicity of magnesium hydroxide filled intumescent fire retardant polypropylene', *J. Appl. Poly. Sci.* 1998, **67** 989–95.

47 Midgett S E and Schultz R E. 'Mechanically ground ATH in XL polyolefins for extrusion and injection moulding', *J. Vinyl and Additive Tech.* 1995, **1**(3) 159–61.

48 Hausmann K and Flaris V. 'Polymeric coupling agents as property enhancers in highly filled polymer systems', *Polym. & Polym. Comp.* 1997, **5**(2) 113–19.

49 Darcy H. *Les fontaines publiques de ville de Dijon*, 1856, 590.

50 Batchelor G K. *An introduction to fluid dynamics.* Cambridge University Press 1987 §4.8.

51 Kanury A M. 'Thermal decomposition kinetics of wood pyrolysis', *Comb. Flame* 1972, **18** 75–83.

52 Kung H C. 'A mathematical model of wood pyrolysis', *Comb. Flame* 1972, **18** 185–95.

53 Kansa E J, Perlee H E and Chaiken R F. 'Mathematical model of wood pyrolysis including internal forced convection', *Comb. Flame* 1977, **29** 311–24.

54 Kkalturinskii N A and Berlin A A. 'On the reduction of combustibility of polymeric materials', *Int. J. Polym. Mater.* 1990, **14** 109–25.

55 Di Blasi C and Wichman I S. 'Effects of solid-phase properties on flames spreading over composite materials', *Comb. Flame* 1995, **102** 229.

56 Staggs J E J. 'A simplified mathematical model for the pyrolysis of polymers with inert additives', *Fire Safety J.* 1999, **32** 221–240.

57 Goodman T R. 'The heat-balance integral and its applications to problems involving a change of phase', *J. of Heat Transfer Trans. ASME*, 1958, **80** 335.

58 Staggs J E J. 'Modelling pyrolysis of char-forming polymers', *Proc. Interflam. '99,* Interscience Comms. London, 1999, 167–79.

59 Antal M J and Varhegyi G. 'Cellulose pyrolysis kinetics – the current state of knowledge', *Ind. Eng. Chem. Res.* 1995, **34** 703.

60 Nelson M I and Brindley J. 'Modelling char formation in isothermal and nonisothermal thermogravimetric experiments', *Thermochimica Acta* 1995, **258** 175.

61 Anthony G M. 'Kinetic and chemical studies of polymer cross-linking using

thermal gravimetry and hyphenated methods', Degradation of polyvinylchloride', *Polym. Degrad. Stab.* 1999, **64** 353–7.

62 Agrawal R K. 'Kinetics of reactions involved in the pyrolysis of cellulose. I. The three reaction model', *Can. J. Chem. Eng.* 1988, **66** 403.

63 Agrawal R K. 'Kinetics of reactions involved in the pyrolysis of cellulose. II. The modified Kilzer-Broido model', *Can. J. Chem. Eng.* 1988, **66** 413.

64 Wu C H, Chang C Y, Hor J L, Shih S M, Chen L W and Chang F W. 'Two-stage pyrolysis model of PVC', *Can. J. Chem. Eng.* 1990, **72** 644.

65 Bockhorn H, Hornung A, Hornung U, Teepe S and Weichmann J. 'Investigation of the kinetics of thermal degradation of commodity plastics', *Comb. Sci. Tech.* 1996, **116–117** 129.

66 Bockhorn H, Hornung A and Hornung U. 'Stepwise pyrolysis for raw material recovery from plastic waste', *J. Anal. and App. Pyrol.* 1998, **46** 1.

67 Bockhorn H, Hornung A, Hornung U and Jakoströer P. 'New mechanistic aspects of the dehydrochlorination of PVC- application of dehydrochlorination to plastic mixtures and electronic scrap', *Comb. Sci. Tech.* 1998, **134** 7.

68 Bockhorn H, Hornung A and Hornung U. 'Gasification of polystyrene as initial step in incineration, fires or smoldering of plastics', *27th. Int. Symp. on Combustion*, Boulder, CO, USA. 1998, Paper no. P079.

69 Leung C H. *Modelling of polymer ablation.* Ph. D. Thesis, Department of Fuel and Energy, University of Leeds, Leeds LS2 9JT, UK. (1999).

70 Nelson M I. 'Everything you wanted to know about the ignition of polymers', *Behaviour of Polymeric Materials in Fire.* (Short Course at Leeds University, 13–14 Oct. 1997) (1997).

71 Sauer F M. 'The charring of wood during exposure to radiation: correlation analysis for semi-infinite solids', *Interim Tech. Report AFSWP-838.* Forest Service, US Dept. Ag., Pacific Southwest Forest and Range Experiment Station, Berkley, CA (1956).

72 Martin S. 'Diffusion-controlled ignition of cellulosic materials by intense radiant energy', *10th. Symp. (Int.) on Combustion*, The Combustion Inst. 1965, 877.

73 Deverall L I and Lai W. 'A criterion for thermal ignition of cellulosic materials', *Comb. Flame* 1969, **13** 8.

74 Atreya A and Wichman I S. 'Heat and mass transfer during piloted ignition of cellulosic solids', *ASME J. Heat Trans.* 1989, **111**(3) 719.

75 Rasbash D J, Drysdale D D and Deepak D. 'Critical heat and mass transfer at pilot ignition and extinction of a material', *Fire Safety J.* 1986, **10** 1.

76 Deepak D and Drysdale D D. 'Flammability of solids: an apparatus to measure the critical mass flux at the firepoint', *Fire Safety J.* 1983, **5** 167.

77 Gallagher P. 'An investigation into the instantaneous mass loss rate at the point of ignition for polymer materials with flame retardant fillers', *Final Year Project Report, May 1998, Dept. Fuel & Energy, University of Leeds* (1998).

78 Nelson M I, Brindley J and McIntosh A C. 'The dependence of critical heat-flux on fuel and additive properties – a critical mass flux model', *Fire Safety J.* 1995, **24**(2) 107.

79 Nelson M I, Brindley J and McIntosh A C. 'Polymer ignition', *Math. and Comp. Modelling* 1996, **24**(8) 39.

80 Nelson M I, Brindley J and McIntosh A C. 'Ignition properties of thermally thin

thermoplastic – the effectiveness of inert additives in reducing flammability', *Polym. Degrad. Stab.* 1996, **54**(2–3) 255.

81 Nelson M I, Brindley J and McIntosh A C. 'The effect of heat sink additives on the ignition and heat release properties of thermally thin thermoplastics', *Fire Safety J.* 1997, **28**(1) 67.

82 Nelson M I, Brindley J and McIntosh A C. 'Ignition mechanisms of thermally thin thermoplastics in the cone calorimeter', *Proc. Roy. Soc. A* 1998, **454**(1971) 789.

83 Rychly J and Costa L. 'Modelling of polymer ignition and burning adopted for cone calorimeter measurements: the correlation between the rate of heat release and oxygen index', *Fire Mat.* 1995, **19** 215.

84 Amos B and Fernandez-Pello A C. 'Model of ignition and flame development on a vaporising combustible surface in a stagnation point flow: ignition by vapour fuel absorption', *Comb. Sci. Tech.* 1988, **62** 331.

85 Park S H and Tien C L. 'Radiative induced ignition of solid fuels', *Int. J. Heat and Mass Trans.* 1990, **33** 1511.

86 Di Blasi C, Crescitelli S, Russo G and Cinque G. 'Numerical model of ignition processes of polymeric materials including gas-phase absorption of radiation', *Comb. Flame* 1991, **83** 333.

87 Durbetaki P and Phuoc T X. 'Numerical study of laser-induced ignition of coals', in R W Lewis (ed.), *Numerical methods in thermal problems* Vol. 8, Pineridge Press, Swansea, 1993, 1310.

88 Durbetaki P, Phuoc T X, Mathur M P and Ekmann J M. 'Numerical study of radiative ignition of pyrolysing solid fuels', *Int. J. for Num. Methods in Fluids* 1995, **20** 507.

Index